Quick Finite Elements for Electromagnetic Waves

Second Edition

DISCLAIMER OF WARRANTY

For a list of recent titles in the
Artech House Electromagnetic Analysis Series,
turn to the back of this book.

Quick Finite Elements for Electromagnetic Waves

Second Edition

Giuseppe Pelosi
Roberto Coccioli
Stefano Selleri

ARTECH HOUSE
BOSTON | LONDON
artechhouse.com

Library of Congress Cataloging-in-Publication Data
A catalog record for this book is available from the U.S. Library of Congress.

British Library Cataloguing in Publication Data
A catalogue record for this book is available from the British Library.

ISBN-13: 978-1-59693-345-3

Cover design by Igor Valdman.

685 Canton Street
Norwood, MA 02062

10 9 8 7 6 5 4 3 2 1

To Peter P. Silvester

Contents

Preface

Ten years have passed since the publication of the first edition of this book. The enormous growth in computing power and the full establishment of high-level mathematical computing languages has revolutionized the world of numerical electromagnetics, together with the presence on the market of very high level finite element-based commercial software.

There is, nevertheless, the clear need for an educational approach to finite elements, valid both at an undergraduate and a graduate student level, allowing for a rapid and effective hands-on approach. This was the aim of the first edition of this book, whose philosophy, sketched in the preface of the 1998 edition, has not changed.

What has changed are the expectations of students and researchers. FORTRAN, which is still a leading language for numerical computation has lost, due to its intrinsic architecture, some of its appeal, especially in undergraduate and graduate student courses, with The MathWorks Matlab, just to cite one of the currently available high-level mathematical languages, taking its place.

Furthermore, if ten years ago a two-dimensional working FEM code was appealing as common personal computers were not able to provide the sheer amount of memory and computing power needed to three-dimensional finite elements implementations, nowadays this computing power is commonly available, so three-dimensional codes are appealing also in an educational framework.

For the reasons expressed above, while maintaining the fundamental philosophy of the first edition, this second edition provides the reader with Part I, which is an upgrade of the 1998 edition, containing the original bidimensional codes in FORTRAN and their newly developed Matlab translation. A complete new set of tools for preprocessing and postprocessing exploiting Matlab graphical capabilities has also been developed.

Furthermore, a brand-new Part II, presenting three chapters devoted to three-dimensional FEM problems has been included. This part presents codes developed in FORTRAN, for speed and efficiency reasons, and presents a three-dimensional tool library as well as two codes, one for resonant cavity problems and one for solving waveguide devices. The development of this second part was carried out with the substantial help of Dr. Giacomo Guarnieri, former Ph.D. student at the University of Florence and now with Galileo Avionica Co., Campi Bisenzio (Florence, Italy).

Finally Part III includes an up-to-date annotated bibliography of the main books and articles published on finite elements for electromagnetics.

The CD-ROM bounded to this book contains the complete software described in the text, in FORTRAN 77 for the whole book and in Matlab for Part I. The software has been designed to be as platform independent as possible and the

Matlab code to be as version independent as possible. While Matlab is of course a commercial code, the codes provided make use, when possible, of some freely available mathematical, mesh generators, and graphical software packages that can be retrieved from public repositories accessible through the Internet. For the convenience of the readers, those packages for which permission was granted are included in the CD-ROM.

This book, as it was already in the first edition, is dedicated to the memory of P. P. Silvester, to whose teachings and expertise is due a great portion of the growth of the scientific activity in the field of numerical analysis at the University of Florence in the nineties. The authors are also indebted to many colleagues for useful advices which lead to noticeable improvements for this second edition.

Giuseppe Pelosi
Florence, Italy
Roberto Coccioli
Los Angeles, California
Stefano Selleri
Florence, Italy
May 2009

Preface to the First Edition

The first complete and clear application of the finite element method (FEM) to applied electromagnetism dates back to 30[1] years ago. At the beginning of 1969, a paper from P. P. Silvester was published in the Italian technical journal *Alta Frequenza*, in which a detailed finite element formulation of the classic problem of guided propagation was presented. Although the problem treated in that paper was rather simple, the author clearly showed and pointed out the promising potentialities of the method. Since then FEM evolved quickly: many improvements in the formulation have widely expanded its capability and flexibility so that applications to electromagnetic problems spawned largely. Currently, FEM has gained a fundamental role for the analysis of electromagnetic problems and is one of the most widely used CAD tools for microwave components ranging from simple guiding structures to circuits, antennas, and scattering problems.

The implementation of finite element methods is not simple though. In comparison to other popular numerical techniques, most notably the finite difference method, it requires a finer analytical development of the formulation before going to the implementation, a deeper knowledge of linear algebra methods, as well as a more involved preprocessing phase. Probably for these reasons, the usage and full knowledge of FEM is still limited to a relatively small number of researchers and microwave engineers despite its enormous possibilities and versatility.

Textbooks on finite element methods are still few in number, even if new ones are published constantly, and the method still lacks a full citizenship in university courses, both at undergraduate and graduate levels. Most of the books on finite elements cover the theory behind the method, rarely providing the reader with codes to experiment with, even very simple ones indeed. This implies that, after having studied the book, the reader needs some time to develop his first working FEM code and, most often, this time is too long with respect to that allocated to finite elements in university courses on numerical methods, or available to practitioners. Consequently, it is very difficult for those approaching the method for the first time to fully appreciate its power without quickly getting some numerical results from an FEM code.

The main goal of this book, as suggested by its title, is that of shortening the learning curve of finite element methods as well as to cut down the implementation time. *Quick FEM* is aimed to provide the user with the essential theory behind FEM, limiting the analytical development to that strictly needed to get started, and to guide the reader to implement the method by describing in detail some applications and the related codes. This allows understanding FEM and writing the very first working FEM code just after having read only the first chapter only. The interested reader may further investigate the theoretical aspects of the method by referring to other available books, such as the classic and well-known text

1. This preface was written in 1998. Now it dates back 40 years.

written by P. P. Silvester and R. L. Ferrari, now in its third revised and expanded edition.

The CD-ROM bounded to *Quick FEM* contains the complete software described in the text and uses the practical, implementation-oriented attitude of the book. The source codes have been written in FORTRAN 77 and are designed to be as platform independent as possible. The class of problems that they can solve ranges from waveguide characterization to scattering and radiation, and can be easily expanded and customized by the reader introducing new modules to treat the problem at hand. The codes provided make use of some freely available mathematical and graphical software packages that can be retrieved from public repositories accessible through the Internet. For the convenience of those who cannot easily access the Internet, those packages for which permission was granted are included in the CD-ROM.

This book is dedicated to P. P. Silvester, to whose teachings and expertise is due a great portion of the growth of the scientific activity in the field of numerical analysis at the University of Florence in the past few years. The authors are also indebted to many colleagues for their useful advice.

Giuseppe Pelosi
Florence, Italy
Roberto Coccioli
Los Angeles, California
Stefano Selleri
Nice, France
1998

How to Use Quick FEM

This book has been organized so that each reader can hopefully benefit the most in the shortest possible time by choosing the chapters to read on the basis of his or her own interests and knowledge.

The book is divided into three parts. The first devoted to two-dimensional problems, the second dealing with three-dimensional problems, the last containing a bibliography referencing key papers and books on finite elements.

In Part I the first chapter describes an extremely simple application to a guided propagation problem, using the quasistatic approach. It gives a brief introduction to FEM and to point out the various phases of the implementation of an FEM code, thus it can be skipped by those already acquainted with the method. Chapter 2, on the other hand, describes a set of software tools and libraries employed by all codes described in the remaining chapters, thus it should be of interest for all those wishing to understand the codes and to modify them. In particular, it describes the various preprocessing and postprocessing tools, as well as some utilities and a library of subroutines for generations of the elemental matrices.

In this second edition all software in Part I is provided both in FORTRAN 77 and in Matlab languages, with an almost 1 to 1 correspondence between implemented functions. It is important to note how Matlab is excellent for prototyping and class exercises, due to its high level and comparable simplicity, whereas FORTRAN codes are more interesting in a research and production environment, due to their general faster execution times and the possibility of handling larger problems on a given machine.

Chapters 3 to 6 describe the finite element formulation and the relative implementation to solve fairly general problems ranging from guided propagation in homogeneous and inhomogeneous waveguides (Chapter 3), scattering from rectangular waveguide devices (Chapter 4), scattering from infinitely extending periodic structures (Chapter 5), and both scattering from isolated objects and radiation from apertures (Chapter 6). All these chapters are self-contained and can be read in any order depending on the interest of the reader. The codes and the examples described cover topics of current research interest, such as complex modes in lossless waveguides and scattering from photonic bandgap structures.

Part II is articulated into three chapters and deals with three-dimensional problems. Chapter 7 is analogous to Chapter 2 inasmuch as it presents the basic elements and tools for three-dimensional finite elements in electromagnetics. In this part, for efficiency reasons, only FORTRAN codes are provided even if some Matlab pre- and postprocessing tools are given. Chapters 8 and 9 then deal with typical waveguide problems; in particular Chapter 8 presents the FEM solution for resonant cavities, while Chapter 9 presents a full three-dimensional code for microwave passive devices in rectangular waveguide, greatly extending the simpler code of Chapter 4.

Finally, Part III contains just a single Chapter 10, presenting a wide and structured bibliography where material to further investigate specific topics of interest, or to read about recent development in FEM applications to electromagnetic engineering can be found.

Some words are due also on the content and usage of the enclosed CD-ROM, as well as on how to retrieve from the Internet the software packages not developed by the authors. The software referenced in this book is subdivided into four categories:

- Tools;
- Electromagnetic analysis;
- Linear algebra solvers;
- Graphical rendering.

The codes belonging to the second category, most of those belonging to the first and to the fourth has been developed expressly for this book and are described in the text. The other essential codes, when permission was granted, are also included into the accompanying CD-ROM (Figure 1). In any case, these freely available codes are available from the network and the reader is encouraged to seek the latest versions.

Except for the first code described in Chapter 1, which is self-contained, all the other codes in this book need linear algebra solvers, sometimes pretty sophisticated ones, as in the case of Chapter 3. For what concerns Matlab codes, Matlab internal solvers are used. For FORTRAN codes any mathematical library is appropriate, and the reader should not encounter any problems in customizing the codes for his own use. The linear algebra package (LAPACK), which is a very widespread mathematical library usually natively available on any LINUX system, has been used.

The main Web site from which the LAPACK library can be downloaded is *http://www.netlib.org*, but European users may find the site *http://www.netlib.no/* more convenient. LAPACK routines are self-explanatory, and the way they are called within the codes here provided should be clear enough. A manual edited by LAPACK authors is available from SIAM. Ordering instructions are on the above WWW addresses.

LAPACK subroutines make ample use of basic linear algebra subprograms (BLAS), a collection of subroutines able to perform standard linear algebra operations. This package is included in the CD-ROM too, and it can also be downloaded from the same sites as the LAPACK. However, BLAS routines vendor optimized for the platform used should be employed whenever available.

Graphical visualization of computed data can be achieved exploiting a great variety of software packages. The commercial packages are numberless but, of course, Matlab is used throughout this book. Among the various freely available packages, for the readers unwilling to use Matlab, we prefer GNUPLOT, which again can be freely downloaded from the site *http://www.gnuplot.info/*.

Mesh visualization is a demanding task, which is well managed via the Matlab codes provided in this book. Otherwise, a visualization program has been devel-

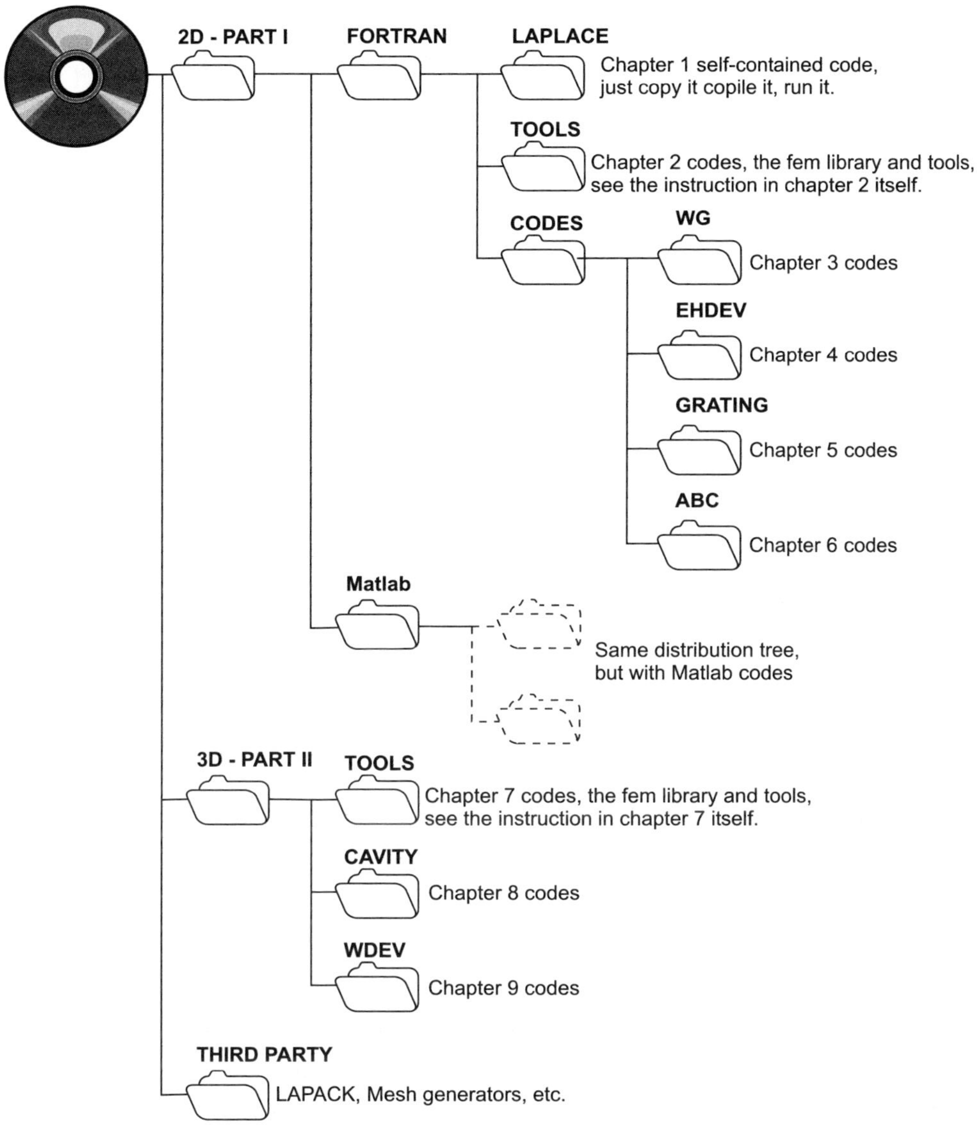

Figure 1 Accompanying CD-ROM tree.

oped in FORTRAN, converting mesh data into PostScript files. These latter can be visualized by resorting to the GhostScript/GhostView packages, whose main site is *http://pages.cs.wisc.edu/ ghost/.*

As for the first edition, a Web site located at *http://www.cem.unifi.it/* is available, publishing news and patches for the codes developed by the authors and accepting comments, suggestions, and bug reports.

PART I

Two Dimensions

CHAPTER 1

Getting Started: Shielded Microstrip Lines

The aim of this chapter is to introduce the finite element method (FEM) by means of a simple example: the evaluation of primary and secondary constants of a microstrip line shielded by a metallic box, indefinitely extending along the axial direction z (Figure 1.1).

The assumption of uniformity effectively makes the problem two-dimensional so that it suffices to analyze the structure in a transverse plane [Figure 1.2(a)].

Electromagnetic propagation in a transmission line can be fully described once its secondary constants Z (characteristic impedance) and β (propagation constant) are known. In general these can be obtained under the hypothesis of TEM or quasi-TEM propagation and lossless substrate and conductors, from the line's primary constants L (inductance per unit length) and C (capacitance per unit length) with the formulas:

$$Z = \sqrt{\frac{L}{C}} = v_p L = \frac{1}{v_p C}, \quad \beta = \omega\sqrt{LC} \tag{1.1}$$

where $v_p = \omega/\beta$ is the phase velocity of the guided wave.

Removing the dielectric substrate [Figure 1.2(b)] allows the structure to support a TEM-guided wave, which, since the medium is air, propagates with a phase velocity $v_p = c$, being c the velocity of light in air ($\simeq 3 \cdot 10^8$m/s). For this new transmission line, the following relation holds:

$$Z^{air} = \sqrt{\frac{L}{C^{air}}} = cL = \frac{1}{cC^{air}} \tag{1.2}$$

where C^{air} is the capacitance per unit length of the equivalent air-spaced shielded line. The inductance per unit length $L = L^{air}$ is unchanged since the relative permeability is $\mu_r = 1$ on the entire cross-section for both lines. Hence:

$$L = \frac{Z_0^{air}}{c} = \frac{1}{c^2 C^{air}} \tag{1.3}$$

Combining (1.1), (1.2), and (1.3) the characteristic impedance of the original shielded microstrip line can be expressed in terms of the capacitance per unit length of the two structures, with and without the dielectric slab:

$$Z = \sqrt{\frac{L}{C}} = \frac{1}{c\sqrt{C^{air}C}} \tag{1.4}$$

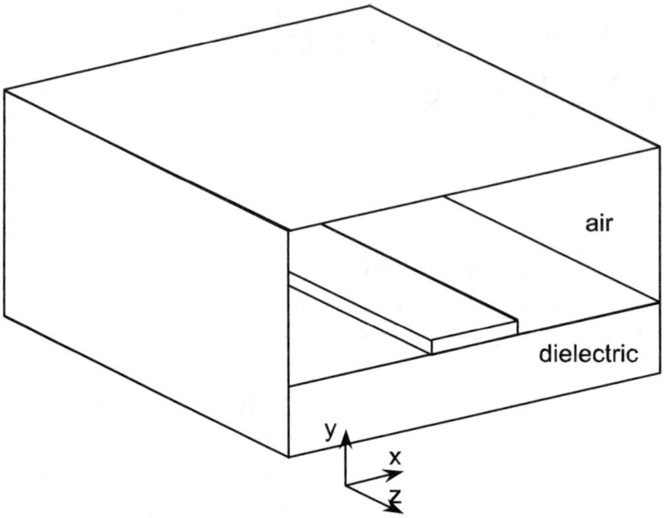

Figure 1.1 Shielded microstrip line.

and, evidently,

$$\beta = \omega\sqrt{LC} = \frac{\omega}{c}\sqrt{\frac{C}{C^{air}}} \tag{1.5}$$

Equations (1.4) and (1.5) can be in general applied to any two-conductor transmission line and extended also to the case of multiconductor lines.

Computation of the primary and secondary constants of the guiding structure is thus reduced to the computation of C and C^{air} which can be easily performed exploiting the flexibility of the finite element method. An FEM-based code can be implemented to evaluate the electrostatic energy U and U^{air} stored per unit length in the structures, from which C and C^{air} can be derived as:

$$C = \frac{2U}{(\Delta V)^2}, \quad C^{air} = \frac{2U^{air}}{(\Delta V)^2} \tag{1.6}$$

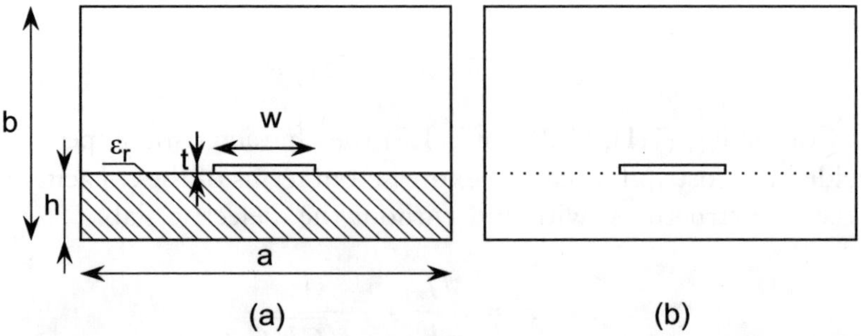

Figure 1.2 Cross-section of the microstrip line. With dielectric loading (a) and air-spaced (b).

with ΔV being the difference of potential between the two conductors. Due to the symmetry of the shielded microstrip line, it is possible to reduce the computational domain to only one-half of the guiding structure cross-section (Figure 1.3), thus saving memory and computation time. To derive a mathematical formulation of the physical problem let us consider the following functional:

$$Q(u) = W(u) = \frac{\epsilon_0}{2}\int_{\Omega_1} |\nabla_t u|^2 d\Omega + \frac{\epsilon_d}{2}\int_{\Omega_2} |\nabla_t u|^2 d\Omega \tag{1.7}$$

where $\epsilon_d = \epsilon_0\epsilon_r$, and Ω_1 and Ω_2 denote the free space and loaded region inside the metallic box, respectively (Figure 1.3). The boundaries of these regions, which constitute the computational domain boundary, are denoted with $\Gamma = \Gamma_D + \Gamma_N = \partial(\Omega_1 + \Omega_2)$. It is well-known that the function $\bar{u}$ with boundary conditions:

$$\bar{u} = u_0 \text{ at } \Gamma_D \tag{1.8}$$

$$\frac{\partial \bar{u}}{\partial n} = 0 \text{ at } \Gamma_N \tag{1.9}$$

which minimizes the previous functional satisfies also Laplace's equation:

$$\nabla_t^2 \bar{u} = 0 \tag{1.10}$$

with the same boundary conditions [1].

The continuity condition on the normal component of the electric displacement $\mathbf{D} = -\epsilon\nabla_t u$ at the dielectric interface Γ_i:

$$\left.\frac{1}{\epsilon_r\epsilon_0}\frac{\partial \bar{u}}{\partial n}\right|_{\Gamma_{i,2}} = \left.\frac{1}{\epsilon_0}\frac{\partial \bar{u}}{\partial n}\right|_{\Gamma_{i,1}} \tag{1.11}$$

must also be taken into account.

Because $\bar{u}$ is the function that satisfies the Laplace equation with the boundary conditions (1.8) and (1.9), and which minimizes the functional (1.7), it follows that $W(\bar{u})$ represents the electrostatic energy of the whole system.

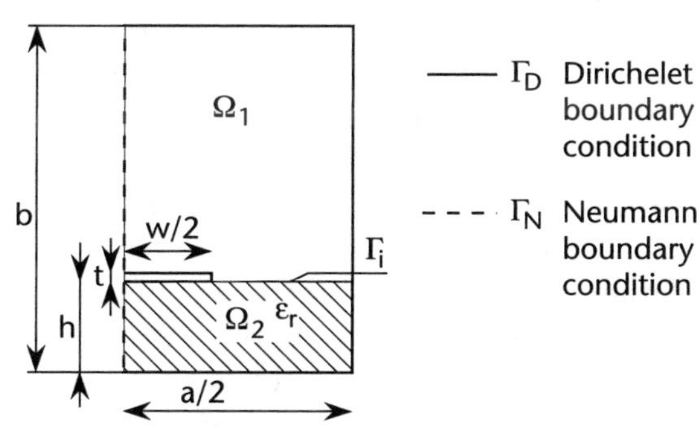

Figure 1.3 Problem to be solved.

Enforcing these boundary conditions is a fundamental step in the finite element solution procedure. Dirichlet-type boundary conditions, such as (1.8), are referred to as *essential boundary conditions* and must be explicitly imposed, as shown later in this chapter. On the other hand, Neumann-type boundary conditions, such as (1.9) and (1.11), are called *natural boundary conditions*, since they automatically arise on boundary nodes without any need for explicit enforcement.

The general FEM solution procedure consists of five steps, which will be outlined in this chapter, more deeply investigated in Chapter 2, and exploited in the subsequent chapters. These steps can be summarized as:

- *Preprocessing*: subdividing the region into finite elements;
- *Building element matrices*: determining the functional on each element;
- *Assembling the global matrix*: summing up the contributions from each element;
- *Minimizing the functional*: solving the system of linear equations;
- *Postprocessing*.

The postprocessing phase, unlike the previous four, may differ widely from problem to problem, according to the quantity sought. It may involve graphical representation of results, as well as their numerical integration or differentiation. In particular, for the problem under examination, this latter phase will comprehend the computation of the capacitance per unit length, in air and with the dielectric, of the shielded microstrip and the computation of its characteristic impedance.

The remaining part of this chapter is devoted to the FEM solution of the shielded microstrip line problem, introducing, step by step, the `LAPLACE` code, which implements a simple but versatile software for solving the Laplace equation on inhomogeneous arbitrarily shaped geometrical domains.

For a quicker approach to FEM, this code is completely self-contained. All the modules here described one by one are contained either in the directory `2D-PART I/MATLAB/LAPLACE` in the CD-ROM or in the file `LAPLACE.F` located in the directory `2D-PART I/FORTRAN/LAPLACE` of the accompanying disk. In both cases all the data files used throughout this chapter are located in the subdirectory `FIRST`; while the subdirectory `STRIP` contains a more refined example, which is also used in Chapter 2 for exemplifying the tools and in Chapter 3, where a more versatile approach to the characterization of guiding structures is presented.

In the following section we will explicitly refer to the Matlab version, but the FORTRAN code is essentially identical except that it is to be compiled and used from the command prompt. The usage is quite simple. After having copied the FORTRAN `LAPLACE` tree onto the hard drive, the program `LAPLACE.F` needs to be compiled. Then it should be run twice using both times the `RAW.FEM` mesh file and the electromagnetic files `RAWA.AUX` first then `RAWD.AUX` later. The first run provides the value of C^{air}, the second C.

For the Matlab code you don't need to copy it on the hard disk, you can run the code directly from the CD-ROM. In this case `Laplace.m` is a function requiring the two input file names as parameters and returning the capacity value. To immediately see some results just launch `raw.m` which is a short M-file launching `Laplace.m` twice and performing secondary line constants computations.

1.1 First Step: Preprocessing

The preprocessing step requires an important tool: the automatic mesh generator. The mesh generator subdivides the region under study into a set of nonoverlapping elements, usually triangles or rectangles, with the former more suitable to fit a generic two-dimensional geometrical shape.

To explain how meshes are described and used in the FEM procedure, the very simple handmade mesh of the shielded microstrip cross-section shown in Figure 1.4 is considered. Although this mesh has nodes spaced too far apart to give results with reasonable accuracy, it is used for the sake of clarity. The program described in the following sections, which uses this mesh, is fully functional and serves as a first example. A simple but flexible automatic mesh generator as well as guidelines to build meshes with an appropriate number of nodes is provided in Chapter 2.

The mesh generator creates a file containing the following information about the mesh:

- Coordinates of the element vertices, which are also called nodes;
- Description of the mesh topology, through a connection matrix;
- Labels marking special nodes and elements.

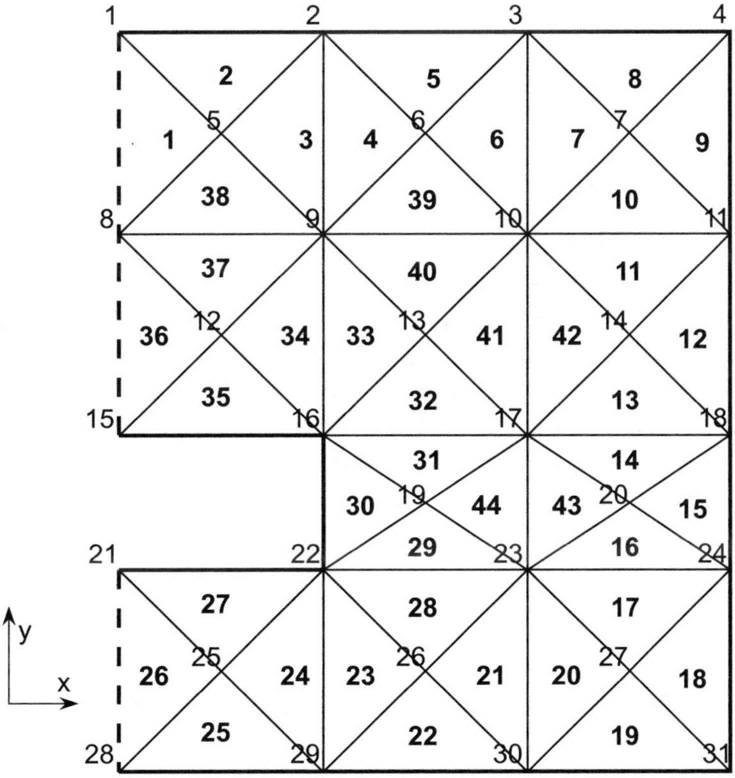

Figure 1.4 A very rough mesh of the region of interest. Node numbers (plain) and element numbers (bold) are shown.

Table 1.1 Node Coordinates

Node n°	*x*	*y*	*Node n°*	*x*	*y*	*Node n°*	*x*	*y*
1	0.000	1.000	12	0.167	0.580	23	0.667	0.280
2	0.333	1.000	13	0.500	0.580	24	1.000	0.280
3	0.667	1.000	14	0.833	0.580	25	0.167	0.140
4	1.000	1.000	15	0.000	0.440	26	0.500	0.140
5	0.167	0.860	16	0.333	0.440	27	0.833	0.140
6	0.500	0.860	17	0.667	0.440	28	0.000	0.000
7	0.833	0.860	18	1.000	0.440	29	0.333	0.000
8	0.000	0.720	19	0.500	0.360	30	0.667	0.000
9	0.333	0.720	20	0.833	0.360	31	1.000	0.000
10	0.667	0.720	21	0.000	0.280			
11	1.000	0.720	22	0.333	0.280			

All of the mesh data required by the program LAPLACE to properly solve the problem are contained in Table 1.1, Table 1.2, and Table 1.3. The first table lists the node coordinates, while the second lists the nodes where Dirichlet boundary conditions are to be enforced, their label, and the value u_0 of their fixed potential. In this example a label of 1 is associated to the nodes lying on the metallic shield ($u_0 = 0$), while nodes on the inner strip ($u_0 = 1$) have a label value of 2.

Table 1.3 lists all of the elements and describes which nodes belongs to each of them. This association is performed by assigning a local counterclockwise numbering scheme to the nodes of each element and a global numbering scheme to the nodes of the whole mesh. From Table 1.3 we can infer, for example, that the element number 1 of the mesh has three nodes, locally numbered 1, 2, 3, which, in the global scheme, are numbered 5, 1, 8, respectively. Of course, a node can belong to more than one element, for instance node number 1 is the second node of element 1 and the third node of element 2.

Elements too may have labels, here used to distinguish elements belonging to region Ω_1 (free space) from those belonging to region Ω_2 (dielectric slab). In this example a label of 1 is associated with elements in free space, while a label of 2 is associated with elements in the dielectric slab.

The mesh generator provided with the software accompanying this book stores the element topology and labels as well as the node coordinates and labels in a file, henceforth referred to as the *output mesh description file*, that, for the mesh of Figure 1.4 is shown in Example 1.3. The mesh generator and the format of the output mesh description file are fully described in Chapter 2.

Table 1.2 Dirichlet Boundary Conditions

Node n°	*Label*	*Prescribed Potential*	*Node n°*	*Label*	*Prescribed Potential*
1	1	0.000	21	2	1.000
2	1	0.000	22	2	1.000
3	1	0.000	24	1	0.000
4	1	0.000	28	1	0.000
11	1	0.000	29	1	0.000
15	2	1.000	30	1	0.000
16	2	1.000	31	1	0.000
18	1	0.000			

Table 1.3 Connection Between Global and Local Numbering Schemes

el.	*Local Node n°*			*lb.*	*el.*	*Local Node n°*			*lb.*	*el.*	*Local Node n°*			*lb.*
n°	*1*	*2*	*3*		*n°*	*1*	*2*	*3*		*n°*	*1*	*2*	*3*	*1*
1	5	1	8	1	16	20	23	24	1	31	16	19	17	1
2	5	2	1	1	17	23	27	24	2	32	13	16	17	1
3	5	9	2	1	18	24	27	31	2	33	9	16	13	1
4	6	2	9	1	19	27	30	31	2	34	9	12	16	1
5	3	2	6	1	20	23	30	27	2	35	12	15	16	1
6	6	10	3	1	21	23	26	30	2	36	8	15	12	1
7	7	3	10	1	22	26	29	30	2	37	8	12	9	1
8	7	4	3	1	23	22	29	26	2	38	5	8	9	1
9	7	11	4	1	24	22	25	29	2	39	6	9	10	1
10	7	10	11	1	25	25	28	29	2	40	9	13	10	1
11	14	11	10	1	26	21	28	25	2	41	10	13	17	1
12	14	18	11	1	27	21	25	22	2	42	10	17	14	1
13	14	17	18	1	28	22	26	23	1	43	17	23	20	1
14	17	20	18	1	29	19	22	23	1	44	17	19	23	1
15	18	20	24	1	30	16	22	19	1					

This latter file is read by the function `ReadFemNFile.m` (Example 1.1) (or FORTRAN subroutine `READNODALMESH`), that makes the data available to the `Laplace.m` code. The function `ReadElecFile.m` (Example 1.2) reads a second input file that specifies values for fixed potentials and dielectric constants (see `READELECT` subroutine for FORTRAN).

Example 1.1

```
function [xy,ele,nlab,elab]=ReadFemNFile(nomefile)
%%%%%%%%%%%%%%%%%%%%%%%%%%%%%%%%%%%%%%%%%%%%%%%%%%%%%%%%%%%%%%%%%%%
% Reads a Geometry file for what concerns the connection matrix
% and the node coordinates, skipping advanced informations
%
% (C) 1997-2008 PELOSI - COCCIOLI - SELLERI
%
%%%%%%%%%%%%%%%%%%%%%%%%%%%%%%%%%%%%%%%%%%%%%%%%%%%%%%%%%%%%%%%%%%%
%---------------- READ THE INPUT FILE AND SET INTERNAL GEOMETRY STRUCTURES
fh = fopen(nomefile,'r');
% Discards header
dummy = fgetl(fh);
% number of elements
NELE = fscanf(fh,'Elements = % i\n',1);
% number of nodes
NNODE = fscanf(fh,'Nodes = % i\n',1);
% Discards edges
dummy = fgetl(fh);
dummy = fgetl(fh); % Reads word 'Elements'
for i=1:NELE
    aux1 = fgetl(fh);
    aux2 = sscanf(aux1,'% i');
    ele(:,aux2(1)) = aux2(3:(2+aux2(2)))';
    elab(aux2(1)) = aux2(3+aux2(2));
end

dummy = fgetl(fh); % Reads word 'Nodes'
for i=1:NNODE
    aux1 = fgetl(fh);
```

```
    aux2 = sscanf(aux1,'% i % f % f % i');
    xy(1,aux2(1))=aux2(2);
    xy(2,aux2(1))=aux2(3);
    nlab(aux2(1))=aux2(4);
end

% That's all!
fclose(fh);
```

Example 1.2

```
function [fp,de]=ReadElecFile(nomefile)
%%%%%%%%%%%%%%%%%%%%%%%%%%%%%%%%%%%%%%%%%%%%%%%%%%%%%%%%%%%%
% Reads a Electromagnetic input file
%
%  (C) 1997-2008 PELOSI - COCCIOLI - SELLERI
%
%%%%%%%%%%%%%%%%%%%%%%%%%%%%%%%%%%%%%%%%%%%%%%%%%%%%%%%%%%%%
%--------------READ THE INPUT FILE AND SET INTERNAL ELECTROMAGNETIC STRUCTURES
fh = fopen(nomefile,'r');
% number of fixed potentials
NFP = fscanf(fh,'% i',1);
for i=1:NFP
    aux = fscanf(fh,'% i, %f',2);
    fp(aux(1)) = aux(2);
end

% number of materials
NDE = fscanf(fh,'%i',1);
for i=1:NDE
    aux = fscanf(fh,'%i, %f',2);
    de(aux(1)) = aux(2);
end

% Sets to 'air' unassigned labels
for i=1:length(de)
    if de(i)==0
        de(i)=1;
    end
end

% That's all!
fclose(fh);
```

Example 1.3

```
Mesh Geometry file for a shielded microstrip
Elements = 44
Nodes = 31
Edges = 0
Elements
     1    3    5     1    8    1
     2    3    5     2    1    1
     3    3    5     9    2    1
     4    3    6     2    9    1
     5    3    3     2    6    1
     6    3    6    10    3    1
     7    3    7     3   10    1
     8    3    7     4    3    1
     9    3    7    11    4    1
    10    3    7    10   11    1
```

```
11   3   14   11   10   1
12   3   14   18   11   1
13   3   14   17   18   1
14   3   17   20   18   1
15   3   18   20   24   1
16   3   20   23   24   1
17   3   23   27   24   2
18   3   24   27   31   2
19   3   27   30   31   2
20   3   23   30   27   2
21   3   23   26   30   2
22   3   26   29   30   2
23   3   22   29   26   2
24   3   22   25   29   2
25   3   25   28   29   2
26   3   21   28   25   2
27   3   21   25   22   2
28   3   22   26   23   2
29   3   19   22   23   1
30   3   16   22   19   1
31   3   16   19   17   1
32   3   13   16   17   1
33   3    9   16   13   1
34   3    9   12   16   1
35   3   12   15   16   1
36   3    8   15   12   1
37   3    8   12    9   1
38   3    5    8    9   1
39   3    6    9   10   1
40   3    9   13   10   1
41   3   10   13   17   1
42   3   10   17   14   1
43   3   17   23   20   1
44   3   17   19   23   1
Nodes
     1     .000000    1.000000    1
     2     .333333    1.000000    1
     3     .666667    1.000000    1
     4    1.000000    1.000000    1
     5     .166667     .860000    0
     6     .500000     .860000    0
     7     .833333     .860000    0
     8     .000000     .720000    0
     9     .333333     .720000    0
    10     .666667     .720000    0
    11    1.000000     .720000    1
    12     .166667     .580000    0
    13     .500000     .580000    0
    14     .833333     .580000    0
    15     .000000     .440000    2
    16     .333333     .440000    2
    17     .666667     .440000    0
    18    1.000000     .440000    1
    19     .500000     .360000    0
    20     .833333     .360000    0
    21     .000000     .280000    2
    22     .333333     .280000    2
    23     .666667     .280000    0
    24    1.000000     .280000    1
    25     .166667     .140000    0
```

```
26     .500000     .140000    0
27     .833333     .140000    0
28     .000000     .000000    1
29     .333333     .000000    1
30     .666667     .000000    1
31    1.000000     .000000    1
```

Example 1.3 is contained in the disk with filename `raw.fem`. Please note that `ReadFemNFile.m` reads the mesh as it is, the mesh is built considering all dimension in millimeters. Coordinates are appropriately scaled in the main program.

For the microstrip shielded line problem, considering a dielectric material of relative permittivity 8.875 [2], the electromagnetic input file (`rawD.aux` in the disk) has the format reported Example 1.4:

Example 1.4

```
2
1  0.0
2  1.0
1
2  8.875
```

The first line of Example 1.4 specifies that there are two fixed potentials, and the following two lines define the labels of the Dirichlet boundary nodes and the associated potential value (i.e., line two assigns potential zero to nodes labeled one). Then there is a line defining the number of dielectric materials, other than air, present in the computational domain. In our case there is only one $\epsilon_r = 8.875$ dielectric corresponding to elements labeled 2. The program LAPLACE assumes that nodes and elements which have labels not specified in this file or elements whose labels have not been associated to any permittivity value, are to be treated as unlabeled, that is, nodes have unknown potential and elements belong to free space. For instance, the input file 1.4 implicitly specifies that elements marked with label 1 belong to air, while those marked with label 2 belongs to a dielectric material of relative permittivity 8.875.

It is worth mentioning that this problem, as well as all the others presented in the book, can be preprocessed using any other commercial, shareware or home-made mesh generating software, provided that the subroutine READNODALMESH, or either `ReadFemNFile.m`, which reads the input mesh geometry file is modified to comply with the output format of the mesh generator actually employed.

1.2 Second Step: Building Element Matrices

The unknown function $u(x,y)$ is approximated inside each element by a polynomial expression, which in this first example is assumed to be of first degree:

$$u(x,y) = a + bx + cy = \begin{bmatrix} 1 & x & y \end{bmatrix} \begin{bmatrix} a \\ b \\ c \end{bmatrix} \tag{1.12}$$

This approximating function represents a plane in the Cartesian coordinate system. In each triangle node $P_i = (x_i, y_i), i = 1,2,3$ the potential assumes the following value:

$$\begin{cases} u_1(x_1,y_1) = a + bx_1 + cy_1 \\ u_2(x_2,y_2) = a + bx_2 + cy_2 \\ u_3(x_3,y_3) = a + bx_3 + cy_3 \end{cases} \tag{1.13}$$

Equations in (1.13) can be recast as a single one in matrix form:

$$\begin{bmatrix} u_1 \\ u_2 \\ u_3 \end{bmatrix} = \begin{bmatrix} 1 & x_1 & y_1 \\ 1 & x_2 & y_2 \\ 1 & x_3 & y_3 \end{bmatrix} \begin{bmatrix} a \\ b \\ c \end{bmatrix} \tag{1.14}$$

from which it is easy to express a, b, c in terms of u_1, u_2, u_3:

$$\begin{bmatrix} a \\ b \\ c \end{bmatrix} = \begin{bmatrix} 1 & x_1 & y_1 \\ 1 & x_2 & y_2 \\ 1 & x_3 & y_3 \end{bmatrix}^{-1} \begin{bmatrix} u_1 \\ u_2 \\ u_3 \end{bmatrix} \tag{1.15}$$

Introducing (1.15) into (1.12), yields:

$$u(x,y) = \frac{1}{2A} \begin{bmatrix} 1 & x & y \end{bmatrix} \cdot \begin{bmatrix} x_2y_3 - x_3y_2 & x_3y_1 - x_1y_3 & x_1y_2 - x_2y_1 \\ y_2 - y_3 & y_3 - y_1 & y_1 - y_2 \\ x_3 - x_2 & x_1 - x_3 & x_2 - x_1 \end{bmatrix} \cdot \begin{bmatrix} u_1 \\ u_2 \\ u_3 \end{bmatrix} \tag{1.16}$$

where A is the triangle area (positive if the vertices are counterclockwise numbered):

$$A = \frac{1}{2} \begin{vmatrix} 1 & x_1 & y_1 \\ 1 & x_2 & y_2 \\ 1 & x_3 & y_3 \end{vmatrix} = \frac{1}{2} [(x_1y_2 - x_2y_1) + (x_3y_1 - x_1y_3) + (x_2y_3 - x_3y_2)] \tag{1.17}$$

Equation (1.16) can be rewritten, in a more concise manner:

$$u(x,y) = \sum_{i=1}^{3} u_i \alpha_i(x,y) \tag{1.18}$$

where the functions α_i, are referred to as *basis* or *shape functions* and assume an unitary value at node P_i and zero at node P_j, for $j \neq i$. In this example, α_i are linear functions representing the planes passing through the points $(x_i, y_i, 1)$ and $(x_j, y_j, 0)$, for $j \neq i$ (Figure 1.5). It can be shown that linear shape functions on a triangle coincides with the simplex coordinates (or area coordinates) ζ_i [1]:

$$\begin{cases} \alpha_1(x,y) = \zeta_1(x,y) = \dfrac{1}{2A} [(x_2y_3 - x_3y_2) + (y_2 - y_3)x + (x_3 - x_2)y] \\ \alpha_2(x,y) = \zeta_2(x,y) = \dfrac{1}{2A} [(x_3y_1 - x_1y_3) + (y_3 - y_1)x + (x_1 - x_3)y] \\ \alpha_3(x,y) = \zeta_3(x,y) = \dfrac{1}{2A} [(x_1y_2 - x_2y_1) + (y_1 - y_2)x + (x_2 - x_1)y] \end{cases} \tag{1.19}$$

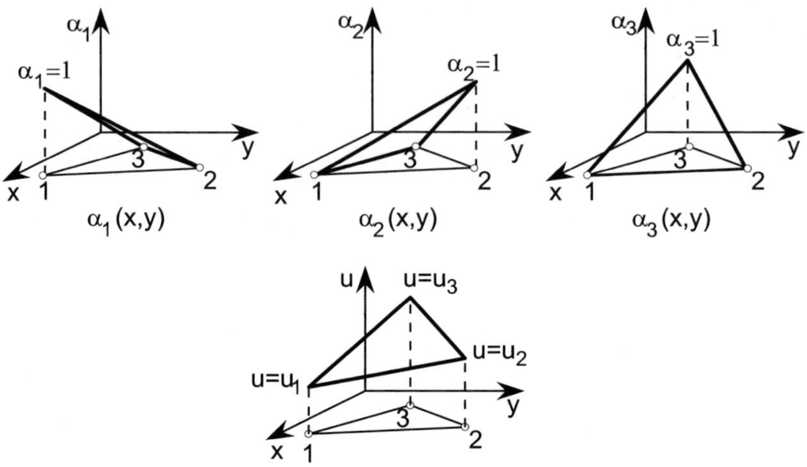

Figure 1.5 Shape functions and linear approximation of the potential inside an element.

shown in Figure 1.5 and have the two following properties:

$$\zeta_i(x_j, y_j) = \begin{cases} 1, \, i = j \\ 0, \, i \neq j \end{cases} \tag{1.20}$$

and:

$$\sum_{i=1}^{3} \zeta_i(x, y) = 1 \tag{1.21}$$

Once the order of the shape functions has been chosen, it is possible to approximate the contribution $W^{(e)}$ of the generic triangle (e) to the functional W by introducing (1.18) into (1.7), and calculating:

$$W^{(e)} = \frac{1}{2}\epsilon^{(e)} \int_{\Delta^{(e)}} |\nabla_t u|^2 dS \tag{1.22}$$

In (1.22), the integration is performed on the element surface $\Delta^{(e)}$ and $\epsilon^{(e)}$ is the permittivity of the material to which the element belongs. Since the gradient of the approximating function u is:

$$\nabla_t u = \sum_{i=1}^{3} u_i \nabla_t \alpha_i \tag{1.23}$$

the functional in each element can be expressed as:

$$W^{(e)} = \frac{1}{2}\epsilon^{(e)} \sum_{i=1}^{3}\sum_{j=1}^{3} u_i u_j \int_{\Delta^{(e)}} \nabla_t \alpha_i \cdot \nabla_t \alpha_j dS \tag{1.24}$$

or, using matrix notation:

$$W^{(e)} = \frac{1}{2}\epsilon^{(e)}[U]^t\left[S^{(e)}\right][U] \tag{1.25}$$

with:

$$[U] = \begin{bmatrix} u_1 \\ u_2 \\ u_3 \end{bmatrix}, \quad \left[S^{(e)}\right] = \begin{bmatrix} S_{11}^{(e)} & S_{12}^{(e)} & S_{13}^{(e)} \\ S_{21}^{(e)} & S_{22}^{(e)} & S_{23}^{(e)} \\ S_{31}^{(e)} & S_{32}^{(e)} & S_{33}^{(e)} \end{bmatrix} \tag{1.26}$$

$$S_{ij}^{(e)} = S_{ji}^{(e)} = \int_{\Delta^{(e)}} \nabla_t \alpha_i \cdot \nabla_t \alpha_j dS \tag{1.27}$$

These latter integrals have an analytical expression [1], whose value for the particular triangle (e) is computed by the function `setri.m` (Example 1.5). The total energy stored per unitary length of the line is then determined by summing up contributions from each triangle.

Example 1.5

```
function [Se] = setri(xy,ele,ie)
%%%%%%%%%%%%%%%%%%%%%%%%%%%%%%%%%%%%%%%%%%%%%%%
% Constructs the element matrix Se for first order
% triangular elements.
%
%   (C) 1997-2008 PELOSI - COCCIOLI - SELLERI
%
%%%%%%%%%%%%%%%%%%%%%%%%%%%%%%%%%%%%%%%%%%%%%%%
%------------------------ Element point definition
idx = ele(:,ie);
area = ((xy(1,idx(2))-xy(1,idx(1)))*(xy(2,idx(3))-xy(2,idx(1)))-...
     (xy(1,idx(3))-xy(1,idx(1)))*(xy(2,idx(2))-xy(2,idx(1))))/2;
dx(1) = xy(1,idx(3))-xy(1,idx(2));
dx(2) = xy(1,idx(1))-xy(1,idx(3));
dx(3) = xy(1,idx(2))-xy(1,idx(1));
dy(1) = xy(2,idx(2))-xy(2,idx(3));
dy(2) = xy(2,idx(3))-xy(2,idx(1));
dy(3) = xy(2,idx(1))-xy(2,idx(2));
%------------------------Compute the matrices entries
Se = (dy'*dy+dx'*dx)/(4*area);
```

1.3 Third Step: Assembling the Global Matrix

Summing up all the contributions from each element given by (1.25) to obtain the value of the functional (1.7), requires appropriate assembling strategies that are able to take into account the fact that adjacent elements share nodes. To illustrate the most widely used assembling technique, let's consider the simple case shown

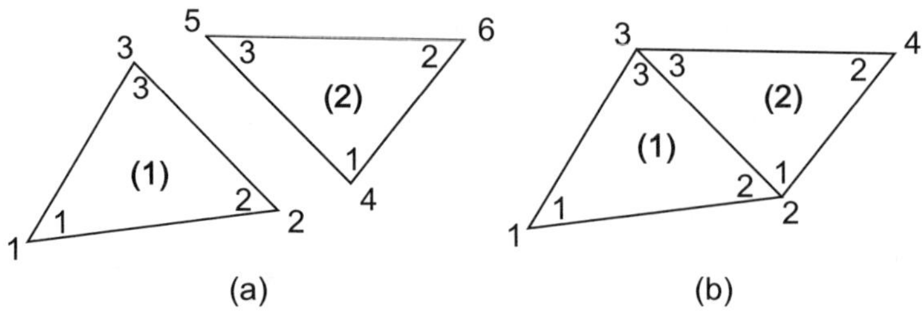

Figure 1.6 (a, b) A two-triangular elements mesh, local node numbering (inside each triangle) and global node numbering (outside).

in Figure 1.6, where the mesh is constituted by only two elements. The value of the total functional W is:

$$W = W^{(1)} + W^{(2)} = \frac{1}{2}\epsilon^{(1)}\left[U^{(1)}\right]^t\left[S^{(1)}\right]\left[U^{(1)}\right] + \frac{1}{2}\epsilon^{(2)}\left[U^{(2)}\right]^t\left[S^{(2)}\right]\left[U^{(2)}\right] \quad (1.28)$$

where $U^{(1)}$ and $U^{(2)}$ are the column vectors of the unknown potential values at the nodes of the two triangles, ordered accordingly to the local numbering scheme. Translating the latter into the global numbering scheme yields:

$$\left[U^{(1)}\right] = \begin{bmatrix} u_1^{(1)} \\ u_2^{(1)} \\ u_3^{(1)} \end{bmatrix} = \begin{bmatrix} u_1 \\ u_2 \\ u_3 \end{bmatrix} = \left[U_g^{(1)}\right], \quad \left[U^{(2)}\right] = \begin{bmatrix} u_1^{(2)} \\ u_2^{(2)} \\ u_3^{(2)} \end{bmatrix} = \begin{bmatrix} u_2 \\ u_4 \\ u_3 \end{bmatrix} = \left[U_g^{(2)}\right] \quad (1.29)$$

where subscript g is used to denote quantities in the global numbering scheme. Considering a new column vector

$$[U] = \begin{bmatrix} u_1 \\ u_2 \\ u_3 \\ u_4 \end{bmatrix} \quad (1.30)$$

containing all the unknowns of the problem, (1.28) can be written as:

$$\begin{aligned} W = W^{(1)} + W^{(2)} &= \frac{1}{2}\epsilon^{(1)}\left[U_g^{(1)}\right]^t\left[S^{(1)}\right]\left[U_g^{(1)}\right] + \\ &+ \frac{1}{2}\epsilon^{(2)}\left[U_g^{(2)}\right]^t\left[S^{(2)}\right]\left[U_g^{(2)}\right] = \frac{1}{2}[U]^t[S][U] \end{aligned} \quad (1.31)$$

where $[S]$ is the global 4×4 global FEM matrix of the problem:

$$[S] = \begin{bmatrix} S_{11} & S_{12} & S_{13} & S_{14} \\ S_{21} & S_{22} & S_{23} & S_{24} \\ S_{31} & S_{32} & S_{33} & S_{34} \\ S_{41} & S_{42} & S_{43} & S_{44} \end{bmatrix} =$$

$$= \begin{bmatrix} \epsilon^{(1)} S_{11}^{(1)} & \epsilon^{(1)} S_{12}^{(1)} & \epsilon^{(1)} S_{13}^{(1)} & 0 \\ \epsilon^{(1)} S_{21}^{(1)} & \epsilon^{(1)} S_{22}^{(1)} + \epsilon^{(2)} S_{11}^{(2)} & \epsilon^{(1)} S_{23}^{(1)} + \epsilon^{(2)} S_{13}^{(2)} & \epsilon^{(2)} S_{12}^{(2)} \\ \epsilon^{(1)} S_{31}^{(1)} & \epsilon^{(1)} S_{32}^{(1)} + \epsilon^{(2)} S_{31}^{(2)} & \epsilon^{(1)} S_{33}^{(1)} + \epsilon^{(2)} S_{33}^{(2)} & \epsilon^{(2)} S_{32}^{(2)} \\ 0 & \epsilon^{(2)} S_{21}^{(2)} & \epsilon^{(1)} S_{23}^{(2)} & \epsilon^{(2)} S_{22}^{(2)} \end{bmatrix} \quad (1.32)$$

The previous equation can be recast in the form:

$$[S] = \epsilon^{(1)} \begin{bmatrix} S_{11}^{(1)} & S_{12}^{(1)} & S_{13}^{(1)} & 0 \\ S_{21}^{(1)} & S_{22}^{(1)} & S_{23}^{(1)} & 0 \\ S_{31}^{(1)} & S_{32}^{(1)} & S_{33}^{(1)} & 0 \\ 0 & 0 & 0 & 0 \end{bmatrix} + \epsilon^{(2)} \begin{bmatrix} 0 & 0 & 0 & 0 \\ 0 & S_{11}^{(1)} & S_{13}^{(1)} & S_{12}^{(1)} \\ 0 & S_{31}^{(1)} & S_{33}^{(1)} & S_{32}^{(1)} \\ 0 & S_{21}^{(1)} & S_{23}^{(1)} & S_{22}^{(1)} \end{bmatrix} \quad (1.33)$$

which shows how to efficiently assemble the global matrix $[S]$ once the local matrices of the elements are known. The generic entry $S_{ij}^{(e)}$ is multiplied by the relative permittivity $\epsilon^{(e)}$ and added to the entry S_{mn} of the global matrix, where m and n are the numbers in the global numbering scheme associated to the local nodes i and j, respectively, of element (e). This assembling scheme is performed by the function `embed.m` (Example 1.6), listed in the following section, which also provides the enforcement of the boundary conditions. Although this latter task is logically part of the global matrix assembling step, it is thoroughly investigated in the following section since it is also deeply tied to the functional minimization process.

It is important to note that the number of nonzero entries in row i of the global FEM matrix is equal to the number of nodes directly connected to node i, that is, the number of all the nodes belonging to the elements which share node i. Hence, the global FEM matrix has a sparse structure reflecting the fact that a local approximation is used for the exact potential. This is one of the most attractive features of FEM since it allows a significant reduction of memory storage requirements and CPU time reduction.

1.4 Fourth Step: Minimizing the Functional

Using the finite element approximation, the functional W is now a function of the N unknown nodal values u_i

$$W = W(u_1, u_2, \ldots, u_N) \quad (1.34)$$

Let's now suppose, for simplicity, that the global node numbering scheme is such that all nodes with free potential are numbered first, and those with prescribed potential last, that is:

$$[U]_{con} = \begin{bmatrix} [U]_f \\ [U]_p \end{bmatrix} \tag{1.35}$$

where the subscripts f and p refer to free and prescribed nodal values, respectively. The process of energy minimization with respect to nodal values free to vary leads to:

$$\frac{\partial}{\partial [U]_f} \left\{ \begin{bmatrix} [U]_f^t & [U]_p^t \end{bmatrix} \begin{bmatrix} [S]_{ff} & [S]_{fp} \\ [S]_{pf} & [S]_{pp} \end{bmatrix} \begin{bmatrix} [U]_f \\ [U]_p \end{bmatrix} \right\} = 0 \tag{1.36}$$

where the superscript t denote the transpose operator. Due to matrix $[S]$ symmetry, the system of linear equations to be solved reduces to:

$$[S]_{ff}[U]_f = -[S]_{fp}[U]_p \tag{1.37}$$

The final square coefficient matrix is of order N_f, being N_f the number of degree of freedom of the problem. Obviously, (1.37) can be written even though the global numbering scheme does not abide to the restriction of numbering the nodes with prescribed potential last. For instance, referring again to Figure 1.6 and assuming that nodes free to vary are those numbered 2 and 3 while nodes numbered 1 and 4 have prescribed potentials $\bar{u}_1$ and $\bar{u}_4$, respectively, the matrix (1.37) becomes:

$$\begin{bmatrix} S_{22} \, S_{23} \\ S_{32} \, S_{33} \end{bmatrix} \begin{bmatrix} u_2 \\ u_3 \end{bmatrix} = - \begin{bmatrix} S_{11} \, S_{14} \\ S_{41} \, S_{44} \end{bmatrix} \begin{bmatrix} \bar{u}_1 \\ \bar{u}_4 \end{bmatrix} \tag{1.38}$$

To avoid unnecessary complex renumbering of nodes to obtain the matrices of (1.38), the usual way of minimizing the functional is to solve the system of linear equations

$$\begin{bmatrix} 1 & 0 & 0 & 0 \\ 0 & S_{22} & S_{23} & 0 \\ 0 & S_{32} & S_{33} & 0 \\ 0 & 0 & 0 & 1 \end{bmatrix} \begin{bmatrix} u_1 \\ u_2 \\ u_3 \\ u_4 \end{bmatrix} = \begin{bmatrix} \bar{u}_1 \\ -S_{11}\bar{u}_1 - S_{14}\bar{u}_4 \\ -S_{41}\bar{u}_1 - S_{44}\bar{u}_4 \\ \bar{u}_4 \end{bmatrix} \tag{1.39}$$

The coefficient matrix and the right-hand side of the above linear system are assembled directly by the function `embed.m` (Example 1.6), which sequentially processes all elements. The system of linear equations built can be solved exploiting appropriate subroutines of the available linear algebra libraries. The solution of a linear system of equation is a trivial one-line task in Matlab, but it is not so straightforward in FORTRAN. To provide the reader with a full-functioning stand-alone program an elementary solver subroutine, SOLVE, is included with the code `LAPLACE.F`. For the more complex problems treated in the rest of the book, the reader is strongly encouraged to use more sophisticated subroutines as, for instance, those included in the freeware libraries provided by *netlib*. The quick reference sheet at the beginning of this book includes instructions to retrieve such libraries.

Example 1.6

```
function [Sg,B] = embed(Sg,B,Se,ele,elab,ie,de,nlab,fp)
%%%%%%%%%%%%%%%%%%%%%%%%%%%%%%%%%%%%%%%%%%%%%%%%%%%%%%%%%%%%
% Embeds local matrix Se of element ie into the global matrix
% Sg or into the RHS term B as appropriate.
%
% (C) 1997-2008 PELOSI - COCCIOLI - SELLERI
%
%%%%%%%%%%%%%%%%%%%%%%%%%%%%%%%%%%%%%%%%%%%%%%%%%%%%%%%%%%%%
iglob = ele(:,ie);
for i=1:3
   % Is the row relative to a fixed potential node?
   if (nlab(iglob(i))>=1)
      Sg(iglob(i),iglob(i)) = 1;
      B(iglob(i)) = fp(nlab(iglob(i)));
   else
       for j=1:3
           if(nlab(iglob(j))>=1)
               % Is the column relative to a fixed potential node?
               B(iglob(i)) = B(iglob(i)) - Se(i,j) * de(elab(ie)) * ...
                   fp(nlab(iglob(j)));
           else
               Sg(iglob(i),iglob(j)) = Sg(iglob(i),iglob(j)) + ...
                   Se(i,j) * de(elab(ie));
           end
       end
   end
end
```

1.5 Fifth Step: Postprocessing

In this example we are mainly concerned with the computation of the stored electrostatic energy and, subsequently, of the capacitance. As already mentioned, the minimum value that the functional (1.7) assume for $u(x,y) = \bar{u}(x,y)$ represents the stored energy-per-unit length. Hence, in FEM approximation the stored energy $\bar{W}$ is given by:

$$\bar{W} = \frac{1}{2}\left[\bar{U}\right]^t [S] \left[\bar{U}\right] \tag{1.40}$$

where $\bar{U}$ is the solution of the linear system (1.37). This quantity is easily evaluated from the global matrix $[S]$ if it is not destroyed in the solution process. Indeed this is the case when boundary conditions are enforced so the stored energy computation is instead performed by summing contributions element by element:

$$\bar{W} = \frac{1}{2}\sum_{i=1}^{M}\left[\bar{U}^{(e)}\right]^t \left[S^{(e)}\right]\left[\bar{U}^{(e)}\right] \tag{1.41}$$

being M the total number of elements.

Result visualization is also an important part of the postprocessing step and it is useful in many instances to gain physical insight into the problem analyzed. To accomplish this task, the nodal values given on the unstructured mesh grid can be interpolated onto a regular rectangular grid and then displayed. All these task are accomplished via standard Matlab functions, as it can be seen at the end of the `Laplace.m` function (Example 1.8), which also draws equipotential lines of the solution.

The aim of the function `Laplace.m` is to evaluate the capacitance of the structure under exam. Two separate runs are needed to compute the line constants. These two runs differs only for the electromagnetic data file, one with dielectric (Example 1.4, file `rawd.aux`), and one without (Example 1.7, file `rawa.aux`), yielding C and C^{air} respectively:

Example 1.7

```
2
1  0.0
2  1.0
1
2  1.0
```

The results for this computation are $C^{air} = 61.22$ pF/m and $\mathrm{C} = 331.4$ pF/m. The secondary constants of the line can then be computed using (1.4) at the beginning of this chapter. The results are $Z = 23.4\Omega$ and $\beta = 2.32\ \omega/\mathrm{c}$.

These values are very rough approximations, since the mesh used is very coarse. The subdirectory STRIP contains another, comparable, set of files, regarding a realistic shielded microstrip. The mesh, much finer, is built with the automatic mesh generator in Chapter 2. The reader is strongly encouraged to read that chapter while experimenting with this second example.

To facilitate the user, two `.m` files are provided, making both computation in sequence. Example 1.9, named `raw.m` on the disk, uses the coarse mesh whereas Example 1.10, named `fine.m` on the disc, uses the fine mesh.

FORTRAN does not provide graphical abilities. For the `LAPLACE.F` file, hence, there is a subroutine `OUTPUT` which writes in a file the nodal values u_i of the computed potential. This file, together with the output geometry file created by the mesh generator, can be converted in a data file suitable for visualization with widely used graphical programs such as, for example, the freeware package GNUPLOT and, of course, Matlab.

Example 1.8

```
function [capacity] = laplace(meshfile,elefile)
%%%%%%%%%%%%%%%%%%%%%%%%%%%%%%%%%%%%%%%%%%%%%%%%%%%%%
%       SOLVES THE ELECTROSTATIC PROBLEM THROUGH AN APPLICATION OF THE
%       FINITE ELEMENT METHOD.
%
%       THE OUTPUT ARE:
%          - THE ELECTROSTATIC ENERGY
%          - THE CAPACITY of the structure. this latter is valid
```

```
%          only for problems with two fixed potentials
%
%     THIS PROGRAM CAN TREAT ONLY FIRST ORDER TRIANGULAR ELEMENTS
%%%%%%%%%%%%%%%%%%%%%%%%%%%%%%%%%%%%%%%%%%%%%%%%%%%%%%%%%%%%%%%%%%%%%%%%%
%     (C) 1997-2008 PELOSI - COCCIOLI - SELLERI
%%%%%%%%%%%%%%%%%%%%%%%%%%%%%%%%%%%%%%%%%%%%%%%%%%%%%%%%%%%%%%%%%%%%%%%%%
%  ---------- Read Input data from FILEFEM
   disp(' [1] -> READING MESH DATA ');
   [xy,ele,nlab,elab]=ReadFemNFile(meshfile);
   nnode = size(xy,2);
   nele = size(ele,2);
%  ---------- Read Input data from FILEELE
   disp(' [2] -> READING ELECTRIC DATA');
   [fp,de]=ReadElecFile(elefile);
%  ---------- BACK TO MKS & Matrix Initialization
    xy=xy/1000;
    Sg = zeros(nnode,nnode);
    B = zeros(nnode,1);
%  ----------  Global Matrix & Right hand side Assembly
   disp(' [3] -> ASSEMBLING GLOBAL MATRIX');
   for ie=1:nele
      Se = setri(xy,ele,ie);
      [Sg,B] = embed(Sg,B,Se,ele,elab,ie,de,nlab,fp);
   end

%  ---------- Solving the System
   disp(' [4] -> SOLVING THE SYSTEM');
   X = Sg\ B;
%  ---------- Computing energy & capacitance
   disp(' [5] -> COMPUTING STORED ENERGY');
   energy = 0;
   for ie=1:nele
      Se = setri(xy,ele,ie);
      energy = energy + X(ele(:,ie))'*Se*X(ele(:,ie))*de(elab(ie));
   end

   % energy = X'*Sg*X
   energy = energy * (8.85*10^(-12)) / 2;
   capacity = 4.*energy*(fp(2)-fp(1))^2;
   disp(' [6] - RESULTS:');
   disp(sprintf(' Energy In the computational box: %e', energy));
   disp(sprintf(' Energy In the complete problem : %e', energy*2));
   disp(sprintf(' Capacity of the system : %e', capacity));
%  ---------- Draws equipotential lines
   Xg = min(xy(1,:)):(max(xy(1,:))-min(xy(1,:)))/10:max(xy(1,:));
   Yg = min(xy(2,:)):(max(xy(2,:))-min(xy(2,:)))/10:max(xy(2,:));
   [XI,YI]=meshgrid(Xg,Yg);
   ZI = griddata(xy(1,:),xy(2,:),X,XI,YI);
   figure;
   contour(XI*1000,YI*1000,ZI);
   title(sprintf('Equipotential lines for %s, %s',meshfile,elefile));
```

Example 1.9

```
c0 = laplace('First/raw.fem','First/rawA.aux');
cd = laplace('First/raw.fem','First/rawD.aux');
disp(' [*] - LINE CHARACTERISTICS:');
disp(sprintf(' Z: %e', 1/(299800000*sqrt(c0*cd))));
disp(sprintf(' beta : %e * omega/c', sqrt(cd/c0)));
```

Example 1.10

```
c0 = laplace('Strip/Strip.fem','Strip/StripA.aux');
cd = laplace('Strip/Strip.fem','Strip/StripD.aux');
disp(' [*] - LINE CHARACTERISTICS:');
    disp(sprintf(' Z: %e', 1/(299800000*sqrt(c0*cd))));
    disp(sprintf(' beta : %e * omega/c', sqrt(cd/c0)));
```

1.6 Variational or Projective?

It is worth noticing that FEM is a variational technique, and this is the approach followed in this chapter and the approach with a more evident physical meaning.

Nevertheless one can arrive at the same FEM formulation by applying a projective approach rather than a variational one. The projective approach is somewhat more general and, for sure, more straightforward to apply to problems where a definition of energy can be less natural than in the one presented here.

The projective approach consists in enforcing the Laplace equation in a weak form. The "strong" form of an equation is that in which the equation itself, in our case $\nabla^2 u = 0$, holds exactly in every point of the whole domain. On the other hand a "weak" form of an equation states that the equation is rather satisfied *on the average* by enforcing that its *weighted integral* is zero. Since this averaging operation is indeed a scalar product in the vector space of the functions, we can also state that the equation is projected onto a subspace of the whole function space, the one generated by the weight functions w used, and that it is the projection of the equation to be zero:

$$\iint_{\Omega} w \nabla^2 u d\Omega = 0 \tag{1.42}$$

If we write (1.42) on a single element we obtain, by applying the two-dimensional Green theorem:

$$\iint_{\Delta^{(e)}} w \nabla^2 u d\Omega = \iint_{\Delta^{(e)}} \nabla w \cdot \nabla u d\Omega + \oint_{\Gamma^{(e)}} w \frac{\partial u}{\partial n} d\Gamma = 0 \tag{1.43}$$

where $\Gamma^{(e)}$ is the boundary of element $\Delta^{(e)}$. By resorting to the expansion on basis functions for the field $u = \sum_{i=1}^{3} u_i \alpha_i$ one obtains:

$$\sum_{i=1}^{3} u_i \left[\iint_{\Delta^{(e)}} \nabla w \cdot \nabla \alpha_i d\Omega + u_i \oint_{\Gamma^{(e)}} w \frac{\partial \alpha_i}{\partial n} d\Gamma \right] = 0 \tag{1.44}$$

If three weighting functions $w_j = \alpha_j$ are chosen then the first term in (1.44) is basically matrix $[S^{(e)}]$, whereas the second term is something new. It is straightforward how, in the assembling process of the contribution of all elements of the first term in (1.44), if only the nodes which are free to vary are considered, leads exactly to the linear system (1.38). For what concerns the second term, its assembly leads to the elimination of the contribution of each *internal* edge of the mesh, since the same edge is considered once for each element sharing it and, in the two cases, the normal vector $\hat{\mathbf{n}}$ assumes two opposite directions.

The contribution of the contour integral is then relevant only on noninternal edges, that is edges on the boundary Γ of the whole domain Ω. The assembled integral is then

$$\oint_{\Gamma} \alpha_j \frac{\partial u}{\partial n} d\Gamma = 0 \tag{1.45}$$

Now Γ is either a Neumann boundary condition, in which case $\frac{\partial u}{\partial n} = 0$, or it is a Dirichlet boundary condition $u = c$. In the first case the integral is of course zero, in the second case the integral is, in general, not zero. Since, on the other hand, the value of the solution is known on the boundary, we are authorized to neglect weighting on the boundary and hence to select weighting functions that are zero on it.

This comes directly from the consideration that bases on the boundary are not relevant to true degrees of freedom of the problem their values being fixed. Hence there is no need to select the corresponding weights.

It can be demonstrated that this weighted residual solution of the problem is equivalent to the variational solution presented in this chapter in a distribution sense, that is $\iint \nabla w \cdot \nabla u$ is *not* zero punctually, but is equal to the null distribution, that is, it is zero almost everywhere on Ω [3].

References

[1] Silvester, P. P. and Ferrari, R. L., *Finite Elements for Electrical Engineers*, Cambridge: Cambridge University Press, 3rd edition, 1996.

[2] Valor, L. and Zapata, J., "Efficient Finite Element Analysis of Waveguides with Lossy Inhomogeneous Anisotropic Materials Characterized by Arbitrary Permittivity and Permeability Tensors," *IEEE Transactions on Microwave Theory and Techniques*, Vol. 43, pp. 2452–2459, 1995.

[3] Quarteroni, A. and Valli, A., *Numerical Approximation of Partial Differential Equations*, New York: Springer-Verlag, 1994.

CHAPTER 2

Tools

The solution of an electromagnetic problem by means of the finite element method (FEM) requires the realization of the five steps described in the previous chapter. Each of these steps can be accomplished using standard routines which constitute the tools required for a straightforward implementation of an FEM code. The purpose of this chapter is to present all the tools necessary for a quick implementation of a general finite element code: some routines can be found on the net, most notably the linear algebra packages, while others, especially those accomplishing the preprocessing and postprocessing steps for two-dimensional geometries, are described here and can be found in the enclosed CD-ROM.

Two sets of tools are provided, a FORTRAN-based set and a Matlab-based set. These sets perform the same operations and are completely interoperable, meaning that you can use Matlab codes for preprocessing, then the FORTRAN code for the EM-analysis, and then the Matlab code again for postprocessing.

As a general rule FORTRAN codes are faster but more difficult to modify and are unable to handle graphical operations, whereas Matlab allows for easy graphical interfaces. Indeed, also for Matlab, the graphical part and the computational part has been kept completely separate, so as to allow a more flexible utilization and avoid possible compatibility issues with different Matlab versions.

2.1 Preprocessing

One of the main features of FEM is that it is applicable to problems with arbitrary domain geometries. Since these domains have to be subdivided into elements, this flexibility is paid in terms of complexity of this first phase. If the problem is not one of the simplest, it is impossible to subdivide the domain manually, as it has been done in Chapter 1, and, even then, the process is at least tedious.

The task of the generation of a suitable FEM mesh, in the following simply referred to as mesh-generation, has been deeply investigated and many—expensive—commercial packages (PATRAN, FEMAP, IDEAS, SIMAIL, etc.) exist to accomplish it as well as some open source or otherwise freeware packages. Among these we might cite `triangle` which is free for personal use but not open source and very similar to the three-dimensional mesh generator described in Chapter 7. Of course commercial FEM codes, one for all Ansoft's High Frequency Structure Simulator (HFSS) embed their own mesh generator.

A very simple and straightforward mesh generator, good enough to solve all the problems discussed in the book and many others, is presented here. This code is basically an improved version of that presented in [1, 2] and can handle:

- First- and second-order triangular nodal elements;
- First- and second-order quadrilateral isoparametric nodal elements;
- First-order triangular and quadrilateral edge elements;
- Meshes containing different kinds of elements;
- Labels assignment for elements, nodes, and edges.

The mesh generation process requires defining the geometrical domain through a set of nonoverlapping quadrilateral *blocks*. Each block is defined by eight *points*, counterclockwise numbered from 1 to 8, and has borders which can be described by second-order curves (Figure 2.1). The blocks are subsequently subdivided by the code into triangular or quadrilateral-type elements, characterized by a set of *nodes* and *edges*, whose number depends on the desired order. The type and order of the elements must be the same within a single block, but may vary between different blocks of the geometrical domain. To take into account the presence of different materials, labels are assigned to elements and nodes. All the elements belonging to the same block share the same label, on the other hand, node labels are assigned only to nodes belonging to the block's boundary. Different node label values can be associated to nodes on different sides of the boundary of the same block. The definition of each block, desired element type and order, and directive for element and node labels assignment are provided to the mesh generator via an input geometry description file.

The output of the mesh generator is a text file containing the number of elements, nodes and, possibly, edges created, the coordinates and labels of the nodes, and the description of each element in terms of its nodes, its edges, and its label.

2.1.1 Input Geometry Description File

The input geometry description file must describe the problem domain once it has been subdivided into quadrilateral blocks. Its syntax is described in Table 2.1, where:

- `Filename` is the name of the output mesh description file to be created;
- `NP` is the total number of points which define the set of blocks;

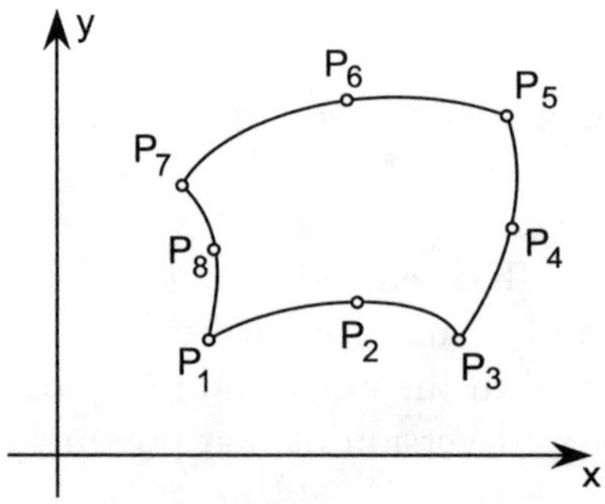

Figure 2.1 Generic quadrilateral block.

Table 2.1 Input Geometry Description File Syntax

Input Data	*Data Type*
`filename`	`Character*256`
`NP NB`	`2 Integers`
⋮	*NB* lines of block definitions, each containing:
`IB ET P1 P2 P3 P4 P5 P6 P7 P8 LE`	`11 Integers`
⋮	
⋮	*NP* lines of point definitions, each containing:
`IP X Y LN`	`1 Integer, 2 Reals, 1 Integer`
⋮	
⋮	*NB* groups of 3 lines; discretization parameters:
`IB N1 N2`	`3 Integers`
`RL1(1) ... RL1(N1)`	`N1 Reals`
`RL2(1) ... RL2(N2)`	`N2 Reals`
⋮	

- `NB` is the number of blocks describing the geometric domain;
- `IB` is the number of the block being defined through:
  - `ET`: element type, (3 = triangle, first order; 4 = quadrilateral, first order; 6 = triangular, second order; 8 = quadrilateral second order);
  - `Pi`: the numbers of the eight points defining the block;
  - `LE`: the label to be assigned to every element belonging to this block.
- `IP` is the number of the point being defined through:
  - `X, Y`: Its Cartesian coordinates;
  - `LN`: the label to be assigned to the nodes belonging to the block boundary section defined by the point. If the point is in the middle of a block side (points 2, 4, 6, 8 in Figure 2.1) the label will be assigned to all the nodes belonging to that side, vertices excluded. If, on the other hand, the point is on a vertex of the block (points 1, 3, 5, 7 in Figure 2.1) the label will be assigned only to the node which will coincide with that point.
- `IB` is the number of the block whose discretization parameters are defined through:
  - `N1`: the number of elements to be created along the block boundary sides 1 and 3 defined by points `P1`, `P2`, `P3` and `P7`, `P6`, `P5`;
  - `N2`: the number of elements to be created along the block boundary sides 2 and 4 defined by points `P3`, `P4`, `P5` and `P1`, `P8`, `P7`;
  - `RL1(i)`: `N1` Reals providing the relative lengths of the element sides along block boundary sides 1 and 3 (left to right);
  - `RL2(i)`: `N2` Reals providing the relative lengths of the element sides along block boundary sides 2 and 4 (bottom to top).

To illustrate further the syntax of the input geometry description file, let's employ the mesh generator to discretize the geometric model of the realistic shielded microstrip line shown in Figure 2.2. The mesh created could be used as input

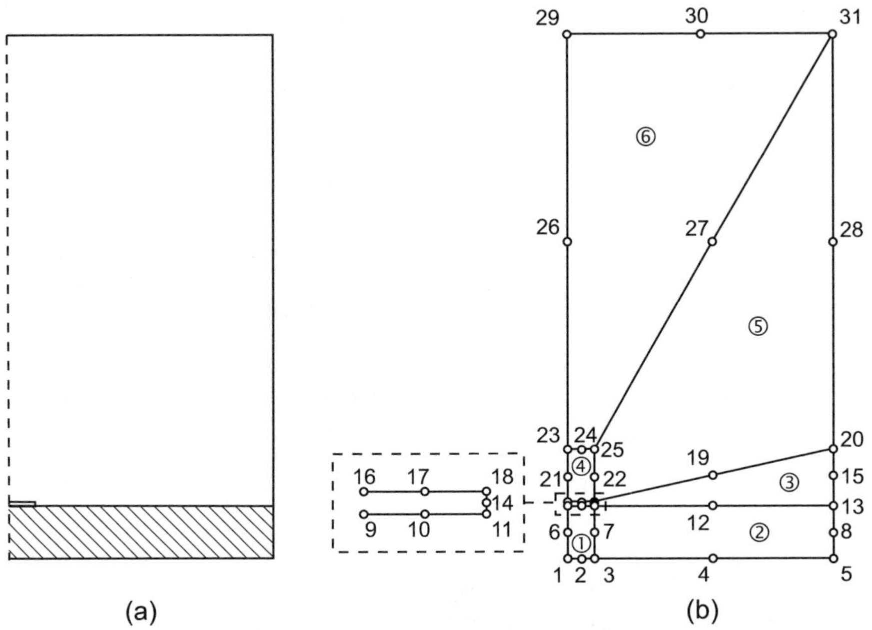

Figure 2.2 Example of mesh generation: (a) geometry of the problem, (b) subdivision in quadrilateral blocks; point numbers and block numbers (circled) are given.

mesh file for the LAPLACE code described in Chapter 1. If the block subdivision in Figure 2.2(b) is chosen, the corresponding input geometry description file for the mesh generator is:

Example 2.1

```
microstr.fem
31 6
1  3   1   2   3   7   11   10    9   6   2
2  3   3   4   5   8   13   12   11   7   2
3  3  11  12  13  15   20   19   18  14   1
4  3  16  17  18  22   26   25   24  21   1
5  3  18  19  20  23   31   28   26  22   1
6  3  24  25  26  28   31   30   29  27   1
1 0.00000 0.00000 1
2 0.31750 0.00000 1
3 0.63500 0.00000 1
4 3.49250 0.00000 1
5 6.35000 0.00000 1
6 0.00000 0.63500 0
7 0.63500 0.63500 0
8 6.35000 0.63500 1
9 0.00000 1.27000 2
10 0.31750 1.27000 2
11 0.63500 1.27000 2
12 3.49250 1.27000 0
13 6.35000 1.27000 1
14 0.63500 1.96850 2
```

```
15 6.35000  1.90500 1
16 0.00000  1.39700 2
17 0.31750  1.39700 2
18 0.63500  1.39700 2
19 3.49250  2.03200 0
20 6.35000  2.66700 1
21 0.00000  2.03200 0
22 0.63500  2.03200 0
23 6.35000  7.68350 1
24 0.00000  2.66700 0
25 0.31750  2.66700 0
26 0.63500  2.66700 0
27 0.00000  7.68350 0
28 3.49250  7.68350 0
29 0.00000 12.7000 1
30 3.17500 12.7000 1
31 6.35000 12.7000 1
1 4 4
1. 1. 1. 1.
1. 1. 1. 1.
2 8 4
.25 .5 .5 1. 1. 1. 1. 1 .
1. 1. 1. 1.
3 8 1
.25 .5 .5 1. 1. 1. 1. 1 .
1.
4 4 4
1. 1. 1. 1.
.25 .5 .5 .5
5 8 4
.25 .5 .5 1. 1. 1. 1. 1 .
.25 .5 .5 .5
6 4 8
1. 1. 1. 1.
.25 .5 .5 1. 1. 1. 1. 1.
```

As it can be easily seen, in Example 2.1, to maintain consistency with the input file for the LAPLACE code used in Chapter 1, the elements belonging to the dielectric slab are assigned a label equal to 2, while the remaining have a label equal to 1. The nodes on the p.e.c. shield walls are assigned a label equal to 1, the nodes on the metallic strip are assigned a label 2, and the nodes on the p.m.c. wall are assigned a label equal to 0, that is, no label.

Some guidelines are useful when choosing suitable block discretization parameters: fine meshing is needed where fields are expected to have rapid spatial variations. In our example, the field is expected to vary more rapidly near the strip (blocks No. 1, 4) and inside the dielectric slab (block No. 2) than near the outside wall (blocks 3, 5 and 6). This consideration is reflected in the choice made for the discretization parameters, both in the number of elements per block and in the relative side length of border elements.

Once the mesh description file is ready the mesh is generated by either the FORTRAN or the Matlab code. The FORTRAN code, `MESH` just asks the name of the input geometry description file and outputs a mesh file, according to the format described in the next section. The Matlab mesh generator `Mesh.m` is a function asking several parameters:

```
Mesh(filename,...
    showGeo,showBand,showBand_ns,showMesh,...
    ElementShow,NodeShow,EdgeShow,...
    dodelaunay,dorenum,period)
```

Besides `filename`, which obviously is a string containing the name of the geometry description file, the other parameters are flags handling options allowing visualization and mesh optimization as described in the following sections. Flag descriptions are contained in the code and can be visualized with

```
help Mesh
```

Code `MeshGUI` opens a Matlab dialog where all these options are presented graphically and can be set interactively in a more intuitive way. Please note that Matlab *is* case sensitive and has a built-in `mesh` function handling two-dimensional graphs.

2.1.2 Output Mesh Description File

The output of the mesh generator is a text file describing the mesh. For accessing this file two subroutine are presented, one for reading and one for writing. It is important to point out that the input geometry description file is indeed a handmade mesh of isoparametric second-order quadrilateral elements, exhibiting some additional information on discretization and lacking edge information, hence within the FORTRAN codes both the geometry and the mesh file data are represented as two similar sets of scalar values and a set of arrays, whereas in all Matlab codes, this same data is grouped in two appropriate structures, `Geo` and `Mesh`. The structure representation in Matlab allows for much simpler data handling and processing.

The output file, even for simple meshes, can be quite big. Its form is given in Table 2.2 where:

- `Header` is a line describing who created the file;
- `NELE` is the number of elements contained in the mesh;
- `NNODE` is the number of nodes contained in the mesh;
- `NSPIG` is the number of edges contained in the mesh;
- `Elements` is a string marking the beginning of the element definition block. Each of the subsequent `NELE` lines defines one element in terms of:
    - `IE`: element number;
    - `NN`: the number of nodes defining the element;
    - `N(i)`: the node numbers, in the global numbering scheme, of the nodes defining the element. These numbers are given in the order defined by the local numbering scheme;
    - `LE`: the element label.
- `Nodes` is a string marking the beginning of the node definition block. Each of the subsequent `NNODE` lines defines one node in terms of:

Table 2.2 Output Mesh Description File Syntax

Input Data	*Data Type*
`Header`	`Character*80`
`Elements = NELE`	`Integer`
`Nodes = NNODE`	`Integer`
`Edges = NSPIG`	`Integer`
`Elements`	a one-line tag of the element definition block
⋮	*NELE* lines of element definition, each containing:
`IE NN N(1) ... N(NN) LE`	`NN + 2 Integers`
⋮	
`Nodes`	a one-line tag of the node definition block
⋮	*NODE* lines of node definition, each containing:
`IN X Y LN`	`1 Integer, 2 Reals, 1 Integer`
⋮	
`Edges`	a one-line tag of the edge definition block
⋮	*NELE* lines of edge definition,
`IE NS N(1) ... N(NS)`	`NS + 1 Integers`
⋮	
`Edge Labels`	a one-line tag of the edge label definition block
⋮	*NSPIG* lines of element label definition, each containing:
`IS LS`	`2 Integers`
⋮	

    - `IN`: node number;
    - `X Y`: the node Cartesian coordinates;
    - `LN`: the node label.
- `Edges` is a string marking the beginning of the edge definition block. This block is completely absent if `NSPIG` is zero. Each of the subsequent `NELE` lines defines one element in terms of:
    - `IE`: element number;
    - `NS`: the number of edges belonging to the element;
    - `N(i)`: the edge numbers, in the global numbering scheme. These numbers are given in the order defined by the local numbering scheme; edge numbers is negative if the direction of the edge in the local scheme is opposite to the direction chosen for that edge in the global scheme.
- `Edge Labels` is a string marking the beginning of the edge label definition block. This block is completely absent if `NSPIG` is zero. Each of the subsequent `NSPIG` lines defines edge labels:
    - `IS`: edge number;
    - `LS`: edge label.

It is often necessary to visualize the mesh described by this output file in order to check its suitability to the problem under examination. This is accomplished by the Matlab functions `DrawGeo.m` and `DrawMesh.m`, which can be used directly from Matlab command line or from within the global Mesh generation function.

For backward compatibility with the first edition of this book a FORTRAN mesh visualization tool, much less flexible than the Matlab function `DrawGeo.m` and `DrawMesh.m`, is provided. Since the software package included with this book was designed with flexibility and portability in mind, providing FORTRAN graphical routines would bind the user to a specific compiler and machine. To overcome this problem, a subroutine which converts the mesh contained in the output mesh description file in a PostScript file is provided. PostScript language is a well-known standard common to every platform and can be directly managed by PostScript printers. Moreover, the freeware package GhostView, available for most platform, provides a simple mean of visualization of the PostScript files and allows the user to print them even on non-PostScript printers. Furthermore, several tools converting PostScript to PDF or Windows metafile formats are available.

Two main problems arise in mesh generation: the elements must be regular, and the node and edge numbers must be smartly assigned. Element regularity implies the generations of triangular and quadrilateral elements which are as close as possible to equilateral triangle and to squares, respectively. For the mesh generator presented here, this depends primarily on how the domain is subdivided into blocks, and how block discretization parameters are set. As already seen in Chapter 1, the application of FEM leads to a sparse global matrix; this matrix can be reduced to a band matrix through appropriate node, or edge, numbering. Sparse and banded matrices allow substantial storage savings and computation time reduction, provided that appropriate solution algorithms are exploited. Indeed the FORTRAN codes will exploit band matrices, whereas Matlab codes will exploit sparse matrices.

Since the mesh generator presented is very simple, it does not provide the best mesh neither for regularity nor for numbering. The next two sections describe two subroutines, included in the CD-ROM, that can be used to improve the quality of the mesh created by the simple mesh generator.

2.1.3 Mesh Regularization

The accuracy of FEM results depends on both element dimension and shape. In regions where the field is expected to have rapid space variations elements must be smaller than where the field is expected to have smooth space variations. The element average dimensions needed to obtain the required accuracy are also a function of the element order. As it will be shown in the examples given throughout this book, accuracy can be increased in two ways: increasing element order (*P-refinement*), or increasing the number of elements used to discretize the computational domain (*H-refinement*). It will soon become evident that P-refinement is more convenient than H-refinement.

A good quality mesh avoids elements with very small internal angles. While there is no special criteria to improve quality of quadrilateral meshes besides moving nodes, the Delaunay criterion can be applied to triangular meshes with a fixed node distribution. Its application leads to a mesh with triangles as close as possible to equilateral. Its essence is that, take a given element and build the circumscribing circle, there must be no other nodes inside the circumference beside those of the element itself. If this is not the case [Figure 2.3(a)], the couple of elements that do

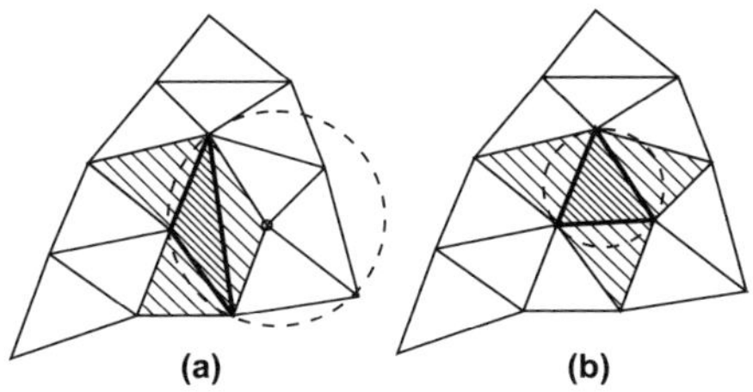

Figure 2.3 Delaunay criterium: (a) unsatisfied, (b) satisfied, after side-swapping.

not satisfy the criterion are transformed through a side-swapping [Figure 2.3(b)]. For completeness a Delaunay regularizator is presented. This subroutine can be included inside the mesh generator or used as a stand-alone program. The effects of the regularization on the mesh previously created (Figure 2.2) can be seen in Figure 2.4. For what concerns FORTRAN the `DELAUNAY` function is called by a stand-alone `DO_DEL` main. This just asks for the name of the mesh file to read and the name of the regularized mesh file to write. In Matlab the regularization is embedded in the mesh generator and can be activated via an input flag.

2.1.4 Numbering Optimization

The numbering scheme used in the mesh to identify either the node or the edges affects the structure of the coefficient matrix of the system of linear equations to be solved. Since many freely available mathematical routines exist for banded matrices in band storage mode, it is often useful to render this global sparse matrix banded. Band storage mode (Section 2.3.1) is very simple to handle and allows storing a $N \times N$ matrix in a $(kl + ku + 1) \times N$ array, where kl and ku are the number of

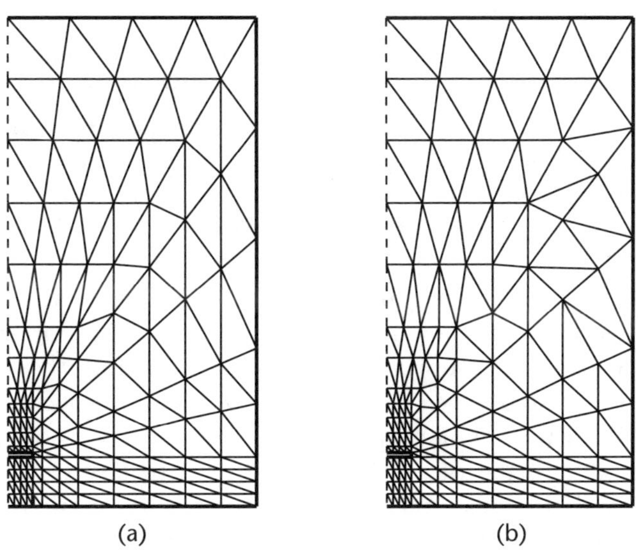

Figure 2.4 The mesh created with the mesh generator provided (a), and the same mesh after regularization (b).

lower and upper codiagonals, respectively. All FORTRAN codes provided in this book will rely on banded storage.

On the other hand, routines able to handle sparse matrices can deal with FEM problems very efficiently without renumbering. Since Matlab allows for a very easy and user-transparent handling of sparse matrices and solvers the Matlab codes will resort to sparse storage (Section 2.3.2). Sparse storage is generally more efficient for memory usage but less efficient for access time to matrix elements, so the decision between the two possible solutions is a matter of balancing the two requirements.

Very few mesh generators provide optimum node or edge numbering for bandwidth minimization, as is the case with the mesh generator presented here. But its performances may be improved by exploiting the simple subroutine RENUMBER which optimizes node and, where needed, edge numbering. This subroutine can be easily embedded in the mesh generator or can be made as a stand-alone program for use with mesh generated by other software.

To minimize the matrix band, the global numbering scheme for nodes (and edges) of a mesh must be chosen so as to minimize the difference between the number of each node (and edge) in the global scheme, within each element of the mesh. This requirement comes from the fact that the global FEM coefficient matrix entry (i,j) is nonzero only if node i and node j belong to the same element. Minimum band is achieved by renumbering the elements first. The code starts from the element which has the fewest number of neighboring elements, which becomes element number 1, and renumbers all the other elements starting from this and proceeding in concentric layers of neighboring elements. In this way the maximum difference between the number of adjacent elements is fairly well minimized.

A straight renumbering of the nodes and of the edges on the basis of this new element numbering scheme provides fairly narrow bandwidth.

Note that when FEM is to be applied to periodic structures, as those in Chapter 5, the concept of adjacent elements becomes slightly different. The subroutine RENUMBER provided thus distinguishes the periodic case from the nonperiodic case to provide the best numbering scheme.

For the mesh analyzed throughout this section [Figure 2.4(b)], the nonzero entries in the global matrix are shown in Figure 2.5. The number of unknowns is $n = 162$; this matrix, in full matrix real, single precision, storage mode, would occupy 102.5 kB. The nonoptimized sparse matrix can be stored in band storage mode, although this is not overly efficient, since $ku = kl = 37$, but memory occupation would drop anyway to 52.5 kB. After having optimized the node numbering the bandwidth becomes $ku = kl = 19$, and memory occupation is further reduced to 31 kB. An even larger gain occurs if edge elements are considered. In this case the bandwidth drops from $ku = kl = 107$ for the nonoptimized matrix to $kd = 28$.

The FORTRAN subroutine RENUMBER, and the corresponding Matlab `Renumber.m` works well and can easily take into account periodic structures, but it might lead to suboptimal results. Matlab has its own renumbering scheme to minimize the band of a matrix based on a reverse Cuthill-McKee ordering for the rows and columns of a sparse matrix. On the basis of the permutation vector provided by Matlab, the element, edges, and nodes can be easily renumbered. Since this routine is highly optimized it usually perform faster than the simple code provided with the book.

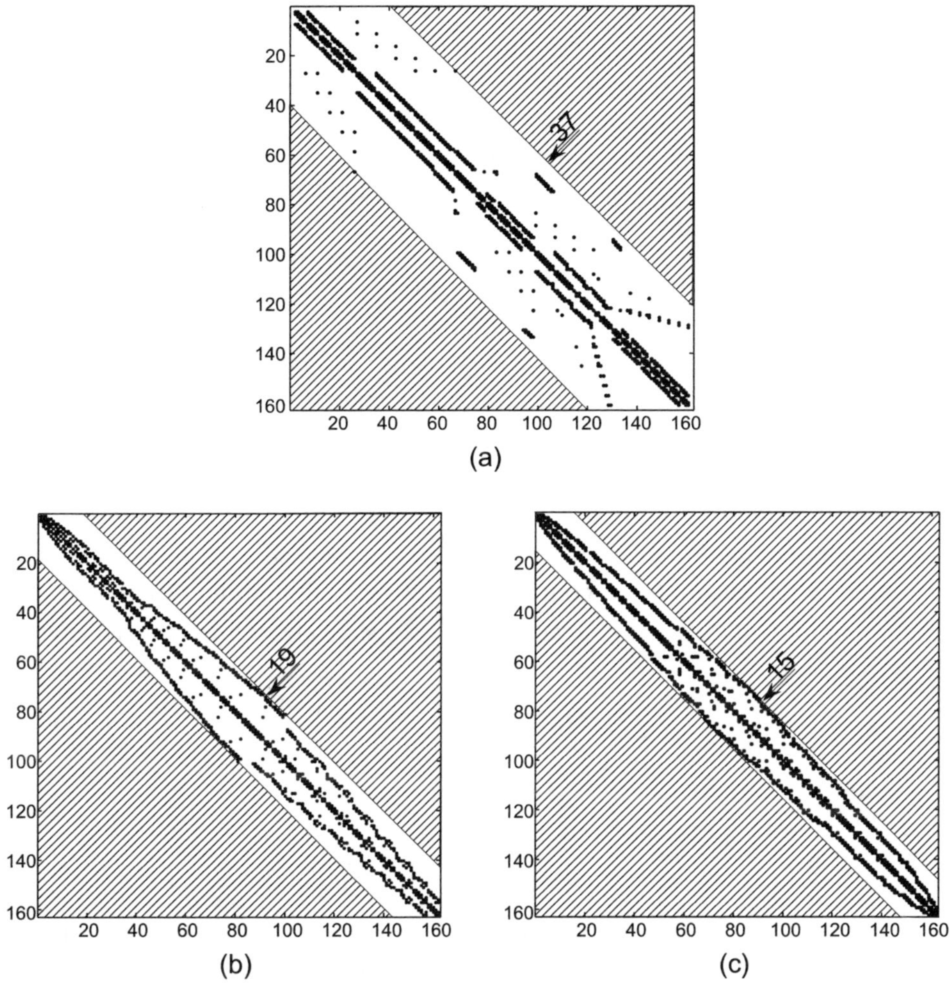

Figure 2.5 Global matrices for nonoptimized (a) numbering scheme. Optimized numbering schemes are (b) with the algorithm described in the book, (c) with internal Matlab renumbering scheme.

A Matlab routine, being unmodifiable, does not take into account periodic conditions, so, even if it leads to usually narrower bands it cannot be used in all problems of this book. The BANDEX function allows for a band occupation estimate. Try it on a mesh before and after renumbering optimization.

2.2 Element Matrices

After the preprocessing phase, the solution procedure requires the global coefficient matrix to be constructed, which, as stated in Chapter 1, is made up of contributions coming from each single element in the form of one ore more local matrices. The first step of the solution process though is to prepare the local matrices needed for the elements we intend to use.

There are two classes of elements: nodal elements and vector elements. The first class makes use of scalar basis function to approximate the unknown on each finite element and it is thus suitable to approximate scalar fields such as the potential in the example given in Chapter 1. It is also used to model single components of vectorial fields such as the longitudinal component (H_z or E_z) in TE_z or TM_z problems. Nodal elements have been used for a long time to fully simulate a vector field by separately approximating each field component. This approach presents numerical problems that can be overcome in many ways. The most popular and simplist is probably resorting to vectorial elements. This second class of elements makes use of vectorial basis functions which are more appropriate than scalar basis functions to approximate a vector field and allow enforcing boundary conditions in a straightforward manner.

Albeit in the very recent years many different kind of elements have been proposed, specifically designed for electromagnetic wave problems [3–5] the few presented in the following are the simplest to understand, implement, and use.

2.2.1 Nodal Elements

In electromagnetic analysis at microwave frequencies, the functionals or equations more often encountered contains the terms $\nabla^2\Phi$ and Φ, which leads, on element (e) to local matrices $\left[S^{(e)}\right]$ and $\left[T^{(e)}\right]$, respectively, whose entries have the form:

$$S_{ij} = \int_{\Omega_e} \nabla\alpha_i \cdot \nabla\alpha_j d\Omega \tag{2.1}$$

and

$$T_{ij} = \int_{\Omega_e} \alpha_i \alpha_j d\Omega \tag{2.2}$$

where Ω_e is the surface of the element and α_i is the ith scalar shape function of the element. The formulas to compute these entries are given in the following paragraphs.

First-order triangular elements: Given a first order triangular element of vertices $P_i = (x_i, y_i)$ [Figure 2.6(a)] the basis functions are simply the simplex coordinates (1.19):

$$\alpha_i = \zeta_i \tag{2.3}$$

The integrals (2.1), (2.2) can be analytically evaluated, leading to the simple expressions:

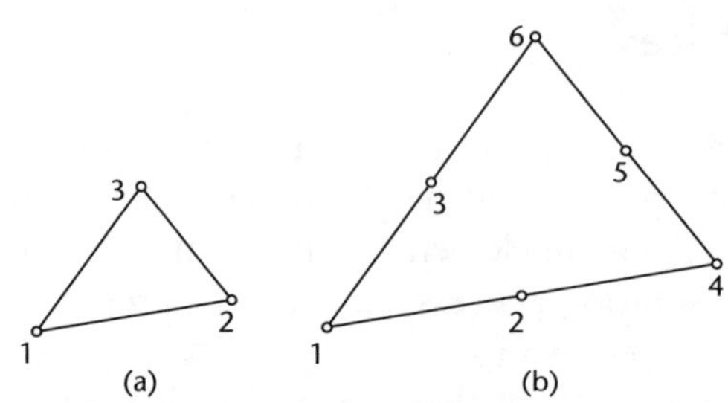

Figure 2.6 Numbering schemes for local nodes in order 1 and 2 triangular elements.

$$[S^{(e)}] = \sum_{i=1}^{3} \cot\theta_i [Q]_i \tag{2.4}$$

$$[T^{(e)}] = \frac{1}{12}\begin{bmatrix} 2 & 1 & 1 \\ 1 & 2 & 1 \\ 1 & 1 & 2 \end{bmatrix} \tag{2.5}$$

where θ_i is the angle at vertex i (Figure 2.7) and its cotangent can be easily computed from the knowledge of the triangle area $A = \frac{(x_2-x_1)(y_3-y_1)-(y_2-y_1)(x_3-x_1)}{2}$ in the following way:

$$\begin{aligned} \cot\theta_1 &= -\frac{(y_3-y_1)(y_1-y_2)+(x_1-x_3)(x_2-x_1)}{2A} \\ \cot\theta_2 &= -\frac{(y_1-y_2)(y_2-y_3)+(x_2-x_1)(x_3-x_2)}{2A} \\ \cot\theta_3 &= -\frac{(y_2-y_3)(y_3-y_1)+(x_3-x_2)(x_1-x_3)}{2A} \end{aligned} \tag{2.6}$$

The constant matrices $[Q]_i$ are given by:

$$\begin{aligned} [Q]_1 &= \frac{1}{2}\begin{bmatrix} 0 & 0 & 0 \\ 0 & 1 & -1 \\ 0 & -1 & 1 \end{bmatrix}, \quad [Q]_2 = \frac{1}{2}\begin{bmatrix} 1 & 0 & -1 \\ 0 & 0 & 0 \\ -1 & 0 & 1 \end{bmatrix} \\ [Q]_3 &= \frac{1}{2}\begin{bmatrix} 1 & -1 & 0 \\ -1 & 1 & 0 \\ 0 & 0 & 0 \end{bmatrix} \end{aligned} \tag{2.7}$$

Second-order triangular elements: Solution accuracy can be increased by introducing more nodes on each element and using polynomial shape functions of higher order (p refinement). For an n-th order element, $(n+1)(n+2)/2$ nodes and shapes functions are required to have a complete family of polynomials, since $\mathcal{P}^n$, the linear space of polynomials of order n in two variables, has dimension $(n+1)(n+2)/2$. This can be easily seen by considering the space of zero-order polynomials, $\mathcal{P}^0$ whose dimension is 1 and whose basis is $\beta = \{1\}$. To build the basis for $\mathcal{P}^n$, once the basis for $\mathcal{P}^{n-1}$ is known, it is necessary

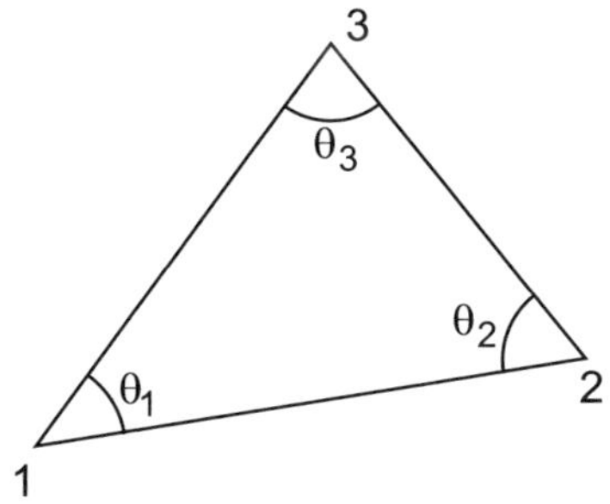

Figure 2.7 Geometry of a triangular element.

to extend this latter with all the monomials of order n. In our case of two variables polynomials, these monomials are of the form $x^{n-m}y^m, m = 0, \ldots, n$. Since $n+1$ elements are added to form the new basis, this implies that $\dim(\mathcal{P}^n) = \dim(\mathcal{P}^{n-1}) + (n+1)$. From the induction process here outlined is thus clear that $\dim(\mathcal{P}^n) = \sum_{i=1}^{n+1} i = (n+1)(n+2)/2$.

Then, to have a complete basis for the set of polynomials of order n on a generic element, $(n+1)(n+2)/2$ linearly independent shape functions must be introduced. To define this basis polynomials the local simplex coordinates ζ_i are preferred, since they may be easily integrated on a triangle regardless of actual vertices coordinates.

The first step is to define the following family R of polynomials in ζ:

$$\begin{cases} R_m(n,\zeta) = \frac{1}{m!}\prod_{k=0}^{(m-1)}(n\zeta - k) & m > 0 \\ R_0(n,\zeta) = 1 \end{cases} \tag{2.8}$$

If the element nodes are locally numbered as shown in Figure 2.6 for first- and second-order elements and in Figure 2.8 for third- and fourth-order elements, the basis function α_i for a triangular element of order n can then be defined as:

$$\alpha_i(\zeta_1, \zeta_2, \zeta_3) = R_{m_1}(n, \zeta_1) R_{m_2}(n, \zeta_2) R_{m_3}(n, \zeta_3) \tag{2.9}$$

In the above equation, $m_1 + m_2 + m_3 = n$ and, if the quantity $S_m = \sum_{i=0}^{m} i$ is defined, the three coefficients m_1, m_2, and m_3 are obtained from the node number i applying:

$$\begin{cases} m_1 \text{ integer which satisfies} & S_{n-m_1} + 1 \leq i \leq S_{n-m_1+1} \\ m_2 \text{ is calculated with} & m_2 = S_{n-m_1} - i + 1 \\ m_3 \text{ is calculated with} & m_3 = n - m_2 - m_1 \end{cases} \tag{2.10}$$

The polynomial-shape function α_i so constructed assumes value 1 in node i and zero in every other node. The first-order shape function introduced in the previous

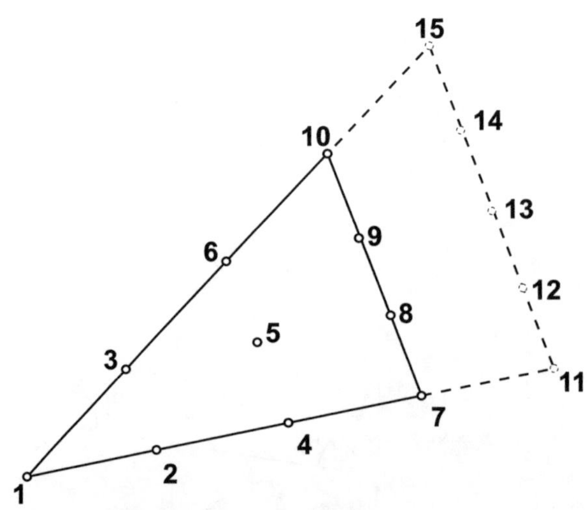

Figure 2.8 Local nodes numbering for high-order triangular elements.

section can be easily obtained using the procedure here outlined. For second-order elements, the shape functions (two of which are drawn in Figure 2.9) have the expressions:

$$\begin{aligned}
\alpha_1 &= (2\zeta_1 - 1)\zeta_1\\
\alpha_2 &= 4\zeta_1\zeta_2\\
\alpha_3 &= 4\zeta_3\zeta_1\\
\alpha_4 &= (2\zeta_2 - 1)\zeta_2\\
\alpha_5 &= 4\zeta_3\zeta_3\\
\alpha_6 &= (2\zeta_2 - 1)\zeta_3
\end{aligned} \tag{2.11}$$

Matrix $[S^{(e)}]$ is still found through application of (2.4), but the constant matrices $[Q]_i$ and $[T^{(e)}]$ become now:

$$[Q]_1 = \frac{1}{6}\begin{bmatrix} 0 & 0 & 0 & 0 & 0 & 0\\ 0 & 8 & -8 & 0 & 0 & 0\\ 0 & -8 & 8 & 0 & 0 & 0\\ 0 & 0 & 0 & 3 & -4 & 1\\ 0 & 0 & 0 & -4 & 8 & -4\\ 0 & 0 & 0 & 1 & -4 & 3 \end{bmatrix} \quad [Q]_2 = \frac{1}{6}\begin{bmatrix} 3 & 0 & -4 & 0 & 0 & 1\\ 0 & 8 & 0 & 0 & -8 & 0\\ -4 & 0 & 8 & 0 & 0 & -4\\ 0 & 0 & 0 & 0 & 0 & 0\\ 0 & -8 & 0 & 0 & 8 & 0\\ 1 & 0 & -4 & 0 & 0 & 3 \end{bmatrix}$$

$$[Q]_3 = \frac{1}{6}\begin{bmatrix} 3 & -4 & 0 & 1 & 0 & 0\\ -4 & 8 & 0 & -4 & 0 & 0\\ 0 & 0 & 8 & 0 & -8 & 0\\ 1 & -4 & 0 & 3 & 0 & 0\\ 0 & 0 & -8 & 0 & 8 & 0\\ 0 & 0 & 0 & 0 & 0 & 0 \end{bmatrix} \tag{2.12}$$

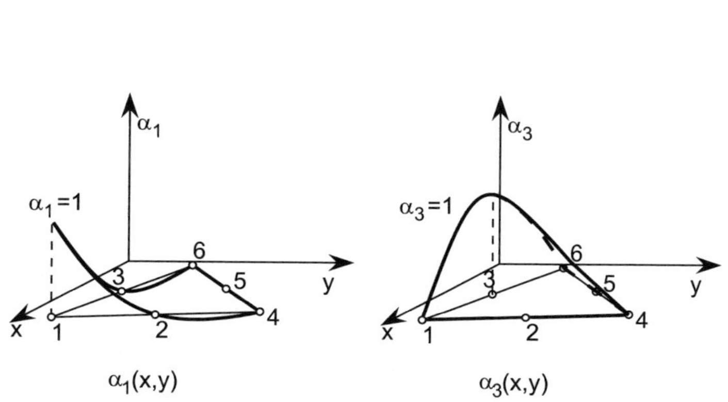

Figure 2.9 Basis function α_1 and α_3 for a second-order triangular elements. The other four basic function are basically a rotation of these. α_4 and α_6 are similar to α_1 and α_2 and α_5 are similar to α_3.

$$[T^{(e)}] = \frac{1}{180}\begin{bmatrix} 6 & 0 & 0 & -1 & -4 & -1 \\ 0 & 32 & 16 & 0 & 16 & -4 \\ 0 & 16 & 32 & -4 & 16 & 0 \\ -1 & 0 & -4 & 6 & 0 & -1 \\ -4 & 16 & 16 & 0 & 32 & 0 \\ -1 & -4 & 0 & -1 & 0 & 6 \end{bmatrix} \tag{2.13}$$

First-order quadrilateral elements: A generic quadrilateral element in the plane x, y (Figure 2.10), can be obtained from the square element with vertices $(-1,-1)$, $(1,-1)$, $(1,1)$, $(-1,1)$ in the plane u, v by means of a coordinates transformation (Figure 2.11). For this reason, let's consider first this special square element, on which the following shape functions can be defined

$$\begin{aligned} \alpha_1 &= \frac{1}{4}(u-1)(v-1) \\ \alpha_2 &= -\frac{1}{4}(u+1)(v-1) \\ \alpha_3 &= \frac{1}{4}(u+1)(v+1) \\ \alpha_4 &= -\frac{1}{4}(u-1)(v+1) \end{aligned} \tag{2.14}$$

with α_i assuming unitary value at node i and zero at all the other nodes. Even though the above shape functions contain second-order monomials in u, v, it is worth noting that they do not constitute a complete basis of $\mathcal{P}^2$ since they do not contain all six possible monomial terms up to order 2, but indeed only the terms $\{1, u, v, uv\}$. Therefore, the accuracy they provide is comparable to that provided by first-order triangular elements.

However, the presence of the term uv makes this element actually able to approximate the field with a second-order polynomial in the directions $v = \pm u$, while only a linear approximation is achieved along the directions $v = const$ and $u = const$. This implies that the accuracy provided by this element depends also on the orientation it has after its mapping in the xy plane, while elements with shape functions forming a complete set for polynomials of order n exhibit the same accuracy in every direction, thus presenting a sort of geometric isotropy.

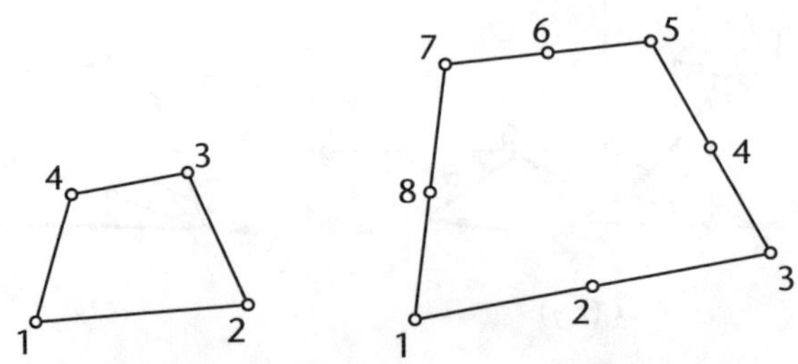

Figure 2.10 Local node numbering for order 1 and 2 quadrilateral elements.

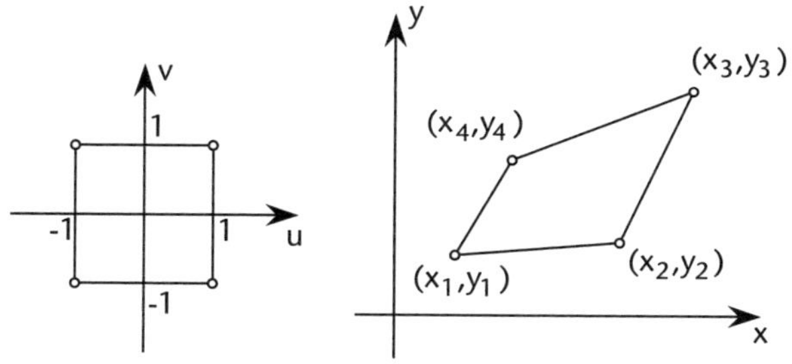

Figure 2.11 Quadrilateral mapping.

The square element is used to build more general quadrilateral elements in the xy plane by resorting to a simple bilinear coordinate transformation. To obtain a general quadrilateral element of vertices (x_1,y_1), (x_2,y_2), (x_3,y_3), (x_4,y_4) (Figure 2.11), the coordinate transformation required is:

$$\begin{aligned} x = &\frac{1}{4}(x_1 - x_2 + x_3 - x_4)uv + \frac{1}{4}(-x_1 + x_2 + x_3 - x_4)u+ \\ &+ \frac{1}{4}(-x_1 - x_2 + x_3 + x_4)v + \frac{1}{4}(x_1 + x_2 + x_3 + x_4) \end{aligned} \tag{2.15}$$

$$\begin{aligned} y = &\frac{1}{4}(y_1 - y_2 + y_3 - y_4)uv + \frac{1}{4}(-y_1 + y_2 + y_3 - y_4)u+ \\ &+ \frac{1}{4}(-y_1 - y_2 + y_3 + y_4)v + \frac{1}{4}(y_1 + y_2 + y_3 + y_4) \end{aligned} \tag{2.16}$$

The entries of the element matrices $[S^{(e)}]$ and $[T^{(e)}]$ for the quadrilateral first-order element in the xy plane cannot be easily derived analytically, and the resulting expressions are quite complex. It is thus convenient resorting to numerical integration. The $[T^{(e)}]$ matrix is the easiest to compute now, and it yields:

$$T_{ij} = \int\int_{\Omega_e} \bar{\alpha}_i \bar{\alpha} j \; dxdy = \int_{-1}^{1}\int_{-1}^{1} \alpha_i \alpha_j |\mathbf{J}| dudv \tag{2.17}$$

where $\bar{\alpha}_i$ is the generic shape function of the quadrilateral element in the xy plane, while α_i is the one for the corresponding square element (2.14) and $|\mathbf{J}|$ is the determinant of the Jacobian $\mathbf{J}$ of the transformation:

$$\mathbf{J} = \begin{bmatrix} \dfrac{\partial x}{\partial u} & \dfrac{\partial y}{\partial u} \\ \dfrac{\partial x}{\partial v} & \dfrac{\partial y}{\partial v} \end{bmatrix} \tag{2.18}$$

And, for transformation (2.17):

$$\mathbf{J} = \frac{1}{4}\begin{bmatrix} \begin{matrix}(x_1 - x_2 + x_3 - x_4)v+ \\ +(-x_1 + x_2 + x_3 - x_4)\end{matrix} & \begin{matrix}(y_1 - y_2 + y_3 - y_4)v+ \\ +(-y_1 + y_2 + y_3 - y_4)\end{matrix} \\ \begin{matrix}(x_1 - x_2 + x_3 - x_4)u+ \\ +(-x_1 - x_2 + x_3 + x_4)\end{matrix} & \begin{matrix}(y_1 - y_2 + y_3 - y_4)u+ \\ +(-y_1 - y_2 + y_3 + y_4)\end{matrix} \end{bmatrix} \tag{2.19}$$

Computation of the $[S^{(e)}]$ matrix is more complex since it involves the reduction of the gradient of the shape function $\bar{\alpha}_i$ in xy to that of the shape function α_i which is expressed in terms of (u,v). This can be performed with the relation:

$$\nabla_{xy} f = \mathbf{J}^{-1} \nabla_{uv} f \tag{2.20}$$

which leads to:

$$\begin{aligned} S_{ij} &= \int\int_{\Omega_e} \nabla\bar{\alpha}_i \cdot \nabla\bar{\alpha}_j dxdy \\ &= \int_{-1}^{1}\int_{-1}^{1} \left[(\nabla_{uv}\alpha_i)^T \left(\mathbf{J}^{-1}\right)^T\right] \cdot \left[\mathbf{J}^{-1}\nabla_{uv}\alpha_j\right] |\mathbf{J}| dudv \end{aligned} \tag{2.21}$$

Both integrals (2.17) and (2.21) can be analytically derived easily only if $\mathbf{J}$ is constant (i.e., if $x_1 - x_2 = x_4 - x_3$ and $y_1 - y_2 = y_4 - y_3$). Otherwise numerical integration is more convenient.

Integral (2.21) implies the computation of $\mathbf{J}^{-1}$ which is not guaranteed to exist. Moreover $\mathbf{J}^{-1}$ contains polynomial ratios, which are much more difficult to integrate numerically.

Second-order quadrilateral elements: To define second-order quadrilateral elements, we can still start from the simple square in the (u,v) plane used to define the first-order element. The basis functions for the square in the (u,v) plane must now be at least six, and indeed, an almost natural choice consists in placing a new node at the midpoint of each side of the square, obtaining an eight-node element with shape functions:

$$\begin{aligned} \alpha_1 &= \frac{1}{4}(u-1)(-v+1)(u+v+1) \\ \alpha_2 &= \frac{1}{2}(u+1)(-v+1)(-u+1) \\ \alpha_3 &= \frac{1}{4}(u+1)(-v+1)(u-v-1) \\ \alpha_4 &= \frac{1}{2}(u+1)(v+1)(-v+1) \\ \alpha_5 &= \frac{1}{4}(u+1)(v+1)(u+v-1) \\ \alpha_6 &= \frac{1}{2}(u+1)(v+1)(-u+1) \\ \alpha_7 &= \frac{1}{4}(-u+1)(v+1)(-u+v-1) \\ \alpha_8 &= \frac{1}{2}(-u+1)(v+1)(-v+1) \end{aligned} \tag{2.22}$$

Similarly to the first-order case, the above set of shape functions is complete up to the second-order and contains some third-order terms (u^2v, uv^2). Again, this kind of element lacks geometric isotropy for the order of approximating function.

This square can be formed from the (u,v) system to the (x,y) system following either the quadrilateral transformation used for the first-order elements or using a much more flexible transformation which uses, for mapping the (u,v) plane into the (x,y), the same basic function used to approximate the unknown function.

$$x = \sum_i x_i \alpha_i(u,v) \tag{2.23}$$

$$y = \sum_i y_i \alpha_i(u,v) \tag{2.24}$$

This transformation is referred to as *isoparametric* because it employs the same shape functions α_i which are used to approximate the unknown function. As a matter of fact other transformations exist, called *subparametric* or *superparametric*, where the transformation maps the two planes using lower or higher order function, respectively, than those used for the unknown function approximation. For example, the quadrilateral transformation (2.15) and (2.16) is subparametric for the second-order elements and isoparametric for the first, as can be easily seen by introducing (2.14) into (2.23) and (2.24). The second-order isoparametric mapping has the great advantage of being quadratic, thus able to map straight lines into curved line (Figure 2.12) while the order one mapping (2.15) and (2.16) simply changes the orientation and length of a segment. This is particularly useful when curved boundaries are present in the problem, allowing much finer approximation of the geometrical domain than that possible with element with straight sides.

Matrices $\left[T^{(e)}\right]$ and $\left[S^{(e)}\right]$ are again computed through (2.17) and (2.21), where now the Jacobian of the transformation is:

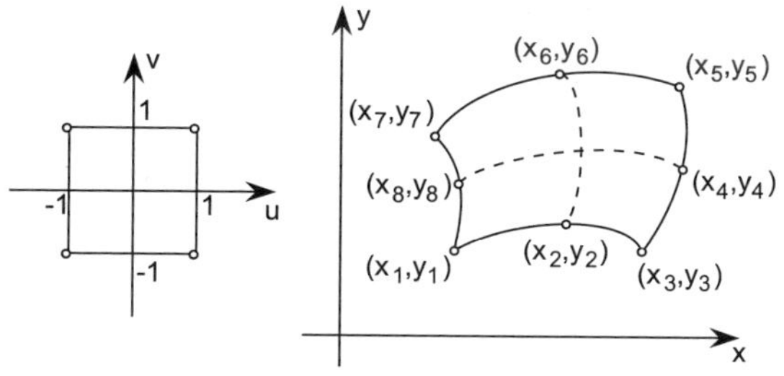

Figure 2.12 Isoparametric mapping.

$$
\mathbf{J} = \begin{bmatrix}
\begin{array}{l}
\left(\frac{x_1-x_3-x_5+x_7}{4}+\frac{x_4-x_8}{2}\right)v^2+ \\
+\left(\frac{x_1+x_3-x_5-x_7}{2}-x_2+x_6\right)uv+ \\
+\left(\frac{x_1-x_3+x_5-x_7}{4}+x_6\right)v+ \\
+\left(\frac{x_1+x_3+x_5+x_7}{2}-x_2-x_6\right)u+ \\
+\frac{x_8}{2}-x_6-\frac{x_4}{2}
\end{array}
&
\begin{array}{l}
\left(\frac{y_1-y_3-y_5+y_7}{4}+\frac{y_4-y_8}{2}\right)v^2+ \\
+\left(\frac{y_1+y_3-y_5-y_7}{2}-y_2+y_6\right)uv+ \\
+\left(\frac{y_1-y_3+y_5-y_7}{4}+y_6\right)v+ \\
+\left(\frac{y_1+y_3+y_5+y_7}{2}-y_2-y_6\right)u+ \\
+\frac{y_8}{2}-y_6-\frac{y_4}{2}
\end{array}
\\
\begin{array}{l}
\left(\frac{-x_1-x_3+x_5+x_7}{4}+\frac{x_2-x_6}{2}\right)u^2+ \\
+\left(\frac{-x_1+x_3+x_5-x_7}{2}-x_4+x_2\right)uv+ \\
+\left(\frac{x_1+x_3+x_5+x_7}{4}+x_4-x_8\right)v+ \\
+\left(\frac{x_1-x_3+x_5-x_7}{2}\right)u+ \\
+\frac{-x_2+x_6}{2}
\end{array}
&
\begin{array}{l}
\left(\frac{-y_1-y_3+y_5+y_7}{4}+\frac{y_2-y_6}{2}\right)u^2+ \\
+\left(\frac{-y_1+y_3+y_5-y_7}{2}-y_4+y_2\right)uv+ \\
+\left(\frac{y_1+y_3+y_5+y_7}{4}+y_4-x_8\right)v+ \\
+\left(\frac{y_1-y_3+y_5-y_7}{2}\right)u+ \\
+\frac{-y_2+x_6}{2}
\end{array}
\end{bmatrix}
\qquad (2.25)
$$

2.2.2 Vector Elements

The use of vector shape functions $\boldsymbol{\alpha}_i$ to approximate vector fields in microwave engineering problems has long been the preferred solution. In this case functionals or equations to be discretized with finite elements contains terms such as $\nabla \times \boldsymbol{\Phi}$ and $\boldsymbol{\Phi}$, which lead to the element matrices $\left[E^{(e)}\right]$ and $\left[F^{(e)}\right]$, respectively, whose entries have the form:

$$E_{ij} = \int_{\Omega_e} (\nabla \times \boldsymbol{\alpha}_i) \cdot (\nabla \times \boldsymbol{\alpha}_j) d\Omega \qquad (2.26)$$

and

$$F_{ij} = \int_{\Omega_e} \boldsymbol{\alpha}_i \cdot \boldsymbol{\alpha}_j d\Omega \qquad (2.27)$$

where Ω_e is the surface of the element. The formulas to compute these matrices will be given in the following paragraphs for triangular and quadrilateral first-order vector elements, also referred to as *edge elements*.

First-order triangular: First-order vector shape functions can be derived from scalar shape functions through the formula

$$\boldsymbol{\alpha}_1 = l_1 (\alpha_1 \nabla \alpha_2 - \alpha_2 \nabla \alpha_1) \qquad (2.28)$$

$$\boldsymbol{\alpha}_2 = l_2 (\alpha_2 \nabla \alpha_3 - \alpha_3 \nabla \alpha_2) \qquad (2.29)$$

$$\boldsymbol{\alpha}_3 = l_3 (\alpha_3 \nabla \alpha_1 - \alpha_1 \nabla \alpha_3) \qquad (2.30)$$

where l_i is the length of edge i (Figure 2.13). These vectorial shape functions have two main attractive features. First, only the shape function $\boldsymbol{\alpha}_i$ has a nonzero tan-

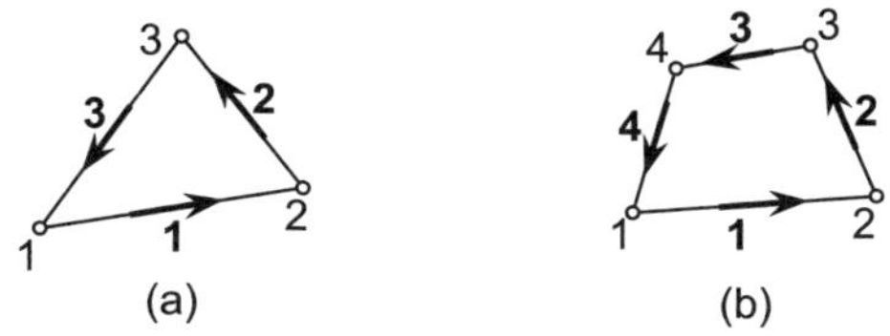

Figure 2.13 Triangular and rectangular edge elements with local numbering schemes.

gential component along edge i, and, for first-order elements, such a tangential component is constant along the entire edge (Figure 2.14). Second, $\nabla \cdot \boldsymbol{\alpha}_i = 0$ inside the triangle. The former feature guarantees continuity of tangential field at element interfaces and allows enforcing boundary conditions in a straightforward manner. The latter is useful since these divergenceless shape functions can approximate the divergenceless electromagnetic fields in a natural way.

The matrices $[E^{(e)}]$ and $[F^{(e)}]$ can be expressed as:

$$[E^{(e)}] = \frac{1}{A}\begin{bmatrix} l_1 & 0 & 0 \\ 0 & l_2 & 0 \\ 0 & 0 & l_3 \end{bmatrix}\begin{bmatrix} 1 & 1 & 1 \\ 1 & 1 & 1 \\ 1 & 1 & 1 \end{bmatrix}\begin{bmatrix} l_1 & 0 & 0 \\ 0 & l_2 & 0 \\ 0 & 0 & l_3 \end{bmatrix} \tag{2.31}$$

$$[F^{(e)}] = \frac{1}{48A}\begin{bmatrix} l_1 & 0 & 0 \\ 0 & l_2 & 0 \\ 0 & 0 & l_3 \end{bmatrix}[G]\begin{bmatrix} l_1 & 0 & 0 \\ 0 & l_2 & 0 \\ 0 & 0 & l_3 \end{bmatrix} \tag{2.32}$$

Where $\left[G^{(e)}\right]$ entries depend on the triangular element geometry:

$$\left[G^{(e)}\right] = \begin{bmatrix} 2(f_{22} - f_{12} - f_{11}) & f_{23} - f_{22} - 2f_{13} + f_{12} & f_{21} - 2f_{23} - f_{11} + f_{13} \\ f_{23} - f_{22} - 2f_{13} + f_{12} & 2(f_{33} - f_{23} - f_{22}) & f_{31} - f_{33} - 2f_{21} + f_{23} \\ f_{21} - 2f_{23} - f_{11} + f_{13} & f_{31} - f_{33} - 2f_{21} + f_{23} & 2(f_{11} - f_{13} - f_{33}) \end{bmatrix} \tag{2.33}$$

being

$$f_{ij} = (y_{i+1} - y_{i-1})(y_{j+1} - y_{j-1}) + (x_{i+1} - x_{i-1})(x_{j+1} - x_{j-1}) \tag{2.34}$$

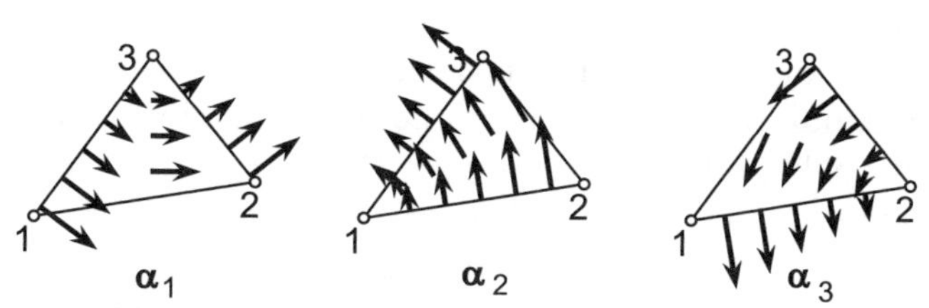

Figure 2.14 Vector basis functions for a triangular element.

where indices i and j always progress modulo 3, that is by cyclically varying as $\overleftarrow{1 \rightarrow 2 \rightarrow 3}$.

Quadrilateral edge elements may be derived in a similar way, and the implementation is similar to that followed for nodal quadrilateral elements. Vector shape functions are derived for a square element in the uv, and the local matrices are computed using the coordinate transformation (2.15) and (2.16). This, unfortunately, results in shape functions which are not divercence-free. Thus, even if they maintain the advantage of being vectorial, and of implicitly enforcing the continuity of the tangential field, they lack the major advantage of zero divergence, unless rectangularity is imposed, thus losing flexibility in domain discretization. Triangular edge elements are thus much more widely used, while quadrilateral ones are used only in conjunction with some sort of control over the divergence or in two-dimensional-and-a-half problems, where the third field component, modeled with scalar shape function, can be held responsible for driving the divergence to zero.

2.3 Global Matrices

Assembling global matrices starting from the local ones is straightforward for nodal elements, as seen in Section 1.4 for first-order elements. All that is required is a look-up table (in the FORTRAN codes of this book stored in the `ELE` array, in the Matlab ones in the `ele` entry of the `Mesh` structure) connecting local and global numbering schemes. For higher order nodal elements that procedure remains unchanged apart from the increased number of nodes of each element.

For edge elements, a slight difference resides in the fact that vectorial shape functions have a direction, thus summing the contributions of two local matrix entries in the appropriate global matrix entry requires checking the relative directions. To handle this, the look-up table connecting local and global edge numbering schemes (in the FORTRAN codes of this book stored in the `SPIG` array, in the Matlab ones in the `spig` entry of the `Mesh` struct) may present negative numbers. An element which contains an edge identified with a positive number has its local edge oriented in the same direction as the global edge. Conversely, a negative number denotes opposite directions. In this latter case the entry of the element matrix relative to that edge is subtracted rather than added to the global matrix entry.

Beside local matrices, boundary conditions also contribute to the assembly of the global matrix. As seen in Chapter 1, Dirichlet-type boundary conditions must be explicitly enforced following the algorithm described in Section 1.4 for first-order triangular nodal elements. The same algorithm can be applied when using edge elements. Neumann boundary conditions, on the other hand, constitute natural boundary conditions and need not be explicitly enforced. Other kinds of boundary conditions often arise in microwave engineering problems, most notably the radiation boundary condition (RBC). The latter condition as well as other useful boundary conditions will be treated in the following applications chapters.

As already noted, global matrices are sparse, hence it is advisable to seek storage methods that are able to minimize memory requirements. Two main methods are suitable to achieve memory savings: sparse matrix storage and banded matrix storage methods. The former approach can be used regardless of the global node numbering adopted and it is particularly convenient when the percentage of nonzero elements is really low. The latter requires the usage of a smart global numbering to effectively reduce the global matrix bandwidth. Although band storage methods perform slightly worse (some zeroes are retained and stored), band storage matrices are faster and easier to access, and there is plenty of freely available routines for their treatment. For such reasons this will be the storage scheme of choice for the FORTRAN codes in this book. On the other hand, Matlab is very easy to use and user-transparent sparse matrices are used in the Matlab codes in this book.

2.3.1 The Band Storage Mode

For memory requirements minimization, it is necessary to minimize the bandwidth of the matrix. The band storage scheme allows storing an $N \times N$ matrix in a $(kl + ku + 1) \times N$ array, being kl and ku the number of subdiagonal and superdiagonal, respectively. If the matrix is also symmetric the storage requirement drops to $(kd + 1) \times N$, being $kd = kl = ku$.

The band storage method described here is the standard used by most existing linear algebra libraries, most notably of all the freely available LAPACK.

Being $[A]$ the full matrix and $[A^{(B)}]$ the array where it will be stored, the general rule is:

$$a_{ij} \rightarrow a^{(b)}_{ku+1+i-j,j} \text{ for } \max(1,j-ku) \leq i \leq \min(n,j+kl) \tag{2.35}$$

for example, the following 5×5 matrix with two subdiagonals and one superdiagonal is stored in a 4×5 array as shown below

$$\begin{bmatrix} a_{11} & a_{12} & & & \\ a_{21} & a_{22} & a_{23} & & \\ a_{31} & a_{32} & a_{33} & a_{34} & \\ & a_{42} & a_{43} & a_{44} & a_{45} \\ & & a_{53} & a_{54} & a_{55} \end{bmatrix} = \begin{bmatrix} * & a_{12} & a_{23} & a_{34} & a_{45} \\ a_{11} & a_{22} & a_{33} & a_{44} & a_{55} \\ a_{21} & a_{32} & a_{43} & a_{54} & * \\ a_{31} & a_{42} & a_{53} & * & * \end{bmatrix} \tag{2.36}$$

Elements marked with an $*$ are unused and need not be set. Of course, much more memory saving is achieved for bigger matrices with a small bandwith. Furthermore, if the matrix is symmetric, as in most FEM problems, the storage can be performed following the rules:

$$a_{ij} \rightarrow a^{(b)}_{kd+1+i-j,j} \text{ for } \max(1,j-kd) \leq i \leq j \tag{2.37}$$

$$a_{ij} \rightarrow a^{(b)}_{1+i-j,j} \text{ for } j \leq i \leq \min(n,j+kd) \tag{2.38}$$

where either (2.37) or (2.38) is used to store the main diagonal and the superdiagonals or the main diagonal and the subdiagonals, respectively.

To ease the usage of LAPACK two simple FORTRAN routines are provided to access matrices stored in band storage mode, according to the standard described above.

2.3.2 The Sparse Storage Mode

There are several possible sparse storage modes, indeed the band storage mode described in the previous section is itself a sparse storage method which assumes that the nonzero entries of the matrix are close to the diagonal.

The most general sparse storage techniques, not assuming any characteristic in the matrix sparsity, are the compressed row and the compressed column techniques. We will briefly describe here the compressed row technique, the compressed column being its dual obtained by exchanging indexes.

Among the advantages of the compressed row technique is that it does not store any unnecessary elements. On the other hand, it is not very time efficient, needing an indirect addressing step for every single scalar operation in a matrix-vector product or preconditioned solver step.

The compressed row storage format keeps the nonzero entries in each matrix row in contiguous memory locations, squeezing out all zero elements. The correct shape of the matrix is retained by using suitable pointer vectors. Let's assume we have a nonsymmetric sparse matrix $[A]$. We need to allocate three vectors: one for floating point reals or complexes, $(\mathbf{v})$, and the other two for integers $[\mathbf{c}^{(idx)}$ and $\mathbf{r}^{(p)}]$. The $\mathbf{v}$ vector stores the nonzero values of matrix $[A]$ as they are read row by row, one after the other. The $\mathbf{r}^{(p)}$ vector stores the location of the $\mathbf{v}$ array where a row starts, that is, if $v_k = a_{i,j}$ then $r_i^{(p)} < k < r_{i+1}^{(p)}$. Then the $\mathbf{c}^{idx}$ vector stores the column indexes of the elements in the $\mathbf{v}$ vector. That is, if $v_k = a_{i,j}$, then $c_k^{(idx)} = j$. By convention, $r_{n+1}^{(p)} = N_{nz} + 1$, where N_{nz} is the number of nonzero entries in matrix $[A]$.

An example will clarify this better, let's assume a generic unsymmetric sparse matrix:

$$[A] = \begin{bmatrix} a_{11} & a_{12} & & & a_{15} \\ a_{21} & a_{22} & & & \\ a_{31} & & a_{33} & a_{34} & \\ & & a_{43} & a_{44} & \\ & & & & a_{55} \end{bmatrix} \tag{2.39}$$

Then it will be

$$\mathbf{v} = [a_{11}, a_{12}, a_{15} | a_{21}, a_{22} | a_{31}, a_{33}, a_{34} | a_{43}, a_{44} | a_{55}] \tag{2.40}$$

$$\mathbf{c}^{(idx)} = [1,2,5|1,2|1,3,4|3,4|5] \tag{2.41}$$

$$\mathbf{r}^{(p)} = [1,4,6,9,11,12] \tag{2.42}$$

where the vertical bars in $\mathbf{v}$ and $\mathbf{c}^{idx}$ have been placed only to help the reader visualize where a row starts and ends.

This storage mode saves significant memory. Instead of storing N^2 elements, we need only $2N_{nz} + N + 1$ storage locations.

Matlab internal storage is not of primary concern here, nevertheless, for completeness, we can briefly report that a sparse matrix is stored in compressed column mode. Each column is stored as a concatenation of sparse vector, and each sparse vector consists of a pair of vectors. A vector of values, containing only nonzero entries, and a vector of row indices. A further vector points to the first element of each sparse vector [6].

The real complexity of Matlab implementation relies on the careful optimization of all sparse matrix operations from the computing time point of view, and the fact that the user does not need to get concerned with the sparsity of the matrix at all.

2.4 Solving the Entire Problem

Once the global matrix is assembled, the last step of the FEM procedure implies solving either a linear system of equations or a generalized eigenvalue problem, depending on the electromagnetic analysis to be performed. Throughout this book many routines belonging to the linear algebra package (LAPACK) will be used. This package is part of the netlib project, a collection of softaware libraries, most of which are written in FORTRAN 77, freely available on the network. Such libraries allow dealing with a wide spectrum of numerical problems ranging from data fitting to optimization.

The LAPACK library aimed to achieve good performance in a portable way on a large class of platforms. Its usage, and indeed that of most of netlib routines, requires implementing the basic linear algebra subroutines (BLAS) and the setting of a small number of tuning parameters, platform dependent, which are already set to reasonable default values. The BLAS itself is available at the netlib Web site. Although LAPACK is written to abide the FORTRAN 77 standard, it contains many routines including double-precision complex data type, which is not part of this standard, so some assumptions about the name of intrinsic functions that do not hold for all machines are made.

Matlab, on its side, presents a variety of built-in solvers, many of which are LAPACK-based but lapack tuning is transparent to the user.

2.5 Postprocessing

The postprocessing phase includes the visualization of the computed approximating function and, when needed, manipulation of results to compute parameters of interest for electromagnetic design, such as scattering parameters, input impedance, or, as in our first example outlined in Chapter 1, the secondary constants of a line.

The importance of result visualization is twofold: in the debugging phase of the code it allows checking its correctness by inspecting field behavior and boundary condition fulfillment; in the subsequent design phase it is of some help in gaining insight in the physical behavior of microwave structures. A qualitative visual inspection of results is an important tool for a first check of code consistency since FEM analysis usually leads to a great amount of data, unevenly scattered in space, which cannot be easily interpreted by the user in any other way. Furthermore, field visualization can be an important feature for microwave engineers in developing new device design by showing critical parts of the structure, such as a zone with too strong a field, or by highlighting undesired effects of discontinuities.

While other numerical methods work on regular grids and the output data generated can be easily plotted, FEM works on highly irregular meshes and obtaining a graph of the fields inside the region of interest requires an appropriate graphical postprocessor. Thus a really complete set of tools for FEM analysis must include some means of field visualization. Since a FORTRAN code with graphics would bound the code to a certain compiler and a certain class of machines, while the aim of this book is to be general, a compromise has been reached providing a FORTRAN program (POSTPRO) which processes the FEM output data to obtain the value of the approximating function on a regularly spaced grid. This will allow the user to see his results on any mathematical CAD or graphical package able to plot surfaces and contour lines. The output of the subroutine POSTPRO is in the format required by the worldwide spread freeware package for surface and contour plotting, and can be used on any other program with minimal modifications.

Of course Matlab codes included in the CD-ROM enclose a complete field visualization tool, as it will be described later on in this chapter.

Besides the mesh geometry description file the POSTPRO program requires a result file containing the value computed by the user's code in the format shown in Table 2.3 where:

- `T` is element type (Number of nodes per element);
- `N` is the number of values contained in the file;
- `NN` is the number of the node;
- `VN` is the value of the computed approximating function at node *NN*.

Example 2.2 shows a typical output data file.

Example 2.2

```
number = 162
       1  .00000
       2  .00000
       3  .22287
       4  .22134
       5  .00000
       6  .45806
       7  .21668
       .   .
       .   .
```

Table 2.3 Results Structure Layout

Input Data	*Data Type*
`Type = T+`	`Integer`
`Number = N`	`Integer`
NN VN	*N* lines, each containing 1 Integer and 1 Real
⋮	

The second important part of the postprocessing phase consists in evaluating field-related parameters. This task usually requires computing stored energy, line or surface integrals, derivatives. Given the wide variety of parameters of interest this part of the postprocessing phase must be implemented by the user to fit his specific needs. Of course, all the problems presented in the applicative chapters of this book do have their specific postprocessing subroutines, as seen for instance in Chapter 1 for the computation of the secondary constants of the line.

2.6 The Matlab Framework

Matlab is a much simpler programming language than FORTRAN 77, able to do complex matrix-vector computations with just one command. It is hence very good for prototyping code and for educational use. This at the cost of being usually slower than a corresponding optimized FORTRAN code.

The quick FEM Matlab framework comprises two main features, a *library* of functions implementing the tools described in this chapter, and others, and a data model for optimizing data exchange between subroutines.

Since Matlab allows for building quite powerful custom graphical user interfaces (GUIs), a few of these have been developed and included in the CD-ROM. These will be described in the following subsection.

Matlab presents several toolboxes, among which the PDE toolbox contains a very good mesh generator. Since these toolboxes are add-ons with their own cost for a license it is quite possible that the computer rooms for numerical computation courses are equipped with the most basic Matlab license.

To allow for the complete utilization of this book with only a basic Matlab license we have included the mesh generator described earlier. However, if the PDE toolbox is available it is suggested to try also the Matlab mesh generator, which has a good GUI and is very robust even though it only handles first-order triangular elements of nodal kind. It outputs three matrices for `point`, `edges`, and `triangles`. These usually are given names `p`, `e`, and `t`, respectively and are created in Matlab global workspace by selecting the option `Mesh->Export Mesh` from the PDE tools GUI. The first matrix is essentially equivalent to the node definition block in the file format of this book, and contains the node coordinates. The latter is essentially equivalent to the element definition block. The `edges` block, despite its name, does not define all the edges of the mesh but only enumerates the edges of the elements which fall on the domain border(s).

The main issues are that no edge definition block is explicitly defined, and that the PDE tool mesh generator assumes that boundary conditions are assigned at the mesh generation level, hence no label assignment is possible.

To overcome these limitations a function is provided in the CD-ROM, `pde2qf.m`, which translates a PDE tools mesh into a quick FEM mesh.

Syntax is:

```
pde2qf(p,e,t,R2E,E2N)
```

where `p`, `e`, and `t` are the matrices exported from PDE tool. `R2E` is a vector mapping the PDE region number to element labels; all the elements belonging to a given region share the region number, but since PDE primitives for drawing are quite few it is often practical to describe a single homogeneous subdomain with more regions which then would have different labels. The same is true for edges, since PDE tools assign a different label to every single edge in a region, comprising internal regions, so `E2N` allows remapping.

A practical example will clarify the procedure. Let's assume that we want to mesh the boxed microstrip problem with PDE tools. It is possible in this case to describe the whole problem with just two regions, one in air and one in the dielectric, but we will use three to exemplify the procedure. Figure 2.15 shows the subdivision of the domain in three quadrilateral regions, and the numeric labels PDE tools automatically assigns to regions and nodes (to show them `Boundary ->Boundary mode`, then check `Show Edge Labels` and `Show Region Labels`).

By pressing `Mesh->Initialize Mesh` and, if needed `Mesh->Refine Mesh` the mesh is generated and shown. You can then use `Mesh->Export Mesh` to obtain the mesh matrices in the global workspace.

Now, from the Matlab command line you can issue the command:

```
Mesh=pde2qf(p,e,t,[2,1,1],[1,1,1,1,1,0,1,1,0,0,0])
```

Here `R2E=[2,1,1]` because region 1 is the dielectric slab, which is labeled 2 in our examples, while regions 2 and 3 contain air, that is a medium of label 1. Then `E2N=[1,2,2,1,2,0,1,1,0,0,0]` because region boundaries 1, 2, 3, 4, 5, 7 and 8 are effectively p.e.c., while 6 and 9 are internal to the domain (no label required) and boundaries 10 and 11 are p.m.c., which does not need a label since p.m.c. conditions are natural.

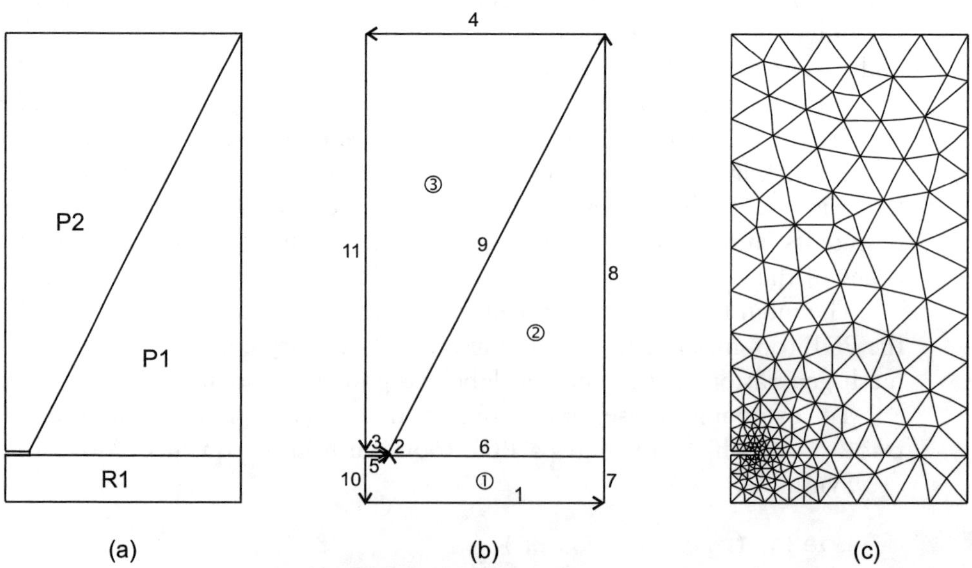

Figure 2.15 The boxed microstrip half-domain: (a) defined via three regions in PDE tools; (b) the numbering PDE tools assigns to regions (circled numbers) and edges; (c) the resulting mesh.

You can now save your mesh issuing the command

```
WriteFemFile(Mesh,'strip.fem')
```

which can be used both by the Chapter 1 electrostatic code and the next chapter full-wave code.

2.6.1 Using the Interface

The approach used in the Matlab quick FEM framework is conceived to allow the largest possible compatibility. Matlab codes are hence developed on three layers.

At the bottom layer there are the basic subroutines, which are the equivalent of the tool library for the FORTRAN codes and are stored in the `TOOLS/LIB` subdirectory. These are usually never called directly by the user but are called by the subroutines of the second layer.

While in the FORTRAN approach libraries are linked at compile time, Matlab is interpreted. This means that the directory holding the library must be added to Matlab PATH, this is done via the `File->Set Path` option. Both the `TOOLS` directory and the `TOOLS/LIB` subdirectory must be explicitly added.

On the second layer there are the main codes and utilities, written in a functional approach, that is, all data is passed to the function as parameters and the programs runs noninteractively. These first two layers are written using the least possible features and should hence run with any Matlab version 6.0 and higher.

The third layer encompasses graphical interfaces. These are merely interfaces allowing interactive setting of the second-layer function parameters. These have been developed in Matlab 2007a.

Functions of every level have been written according to Matlab standards hence the command

```
help functionname
```

will give clear indications of what a function does and how it should be called.

A functional approach has been used also for programs, such as the mesh generator and all the computational codes of the next chapters allows for an easy embedding of the codes in iterative loops to compute material parameter sweeps or optimizations.

Since typing in function parameters might be tedious, especially for complex programs, a GUI is provided for the most-used tools. The GUIs are called just by typing their name in Matlab command window and, different from other codes, consists of *two* files, with the same name but with extensions `.m` and `.fig`, the former containing the code, the latter the graphical layout. The provided GUIs are:

- MeshGUI: It is the interface to the mesh generator `Mesh.m`. It asks for the geometric description files and allows choosing many options:
  - To apply Delaunay regularization;
  - To apply the basic renumbering scheme described in this book;
  - To apply the Matlab internal Cuthill-McKee renumbering;
  - To specify whether the mesh is to be a periodic cell.

- It also visualizes several items:
  The initial geometry;
  The final mesh;
  The bandwidth, this can be done numerically or also graphically, for nodal or edge elements;
  The mesh itself can be shown with several options, which are described in the `SeeMeshGUI` paragraph hereafter.
- SeeMeshGUI: It is the interface to `SeeMesh` and it shows an existing mesh whose filename is asked. Options are:
  - To show element labels as colored symbols, or as numbers, or show the element numbers or none of these;
  - To show node labels as colored symbols, or as numbers, or show the node numbers or none of these;
  - To show edge labels as colors associated to the edge, or numbers next to the edge, or show the edge numbers or none of these.
  - This interface allows again for band estimation.
- SeeResultsGUI: It is an interface to both `SeeFieldMap` and to `SeeContours`. The former, given a mesh and a result file shows either a color map, if results are nodal, or an arrow plot, if results are vectorial. The second works only for nodal elements and shows countour lines rather than a color map. Options are:
  - To show the elements above the results.
  - To select the style: *standard* to use `SeeFieldMap`, or *contours* to use `SeeContours`.
  - To use a given number of interpolation points for the results. Warning: This option is useless for nodal fieldmaps, which are interpolated internally by Matlab, while governs the amount of arrows drawn within each edge element, by giving the number of arrows along each edge, that is 1 is 1, 2 are three arrows in total, 3 are 6 arrows in total and so on. If a contour is requested, it is necessary to convert the data on the unstructured FEM grid over a structured rectangular grid. The grid is chosen so as to have ten times the number selected for the interpolation points along its *shortest* dimension.

2.6.2 The Data Framework

The core difference between FORTRAN codes and Matlab codes relies in how mesh data is represented internally. Notwithstanding that the data is the same for both codes, the FORTRAN mesh data is scattered over six matrices and a few integer values. This makes FORTRAN function call quite long and complex.

To make Matlab programming a lot easier than FORTRAN, the mesh data is stored in a mesh data structure, which is a Matlab feature allowing storage of heterogeneous data in a single container. This container is then passed as a single item to the subroutines and functions.

A Matlab structure data is accessed by adding to the structure name the name of the contained data, separated by a period. For example if `Mesh` is the name of the structure, `Mesh.ele` is the array containing the element definitions, as given

Table 2.4 Mesh Structure Layout

Container	*Field*	*Description*
Mesh		Mesh data structure.
	.ele	Element definition array. It has 3, 4, 6, or 8 rows, depending on element type (all elements are assumed of the same type within the mesh) and as many columns as there are elements in the mesh.
	.elab	Element label vector. It has as many entries as there are elements in the mesh.
	.xy	Node array. It has two rows, one containing x coordinates, one containing y coordinates and with as many columns as there are nodes in the mesh.
	.nlab	Node label vector. It has as many entries as there are nodes in the mesh.
	.spig	Edge definition array. It has three or four rows, and is defined only for first-order elements. It contains one column for each element.
	.slab	Edge label vector. It has as many entries as there are edges in the mesh.

previously in this chapter. Each entry in the element definition array is addressed as `Mesh.ele(i,j)` as for a conventional array.

The full data structure is described in Table 2.4.

As for the outputs, nodal and edge values are stored exactly as in the FORTRAN codes, for compatibility and to allow Matlab GUI tools to also visualize results from FORTRAN codes. Results are themselves arranged in a structure, as sketched in Table 2.5.

2.6.3 How to Code Yourself

Coding in Matlab is easier than coding in FORTRAN. Working with interpreted languages makes debugging much easier and code prototyping faster.

If you want to develop your own codes there are a few library routines you might want to know. First the `[Mesh]=ReadFemMesh(nomefile)` which loads a mesh in memory and returns a mesh structure as defined earlier, then the `varargout = ReadAuxLine(fh,n,flag)` function, which reads numeric data from a file one line by one line. Matlab is more sensitive than FORTRAN on file reading and end-of-line comments are not allowed. Variable length lines too are not easily handled. This function allows writing input files with end-of line-comments and

Table 2.5 Results Structure Layout

Container	*Field*	*Description*
Res		Results data structure.
	.N	Number of results (the length of vector R).
	.Type	Results type (1 = nodal; 2 = edge).
	.R	An array of N elements, containing real numbers, relative to nodal or edge results.

variable line length as it is the standard in this book, plus returning the data both as a single vector or as a list of scalars, depending on your needs.

Then, to build elemental matrices `function [Se,Te] = EleN(Mesh, ie)` outputs matrices T and S for element *ie* in the given mesh. Similarly, `[Ee,Fe] = EleE(Mesh, ie)` computes edge elements elemental matrices. No explicit routine is given for assembling matrices, since the task is fairly trivial, or for solving the system, since Matlab includes many full-featured solvers.

Finally, the routine `WriteFemRes(fnameout,fields,type)` writes either nodal (`type=1`) or edge (`type=2`) results.

2.7 Disc Description and Installation

The accompanying CD-ROM is organized as follows:

2.7.1 FORTRAN Framework

- TOOLS: The codes of this chapter:
  - LIB: The subroutines described in this chapter, and a couple of general subroutines that deserve to be included in the library but are described in the following chapters, where firstly needed:
    BAND.F: The functions BPUT, BGET, CBPUT, and CBGET used to access matrices in band storage mode;
    BANDEX.F: Straightforward band occupation estimator;
    BSOLVE.F: Block solver, described in Chapter 4, is the only library subroutine which exploits LAPACK;
    DELAUNAY.F: The Delaunay regularization subroutine;
    ELEE.F: Edge element matrices;
    ELEN.F: Nodal element matrices;
    FEMPS.F: Subroutine to translate meshes into PostScript files;
    FINDEQN.F: Find equivalent nodes in periodic meshes, as described in Chapter 5;
    PNJEXP.F: Evaluates the integral of the product of the element basis function and a complex exponential, as described in Chapter 4;
    READ.F: Subroutine to read a mesh file;
    READN.F: Subroutine to read a mesh file, skipping edges;
    RENUMBER.F: Subroutine to perform optimization;
    WRITE.F: Subroutine to write a mesh file;
    WRITECR.F: Subroutine to write results in a form usable by the postprocessor.
  - MAINS: The driver programs to exploit part of the library interactively. All these files should be compiled singularly by resorting to the above-mentioned library:
    DO_BEX.F: Driver for the band estimator;
    DO_DEL.F: Driver for the Delunay regularizator;
    DO_F2PS.F: Driver for the translator mesh → PostScript;

DO_REN.F: Driver for the number optimization;
MESH.F: The mesh generator;
POSTPRO.F: The postprocessor.

- CODES: This directory contains all the programs of the following chapters. The single subdirectories will be discussed in the relative chapters, but all share the same organization:
  A top-level subdirectory containing the source code(s);
  A series of subdirectories, one for each example discussed in the text.

2.7.2 Matlab Framework

- TOOLS: The codes of this chapter:
  - LIB: The subroutines described in this chapter, and a couple of general subroutines that deserve to be included in the library but are described in the following chapters, where firstly needed.
    As opposed to what is done for FORTRAN codes, each Matlab subroutine is contained in a file of its own, which makes listing all of them cumbersome here.
- CODES: This directory contains all the programs of the following chapters. The single subdirectories will be discussed in the relative chapters, but all share the same organization:
  - A top-level subdirectory containing the source code(s);
  - A series of subdirectories, one for each example discussed in the text.

References

[1] Hinton, E. and Owen, D. R. J., *An Introduction to Finite Element Computations*, Swansea, UK: Pineridge Press, 1979.

[2] Sadiku, M. N. O., *Numerical Techniques in Electromagnetics*, London, UK: CRC Press, 1992.

[3] Graglia, R. D., Wilton, D. R., and Peterson, A. F., "Higher Order Interpolatory Vector Bases for Computational Electromagnetics," *IEEE Transactions on Antennas and Propagation*, Vol. 45, pp. 329–332, 1997.

[4] Graglia, R. D. and Lombardi, G., "Singular Higher Order Complete Vector Bases for Finite Methods," *IEEE Transactions on Antennas and Propagation*, Vol. 52(7), pp. 1672–1685, July 2004.

[5] Sun, D.-K., Vardapetyan, L., and Cendes, Z., "Dimensional Curl-Conforming Singular Elements for FEM Solutions of Dielectric Waveguide Structures," *IEEE Transactions on Microwave Theory and Techniques*, Vol. 53(3), pp. 984–992, March 2005.

[6] Gilbert, J. R., Moler, C., and Schreiber, R., "Sparse Matrices in MATLAB: Design and Implementation," *SIAM J. Matrix Anal. Appl.*, Vol. 13, pp. 333–356, 1992.

CHAPTER 3

Microwave Guiding Structures: Characterization

Guided propagation problems in general, and the transverse problems subclass in particular, were the first group of electromagnetic phenomena at microwave frequency to be analyzed with finite elements. Typical examples of transverse problems comprise determining the modal structure of homogeneous and inhomogeneous waveguides of arbitrary cross-sections, including optical guides, and finding the equivalent circuit parameters of inhomogeneous multiconductor lines (microstrip, coplanar waveguide, finline, etc.). The problems of this subclass require solving an eigenvalue problem to determine the propagation constant along the waveguide as well as the electromagnetic field distribution in the plane transverse to the direction of propagation, hence a two-dimensional finite element technique can be used.

The second subclass of guided propagation problems includes the analysis and characterization of components and, from a mathematical point of view, differs from the first subclass in the fact that it requires the solution of the deterministic problem of finding the field distribution due to a specific excitation. In general, discontinuities in guiding structures such as transitions and multiport devices, must be given a full three-dimensional treatment. However, in some simple but yet of practical importance cases such as bends, corners, as well as E- and H-plane junctions, the analysis can be carried out on a plane, reducing the computational domain to a two-dimensional one, thus achieving significant savings in memory and CPU time.

In this chapter, the transverse problems subclass is considered. In particular, the simple example of the shielded microstrip line considered in Chapter 1 is reviewed, removing the quasistatic approximation previously introduced to compute the line primary constants and adopting a full-wave analysis to evaluate the modal dispersion of the inhomogeneous waveguide. Two formulations are given: the first, valid for arbitrary cross-section homogeneous waveguides, is based on the fact that for homogeneous guiding structures TE_z ($E_z = 0$) and TM_z ($H_z = 0$) modes can be considered separately, whereas the second is more general and allows dealing with the hybrid modes arising in inhomogeneous structures. The code included in this chapter implements the more general formulation for inhomogeneous waveguides and employs first-order mixed vector and nodal finite elements [1].

3.1 Homogeneous Waveguides

The earliest finite element applications to rectilinear propagation problems concerned closed metallic waveguides with uniform cross-section along the direction of propagation, both homogeneous [2] and inhomogeneous [3, 4] cross-sections have been analyzed employing the longitudinal components of the electric and magnetic fields as unknown. Later on, the method was applied also to the analysis of open structures such as dielectric waveguides and microstrips by combining classical finite elements with either boundary integrals (BI) or absorbing boundary conditions (ABC), to fully account for the open nature of the guiding structure.

For a homogeneous closed waveguide with arbitrary cross-section uniform along the z-direction (Figure 3.1) it is necessary to solve the following eigenvalue problem:

$$\nabla_t^2\Phi + k_t^2\Phi = 0 \tag{3.1}$$

defined into the region Ω representing the waveguide cross-section. In particular, for the TM_z ($H_z = 0$) case, $\Phi(x,y) = E_z(x,y)$ and the solution must fulfill the Dirichlet boundary conditions $\Phi = 0$ at metallic walls and the Neumann boundary conditions $\partial\Phi/\partial n = 0$ at magnetic walls, while for the TE_z ($E_z = 0$) case $\Phi(x,y) = H_z(x,y)$ and the boundary conditions at perfectly conducting and magnetic walls are $\partial\Phi/\partial n = 0$ and $\Phi = 0$, respectively. The solution of (3.1) can be sought by following two different approaches: either the variational formulation or the Galerkin formulation of the weighted residual method (WRM). Both formulations lead to exactly the same generalized eigenvalue problem. Similar to the static problem solved in Chapter 1, the variational approach seeks for the function which minimizes an appropriate functional. It can be shown [5] that the exact solution of the Helmholtz (3.1) minimizes the functional

$$F(\Phi) = \frac{1}{2}\iint_\Omega \left[(\nabla_t\Phi)^2 - k_t^2\Phi\right] d\Omega \tag{3.2}$$

and that Dirichlet boundary conditions are essential (must be explicitly imposed), while homogeneous Neumann boundary conditions are natural (they are automatically satisfied by the function about which the functional is stationary).

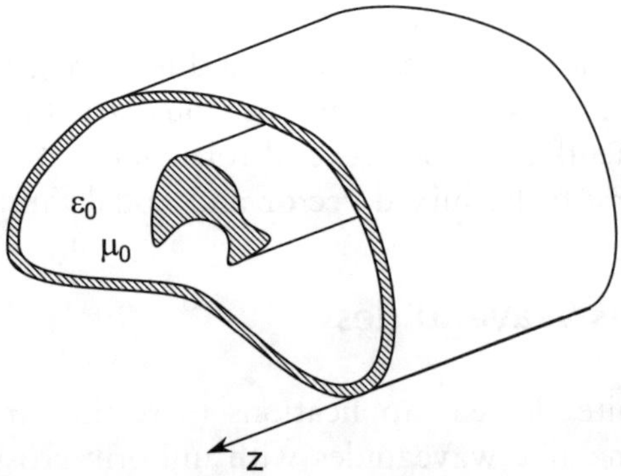

Figure 3.1 Example of homogeneous waveguide with uniform cross-section along the direction of propagation (z).

Applying the minimization procedure described in Chapter 1, the following generalized eigenvalue problem is set up:

$$[S][\Phi] = k_t^2[T][\Phi] \tag{3.3}$$

where $[S]$ and $[T]$ are built by assembling the local matrices $[S^{(e)}]$ and $[T^{(e)}]$ with entries

$$S_{ij}^{(e)} = \iint_{\Delta_e} \nabla_t \alpha_i^{(e)} \cdot \nabla_t \alpha_j^{(e)} dS \tag{3.4}$$

$$T_{ij}^{(e)} = \iint_{\Delta_e} \alpha_i^{(e)} \alpha_j^{(e)} dS \tag{3.5}$$

In (3.4) and (3.5), Δ_e is the area of element (e) and $\alpha_i^{(e)}$ is the i-th shape function defined on (e). Since the derivation of an appropriate functional for practical electromagnetic problems with complex material composition can be lengthy, the weighted residual method in the Galerkin formulation provides a more flexible and useful tool for the numerical solution of boundary value problems. For this reason this approach is outlined here and applied to the solution of the Helmholtz equation (3.1). To describe this procedure let us assume that the scalar Helmholtz equation must be solved in the geometrical domain Ω, with the Dirichlet boundary condition $\Phi = \Phi_0$ at Γ_D, and the Neuman boundary condition $\partial\Phi/\partial n = 0$ at Γ_N, where $\Gamma = \partial\Omega = \Gamma_D \cup \Gamma_N$. By introducing arbitrary weighting functions W, W_D, and W_N defined in Ω, the above boundary value problem can be replaced by:

$$\iint_\Omega W\left[\nabla_t^2\Phi + k_t^2\Phi\right] d\Omega + \int_{\Gamma_D} W_D(\Phi - \Phi_0)d\Gamma + \int_{\Gamma_N} W_N(\nabla_t\Phi \cdot \hat{\mathbf{n}})d\Gamma = 0 \tag{3.6}$$

and applying Green's theorem to the first integral it is possible to reduce the order of the derivatives involved in the above equation:

$$\begin{aligned}\iint_\Omega \nabla_t W \cdot \nabla_t \Phi d\Omega - \iint_\Omega k_t^2 W\Phi d\Omega - \int_\Gamma W(\nabla_t\Phi \cdot \hat{\mathbf{n}})d\Gamma \\ - \int_{\Gamma_D} W_D(\Phi - \Phi_0)d\Gamma \\ - \int_{\Gamma_N} W_N(\nabla_t\Phi \cdot \hat{\mathbf{n}})d\Gamma = 0\end{aligned} \tag{3.7}$$

By choosing, Φ such that it satisfies the Dirichlet condition $\Phi = \Phi_0$ at Γ_D and the weighting functions W and W_N so that $W_N = -W$ at Γ_N, $W = 0$ at Γ_D, the boundary value problem reduces to

$$\iint_\Omega \nabla_t W \cdot \nabla_t \Phi d\Omega - \iint_\Omega k_t^2 W\Phi d\Omega = 0 \tag{3.8}$$

which is often referred to as the weak form of the Helmholtz equation. If the function $\bar{\Phi}$ is an approximation to the exact solution Φ, then the right-hand side of (3.8) is not exactly zero, but it will be equal to a certain quantity called *residue*.

In the WRM, the solution to the boundary value problem (3.6) is sought as the best approximation to the exact solution, that is, as the function which annihilates the right-hand side of (3.8). The application of FEM in the framework of the WRM consists of the following five steps:

1. Subdivide the region Ω into nonoverlapping elements;
2. Approximate the unknown function on each element;
3. Define weighting functions and express the residue on each element;
4. Sum up contributions from all elements to obtain the residue on the whole domain;
5. Annihilate the residue and solve the generalized eigenvalue problem.

The first step can be easily implemented using the software tools described in Chapters 1 and 2, and will not be further investigated here. As for the second step, on each element (e) the solution is sought among the approximating functions $\bar{\Phi}^{(e)}$ of the form:

$$\bar{\Phi}^{(e)}(x,y) = \sum_{j=1}^{N^{(e)}} \bar{\Phi}_j^{(e)} \alpha_j^{(e)}(x,y) \tag{3.9}$$

where $\bar{\Phi}_j^{(e)}$ and $\alpha_j^{(e)}(x,y)$ $(j = 1,\ldots,N^{(e)})$ are unknown coefficients and a set of differentiable basis functions, respectively. Either nodal or vectorial-shape functions like those described in Chapter 2 can be used to this end, depending on the scalar or vectorial nature of the solution sought.

To implement the third step, a set of weighting functions $W_i^{(e)}$ $(i = 1,\ldots,N^{(e)})$ is defined on each element and introduced into the weak form of the Helmholtz equation together with the approximating functions $\bar{\Phi}^{(e)}(x,y)$ (3.9). The Galerkin formulation of the WRM is preferred when applying FEM, that is, the weighting functions on each element are chosen equal to the basis functions: $W_i^{(e)} = \alpha_i^{(e)}$ $(i = 1,\ldots, N^{(e)})$. With this choice, the residue $R_i^{(e)}$ relative to the ith weighting function on element (e) is expressed as:

$$\iint_{\Delta^{(e)}} \nabla_t \alpha_i^{(e)} \cdot \nabla_t \bar{\Phi}^{(e)} d\Omega - \iint_{\Delta^{(e)}} k_t^2 \alpha_i^{(e)} \bar{\Phi}^{(e)} d\Omega = R_i^{(e)} \quad i = 1, N^{(e)} \tag{3.10}$$

or, in matrix form:

$$[S^{(e)}] \cdot [\bar{\Phi}^{(e)}] - k_t^2 [T^{(e)}] \cdot [\bar{\Phi}^{(e)}] = [R^{(e)}] \tag{3.11}$$

The entries of the matrices $[S^{(e)}]$ and $[T^{(e)}]$ are exactly the same as those of the local matrices obtained following the variational approach, while the vector $[R^{(e)}]$ contains the residues $R_i^{(e)}$. Note that if the vertex j of element (e) lies at the Dirichlet boundary Γ_D then the coefficient $\bar{\Phi}_j^{(e)}$ is known (and its value is equal to Φ_0) and no weighting function should be centered at j. This is consistent with the condition $W = 0$ at Γ_D necessary to derive (3.8). Consequently, if an element has N_D nodes at Γ_D, the matrices $[S^{(e)}]$ and $[T^{(e)}]$ are of order $N^{(e)} - N_D$.

In the fourth step of the procedure, the local matrices $[S^{(e)}]$ and $[T^{(e)}]$ are assembled into the global ones $[S]$ and $[T]$, respectively. This corresponds to sum up the residual contributions coming from all the weighting functions centered at the same node, but spanning different elements, and it is equivalent to weighting the Helmholtz equation with pyramidal *global* weighting functions $W_i^{(g)}$, which can be thought of as the connection of all the weighting functions centered at the same node but spanning different elements (Figure 3.2). The assembling strategy is the same as that described in Chapter 1, with the only difference arising on the treatment given to nodes at the Dirichlet boundary Γ_D. In fact, while for deterministic problems (e.g., the solution of the Laplace equation in Chapter 1) Dirichlet boundary conditions are assembled into the linear system of equations to be solved, in the case of eigenvalue problems the homogeneous Dirichlet conditions are enforced by removing from the global system of equations the rows and the columns pertaining to the nodes on Dirichlet boundaries. This is equivalent to set to zero the corresponding FEM coefficients $\bar{\Phi}_j^{(e)}$ and of course has the advantage of reducing the dimension of the eigenvalue problem to be solved.

The final step is the solution of the generalized eigenvalue problem (3.3) obtained assembling the local matrices $[S^{(e)}]$ and $[T^{(e)}]$. This can be accomplished by employing one of the routines in the free mathematical software package LAPACK, available on the Internet or either the `eig` Matlab function. It is worth noticing that Matlab internally uses LAPACK so results, if FORTRAN LAPACK has been correctly compiled for your system, are highly consistent.

The eigenvalues resulting from the computations represent the cutoff frequency k_t of the modes of the homogeneous structure, from which, using the dispersion relation

$$k_z^2 = k_0^2 - k_t^2 \tag{3.12}$$

is easy to compute the propagation and attenuation constant of propagating and evanescent modes at the operating normalized frequency k_0.

The eigenvectors resulting from the computation represent the values of the unknown expansion coefficients in (3.9) and can hence be used to construct field maps as described in Chapter 2.

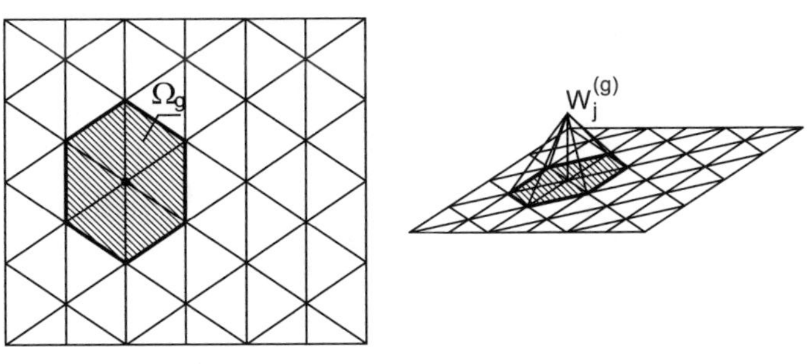

Figure 3.2 Global weighting function $W_j^{(g)}$ obtained by connecting first-order weighting functions defined on adjacent elements and centered at the same global node j.

3.2 Inhomogeneous Waveguides

The characterization of inhomogeneous guiding structures (Figure 3.3) requires a mathematical formulation that includes more than one component of the field. Moreover, using the classical nodal shape functions to approximate either the longitudinal components of both the electric and magnetic fields or the three components of the magnetic or electric fields is rendered difficult by the *spurious modes* problem. These latter are computed solutions that have no physical existence. The presence of these extra modes has plagued the method for a long time and the cause of their presence has been thoroughly discussed in the scientific literature. Three solutions to this problem have been explored during the last thirty years: two of them attack the formulation of the electromagnetic problem, while the other makes use of special elements able to eliminate spurious modes.

The first hypothesis on the cause of the spurious modes was that these modes represent solutions of the curl Maxwell's equations but do not satisfy the divergence equations [6]. Following this reasoning, many researchers have developed ways to include the divergence equation into the equation solved by means of FEM. The first attempt to eliminate spurious modes by amending the mathematical formulation was made by Rahman and Davies [7], who added a penalty term to the functional of the problem. This new term is introduced to minimize the divergence of either the electric or magnetic field (whichever is used as the unknown variable) in a least squares sense. Actually, the introduction of the penalty term does not eliminate the spurious modes, but it shifts the spurious critical frequencies to the higher part of the spectrum removing them from the frequency range of interest. This method has the shortcoming of leading to approximate solutions, for it also shifts the true physical mode spectrum, albeit only slightly, so other technique are preferred.

Various other variational formulations embedding the divergence equation have been proposed, all using more than one field component. Some are physically valid but computationally inconvenient because they do not retain the symmetry and sparsity of the finite element matrix, or put the eigenvalue problem in an unusual form. Fernandez and Lu [8] proposed a functional which does lead to

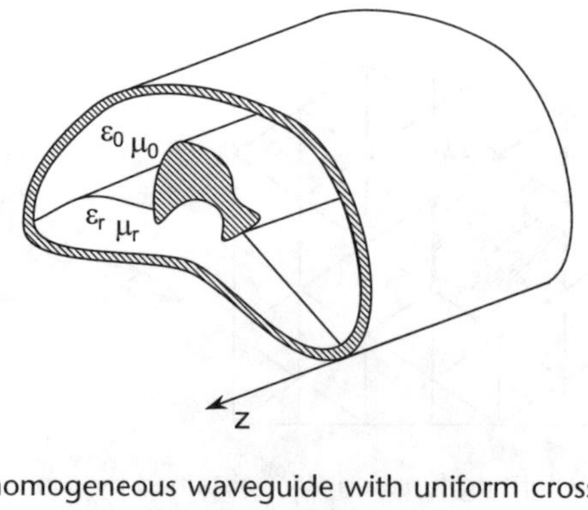

Figure 3.3 Example of inhomogeneous waveguide with uniform cross-section along the direction of propagation (z).

a standard eigenvalue problem. It involves two components of the magnetic field and is able to account for complex modes that may arise in certain inhomogeneous guiding structures.

The second solution developed to eliminate spurious modes starts from the transversalization of the vector Helmholtz equation which is subdivided into a transverse and a longitudinal equation. Then, the divergence equation is used to express the longitudinal component of the electric (magnetic) field in terms of its transverse component and this expression is introduced into the transverse part of the Helmholtz equation. In this way, a modified transverse equation embedding the divergence equation is obtained, where only the transverse part of the electric (magnetic) field appears and it can be solved following a weighted residual procedure similar to that outlined in the previous section for the case of homogeneous waveguides. The fact that now the unknown is a vector rather than a scalar means that at each node of the finite element mesh two unknowns must be defined, associated to the two different transverse Cartesian components of the field. This method was described by Hayata et al. [9] and subsequently elaborated upon [10]. Reviews of the variational and weighted residual formulations available for modal analysis, as well as comparative evaluations of their characteristic and potentialities, have been published by Davies [11], by Rahman, Fernandez, and Davies, [12], and by Dillon and Webb [13].

The third and last approach developed to avoid spurious modes, which proved the best in the long run, makes use of vector shape functions to approximate the solution on each element. These shape functions were initially constructed so as to be divergence free on each finite element of the mesh [14]. In this way the approximated solution is forced to satisfy the divergence condition on each element. Their usage has rapidly gained popularity because they not only free the numerical solution from spurious modes, but also permit to easily enforce Dirichlet or Neumann boundary conditions at sharp metal edges as well as at curved boundaries. For this reason they have been extensively employed also in conjunction with mathematical formulations capable of eliminating spurious modes per se. Since the introduction of the so-called edge elements [14] (vector element of the order zero) many varieties of higher order [15, 16] or mixed order [17, 18] elements have been reported. It must be noted that some of them are not solenoidal but nevertheless do not exhibit spurious modes. This has been recently explained by noting that the divergence equation is already included into the curl equation and thus the cause of spurious modes must be sought somewhere else. Spurious modes have been shown to be poor approximations of static solutions [19, 20] and can be removed by choosing a set of shape functions that correctly model the null space of the curl operator. This is actually what higher order or mixed order vectorial elements accomplish, even though some of them are not solenoidal per se.

3.3 Inhomogeneous Waveguides: Formulation

To analyze electromagnetic wave propagation in an inhomogeneous waveguide, the finite element method is employed in the framework of the Galerkin formulation of the weighted residual method to solve the vector Helmholtz equation.

Spurious modes are avoided by using mixed edge and nodal elements to approximate the transverse and longitudinal components of the electric field, respectively, on the cross-section of the waveguide [1]. To derive the mathematical formulation adopted, let us consider the curl Maxwell's equations:

$$\begin{cases} \nabla \times \mathbf{E} = -j\omega\mu\mathbf{H} \\ \nabla \times \mathbf{H} = +j\omega\bar{\epsilon}\mathbf{E} \end{cases} \tag{3.13}$$

where $\bar{\epsilon} = \epsilon + \sigma/(j\omega)$ with ϵ and σ permittivity and conductivity, respectively, of dielectric materials. Taking the curl of the second equation and substituting it into the first one yields the vector Helmholtz equation:

$$\nabla \times \frac{1}{\mu_r}\nabla \times \mathbf{E} = k_0^2\,\bar{\epsilon}_r\mathbf{E} \tag{3.14}$$

where $k_0^2 = \omega^2\mu_0\epsilon_0$, $\mu_r = \mu/\mu_0$ and $\bar{\epsilon}_r = \bar{\epsilon}/\epsilon_0$. Assuming for all the field components the dependence from the spatial coordinate z of the form $exp(-\gamma z)$, with $\gamma = \alpha + j\beta$ complex propagation constant, and subdividing the electric field into its transverse ($\mathbf{E}_t$) and longitudinal (E_z) parts, we get

$$\mathbf{E}(x,y,z) = \mathbf{E}(x,y)e^{-\gamma z} = [\mathbf{E}_t(x,y) + \hat{\mathbf{z}}E_z(x,y)]e^{-\gamma z} \tag{3.15}$$

the vector Helmholtz equation (3.14) can be split in its transverse and longitudinal parts too:

$$\nabla_t \times \left(\frac{1}{\mu_r}\nabla_t \times \mathbf{E}_t\right) - \gamma\frac{1}{\mu_r}(\nabla_t E_z + \gamma\mathbf{E}_t) = k_0^2\bar{\epsilon}_r\mathbf{E}_t \tag{3.16}$$

$$\nabla_t \times \left[\frac{1}{\mu_r}(\nabla_t E_z + \gamma\mathbf{E}_t) \times \hat{\mathbf{z}}\right] = k_0^2\bar{\epsilon}_r E_z\hat{\mathbf{z}}. \tag{3.17}$$

If this formulation is employed to compute the propagation constant γ when the operating frequency (k_0^2) is fixed, a generalized eigenvalue system is obtained. However, since in such system the eigenvalue γ appears at the first- and second-order power, it is convenient introducing the auxiliary variables

$$\mathbf{e}_t = \gamma\mathbf{E}_t \tag{3.18}$$

$$e_z = E_z \tag{3.19}$$

Employing these variables, (3.16) and (3.17) become

$$\nabla_t \times \left(\frac{1}{\mu_r}\nabla_t \times \mathbf{e}_t\right) - \gamma^2\frac{1}{\mu_r}(\nabla_t e_z + \mathbf{e}_t) = k_0^2\,\bar{\epsilon}_r\mathbf{e}_t \tag{3.20}$$

$$\gamma^2\nabla_t \times \left[\frac{1}{\mu_r}(\nabla_t e_z + \mathbf{e}_t) \times \hat{\mathbf{z}}\right] = \gamma^2 k_0^2\,\bar{\epsilon}_r e_z\hat{\mathbf{z}} \tag{3.21}$$

and must be solved with the boundary conditions:

$$\hat{\mathbf{n}} \times \mathbf{e}_t = 0 \qquad e_z = 0 \tag{3.22}$$

at perfectly conducting material, and

$$(\nabla_t e_z + \mathbf{e}_t) \cdot \hat{\mathbf{n}} = 0 \qquad \nabla_t \times \mathbf{e}_t = 0 \tag{3.23}$$

at magnetic walls. To apply the weighted residual procedure, two sets of basis functions and two corresponding sets of weighting functions have to be defined. Since the Galerkin formulation is adopted, each set of weighting function is equal to the corresponding set of basis functions. The vectorial-shape functions $\boldsymbol{\alpha}_i^{(e)}$ defined in Chapter 2, are used as the set of basis functions to express, on element (e), the approximation $\bar{\mathbf{e}}_t^{(e)}$ to the exact transverse part $\mathbf{e}_t$ of the electric field:

$$\bar{\mathbf{e}}_t^{(e)}(x,y) = \sum_{j=1}^{3} \bar{\mathbf{e}}_{tj}^{(e)} \boldsymbol{\alpha}_i^{(e)}(x,y) \tag{3.24}$$

while the approximation $\bar{e}_z^{(e)}$ to the exact longitudinal component e_z of the electric field on element (e) is expressed in terms of nodal shape functions $\alpha_i^{(e)}$:

$$\bar{e}_z^{(e)}(x,y) = \sum_{j=1}^{3} \bar{e}_{zj}^{(e)} \alpha_j^{(e)}(x,y) \tag{3.25}$$

Accordingly, vectorial and nodal shape functions are also used to weight the transverse and longitudinal (3.20) and (3.21), respectively, obtaining:

$$\begin{aligned} \iint_{\Delta^{(e)}} \boldsymbol{\alpha}_i^{(e)}(x,y) \cdot \left(\nabla_t \times \left(\frac{1}{\mu_r} \nabla_t \times \mathbf{e}_t\right)\right) dS - \frac{\gamma^2}{\mu_r} \iint_{\Delta^{(e)}} \boldsymbol{\alpha}_i^{(e)}(x,y) \cdot (\nabla_t e_z + \mathbf{e}_t)\, dS \\ - k_0^2\, \bar{\epsilon}_r \iint_{\Delta^{(e)}} \boldsymbol{\alpha}_i^{(e)}(x,y) \cdot \mathbf{e}_t\, dS = R_i^{(e)} \quad i = 1,\ldots,3 \end{aligned} \tag{3.26}$$

$$\begin{aligned} \gamma^2 \iint_{\Delta^{(e)}} \hat{\mathbf{z}} \alpha_i^{(e)}(x,y) \;\cdot\; \nabla_t \times \left[\frac{1}{\mu_r} (\nabla_t e_z + \mathbf{e}_t) \times \hat{\mathbf{z}}\right] dS \\ - \gamma^2 k_0^2\, \bar{\epsilon}_r \iint_{\Delta^{(e)}} \hat{\mathbf{z}} \alpha_i^{(e)}(x,y) \\ \cdot\; e_z \hat{\mathbf{z}}\, dS = R_i^{(e)} \quad i = 1,\ldots,3 \end{aligned} \tag{3.27}$$

The order of partial derivatives in the integrals in the left-hand side may be decreased by exploiting both the vector integration by parts formula

$$\iint_{\Omega} \mathbf{A} \cdot (\nabla \times \mathbf{B})\, d\Omega = \iint_{\Omega} (\nabla \times \mathbf{A}) \cdot \mathbf{B}\, d\Omega - \int_{\Gamma} (\mathbf{A} \times \mathbf{B}) \cdot \hat{\mathbf{n}}\, d\Gamma \tag{3.28}$$

where $\Gamma = \partial\Omega$, and the vector form of Green's theorem

$$\begin{aligned} \iint_{\Omega} \mathbf{A} \cdot (\nabla \times \nabla \times \mathbf{B})\, d\Omega = \iint_{\Omega} (\nabla \times \mathbf{A}) \cdot (\nabla \times \mathbf{B})\, d\Omega \\ - \int_{\Gamma} (\mathbf{A} \times \nabla \times \mathbf{B}) \cdot \hat{\mathbf{n}}\, d\Gamma \end{aligned} \tag{3.29}$$

Introducing (3.28) into (3.28) and (3.29) into (3.26) gives, after rearranging terms, the relations:

$$\begin{aligned}\frac{1}{\mu_r}\iint_{\Delta^{(e)}}\left(\nabla_t\times\boldsymbol{\alpha}_i^{(e)}\right)&\cdot(\nabla_t\times\mathbf{e}_t)\,dS-k_0^2\,\bar{\epsilon}_r\iint_{\Delta^{(e)}}\boldsymbol{\alpha}_i^{(e)}\cdot\mathbf{e}_t\,dS\\&-\frac{1}{\mu_r}\int_{\Gamma^{(e)}}\left(\boldsymbol{\alpha}_i^{(e)}\times\nabla_t\times\mathbf{e}_t\right)\cdot\hat{\mathbf{n}}\,d\Gamma\\&-\frac{\gamma^2}{\mu_r}\int_{\Delta^{(e)}}\boldsymbol{\alpha}_i^{(e)}\cdot\nabla_t e_z\,dS\\&-\frac{\gamma^2}{\mu_r}\iint_{\Delta^{(e)}}\boldsymbol{\alpha}_i^{(e)}\cdot\mathbf{e}_t\,dS=R_i^{(e)}\quad i=1,\ldots,3\end{aligned}\tag{3.30}$$

$$\begin{aligned}\frac{\gamma^2}{\mu_r}\iint_{\Delta^{(e)}}\left(\nabla_t\times\hat{\mathbf{z}}\alpha_i^{(e)}\right)&\cdot[(\nabla_t e_z+\mathbf{e}_t)\times\hat{\mathbf{z}}]\,dS\\&-\frac{\gamma^2}{\mu_r}\int_{\Gamma^{(e)}}\hat{\mathbf{z}}\alpha_i^{(e)}\\&\times[(\nabla_t e_z+\mathbf{e}_t)\times\hat{\mathbf{z}}]\,d\Gamma\\&-\gamma^2k_0^2\bar{\epsilon}_r\iint_{\Delta^{(e)}}\hat{\mathbf{z}}\alpha_i^{(e)}\\&\cdot\, e_z\hat{\mathbf{z}}\,dS=R_i^{(e)}\quad i=1,\ldots,3\end{aligned}\tag{3.31}$$

where the line integrals extends to the element $\Delta^{(e)}$ contour $\Gamma^{(e)}$ and need some comments. Three situations may arise: the contour is shared between adjacent elements, it lies at a metallic wall, or at magnetic wall. In all three cases the line integrals vanish and should not be included into the final equation. Let us consider, for instance, the line integral in (3.30) and the relation

$$-\frac{1}{\mu_r}\nabla_t\times\mathbf{e}_t=jk_0Z_0\gamma\mathbf{H}_z\tag{3.32}$$

It is easy to see that contributions coming from adjacent elements annihilate each other because all the quantities in the integral are equal but the outward normal vector, which is opposite. In the case of a perfectly conducting wall the weighting functions vanish at the boundary and finally in the case of a magnetic wall H_z vanish. Similar reasonings hold for the line integral in (3.32). Introducing the finite element expansions of the unknown field on element (e), (3.24), (3.25), into (3.30), (3.32) and annihilating the residue, yields the matrix:

$$\begin{aligned}&\left\{\begin{bmatrix}0 & 0\\ 0 & \frac{1}{\mu_r}\left[S_t^{(e)}\right]-k_0^2\bar{\epsilon}_r\left[T_t^{(e)}\right]\end{bmatrix}\right\}\begin{bmatrix}\left[\bar{E}_z^{(e)}\right]\\ \left[\bar{E}_t^{(e)}\right]\end{bmatrix}=\\&=\gamma^2\left\{\begin{bmatrix}\frac{1}{\mu_r}\left[S_z^{(e)}\right]-k_0^2\bar{\epsilon}_r\left[T_z^{(e)}\right] & \frac{1}{\mu_r}\left[G^{(e)}\right]^T\\ \frac{1}{\mu_r}\left[G^{(e)}\right] & \frac{1}{\mu_r}\left[T_t^{(e)}\right]\end{bmatrix}\right\}\begin{bmatrix}\left[\bar{E}_z^{(e)}\right]\\ \left[\bar{E}_t^{(e)}\right]\end{bmatrix}\end{aligned}\tag{3.33}$$

where the entries of the local matrices are given by:

$$S_{zij}^{(e)} = \iint_{\Delta_e} \nabla_t \alpha_i^{(e)} \cdot \nabla_t \alpha_i^{(e)} dS \tag{3.34}$$

$$T_{zij}^{(e)} = \iint_{\Delta_e} \alpha_i^{(e)} \alpha_j^{(e)} dS \tag{3.35}$$

$$S_{tij}^{(e)} = \iint_{\Delta_e} \left(\nabla_t \times \boldsymbol{\alpha}_i^{(e)}\right) \cdot \left(\nabla_t \times \boldsymbol{\alpha}_j^{(e)}\right) dS \tag{3.36}$$

$$T_{tij}^{(e)} = \iint_{\Delta_e} \boldsymbol{\alpha}_i^{(e)} \cdot \boldsymbol{\alpha}_j^{(e)} dS \tag{3.37}$$

$$G_{ij}^{(e)} = \iint_{\Delta_e} \boldsymbol{\alpha}_i^{(e)} \cdot \nabla_t \alpha_j^{(e)} dS \tag{3.38}$$

Assembling on all elements and zeroing the residuals provides the final generalized eigenvalue problem:

$$\begin{bmatrix} 0 & 0 \\ 0 & [S_t] \end{bmatrix} \begin{bmatrix} [\bar{E}_z] \\ [\bar{E}_t] \end{bmatrix} = \gamma^2 \begin{bmatrix} [S_z] & [G]^T \\ [G] & [T_t] \end{bmatrix} \begin{bmatrix} [\bar{E}_z] \\ [\bar{E}_t] \end{bmatrix} \tag{3.39}$$

Once the normalized operating frequency k_0 is fixed, the above equation allows computing the propagation and attenuation constants of the characteristic modes of the guiding structures, which can be used to plot the dispersion diagram sought.

3.4 Numerical Implementation

The algorithm described in the previous section has been implemented by using a mixed vector-nodal elements that approximate the longitudinal component e_z of the electric field with first-order nodal shape functions and the transverse component $\mathbf{e}_t$ with first-order vectorial shape functions. Both kinds of elements are described in Chapter 2 where also the pertinent element matrices $[S_t^{(e)}]$, $[S_z^{(e)}]$, $[T_t^{(e)}]$, and $[T_z^{(e)}]$ are provided. As for the element matrix $[G^{(e)}]$, it can be computed as follows: first, the gradient $\nabla_t \alpha_j^{(e)}$ that appears in (3.38) is expressed as:

$$\nabla_t \alpha_j^{(e)} = \frac{1}{2A} \begin{bmatrix} y_{j+2} - y_{j+1} \\ x_{j+1} - x_{j+2} \end{bmatrix} \tag{3.40}$$

being, as usual, A the element's area, (x_i, y_i) the vertexes coordinates of element (e), and cycling the index j modulus 3 $\overleftarrow{1 \to 2 \to 3}$. Then, algebraic manipulations lead to the form:

$$
\begin{aligned}
G_{11}^{(e)} &= K_1\left((y_2 - y_3)((y_2 - y_3) - (y_3 - y_1))\right. \\
&\quad \left. + (x_3 - x_2)((x_3 - x_2) - (x_1 - x_3))\right) \\
G_{12}^{(e)} &= -K_1\left((y_3 - y_1)((y_3 - y_1) - (y_2 - y_3))\right. \\
&\quad \left. + (x_1 - x_3)((x_1 - x_3) - (x_3 - x_2))\right) \\
G_{13}^{(e)} &= K_1\left(-y_1(y_3 - y_1) - y_2(y_2 - y_3)y_3(y_1 - y_2)\right. \\
&\quad \left. + x_1(x_1 - x_3) + x_2(x_3 - x_2) + x_3(x_2 - x_1)\right) \\
G_{21}^{(e)} &= K_2\left(-y_1(y_2 - y_3) - y_2(y_1 - y_2) - y_3(y_3 - y_1)\right. \\
&\quad \left. + x_1(x_3 - x_2) + x_2(x_2 - x_1) + x_3(x_1 - x_3)\right) \\
G_{22}^{(e)} &= K_2\left((y_3 - y_1)((y_3 - y_1) - (y_1 - y_2))\right. \\
&\quad \left. + (x_1 - x_3)((x_1 - x_3) - (x_2 - x_1))\right) \\
G_{23}^{(e)} &= -K_2\left((y_1 - y_2)((y_1 - y_2) - (y_3 - y_1))\right. \\
&\quad \left. + (x_2 - x_1)((x_2 - x_1) - (x_1 - x_3))\right) \\
G_{31}^{(e)} &= -K_3\left((y_2 - y_3)((y_2 - y_3) - (y_1 - y_2))\right. \\
&\quad \left. + (x_3 - x_2)((x_3 - x_2) - (x_2 - x_1))\right) \\
G_{32}^{(e)} &= K_3\left(-y_1(y_1 - y_2) - y_2(y_3 - y_1) - y_3(y_2 - y_3)\right. \\
&\quad \left. + x_1(x_2 - x_1) + x_2(x_1 - x_3) + x_3(x_3 - x_2)\right) \\
G_{33}^{(e)} &= K_3\left((y_1 - y_2)((y_1 - y_2) - (y_2 - y_3))\right. \\
&\quad \left. + (x_2 - x_1)((x_2 - x_1) - (x_3 - x_2))\right)
\end{aligned}
\tag{3.41}
$$

with $K_j = \sqrt{(x_j - x_{j+1})^2 + (y_j - y_{j+1})^2}/(12A)$.

Once the local element have been computed, what is only needed is to build the global matrices

$$
[A] = \begin{bmatrix} 0 & 0 \\ 0 & [S_t] \end{bmatrix}; \quad [B] = \begin{bmatrix} [S_z] & [G]^T \\ [G] & [T_t] \end{bmatrix}
\tag{3.42}
$$

and solve the generalized eigenvalue problem $[A][x] = \gamma^2[B][x]$ to find the possible values of the propagation constant γ corresponding to a given value of the normalized frequency k_0^2. If dielectric losses have to be considered, the two matrices are complex, thus memory requirement is doubled with respect to the case of lossless waveguides. Moreover, since the matrix $[A]$ is not positive definite, the LAPACK software library does not include routines suitable to solve the eigenvalue problem (3.39) exploiting also the banded nature of the matrices $[A]$ and $[B]$. The LAPACK routine SGGEV used in the code WG (waveguides) makes use of full matrices thus increasing memory requirements. The local matrices must be assembled into the global one accordingly. To keep memory requirement to a minimum improving also the performance of the code WG, the interested reader should resort to more efficient eigenvalue problem solvers, based for instance on the subspace iteration method [21], that allows fully exploiting the sparse nature of the finite element matrices. The Matlab version of the code, thanks to the Matlab memory model, can indeed treat larger problems, even if more slowly.

Another difficulty encountered in the solution of the generalized eigenvalue problem (3.39), stems from the fact that the matrix $[A]$ presents three blocks which are zero, and this leads to a highly ill-conditioned matrix, thus rendering the solution prone to numerical errors. To obtain accurate results is necessary to balance the two matrices $[A]$ and $[B]$. In the code WG this is accomplished by first scaling the mesh so as to have the average length of an edge equal to 1 (frequencies are of course scaled accordingly). This leads to a low difference between the eigenvalues of $[T_t]$ and those of $[S_z] - k_0^2\bar{\epsilon}_r[T_z]$, thus improving the conditioning of $[B]$. Then the maximum entries of $[A]$ and $[B]$ are sought for, and the matrix scaled accordingly, to get more accurate solutions from the eigenvalue solver. This might be unnecessary in the Matlab version of the code, since the function `eig` does some internal automatic balancing, but it is anyway performed to keep the two code versions consistent.

3.5 The Code WG: Waveguides

The structure of the code WG is better explained by resorting to the solution of a simple example, the analysis of an empty rectangular waveguide with dimensions $a = 18.35$ mm and $b = 9.175$ mm, which can serve both to validate the code and discuss its accuracy. The chosen waveguide has a unimodal band in the frequency range 8.17–16.34, hence can be used with good performances in the 10–15 GHz band, and will be extensively used throughout the book.

The analysis is done three times, on the entire waveguide in Figure 3.4(a), on half the waveguide, with a p.e.c. fictitious boundary in Figure 3.4(b), and on half waveguide with a p.m.c. fictitious boundary, in Figure 3.4(c).

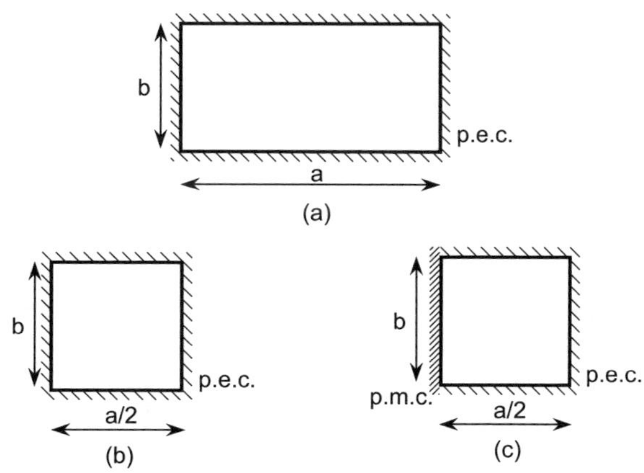

Figure 3.4 Rectangular waveguide in air: (a) full domain; (b) half waveguide (with p.e.c. wall); (c) half waveguide (with p.m.c. wall).

The first task of the analysis is building the geometric model and its finite element discretization using the mesh generator provided. The input file for the mesh generator, in the case of the whole cross-section waveguide Figure 3.4(a), is the following Example 3.1.

Example 3.1

```
wg1.fem
8 1
1 3 1 2 3 5 8 7 6 4 1
1 0.000 0.000 1
2 9.175 0.000 1
3 18.35 0.000 1
4 0.000 4.5875 1
5 18.35 4.5875 1
6 0.000 9.175 1
7 9.175 9.175 1
8 18.35 9.175 1
1 12 6
1 1 1 1 1 1 1 1 1 1 1 1
1 1 1 1 1 1
```

The CD-ROM contains the input files for the other two configurations, which can anyway easily derived from that shown above. It is important noting that dimensions must be given in millimeters. The geometric data file created by the mesh generator is the first input file needed by the code WG, the second one is a file containing the electromagnetic data for the analysis. This latter file can be created using a text editor of reader's choice and must have the format described in Example 3.2.

Example 3.2

```
5, 30, 26        !Frequencies: Start, End, Steps
9                !Modes
1                !Materials
1                !Material Label(s)
1.               !Epsilon(s)
1                !PEC Label
```

Please note that the !-beginning comments are placed here for clarity and are unnecessary, and that the separator between numbers on the same line must be a comma.

The first line defines the frequency range which must be investigated in the analysis, providing the initial and final frequency, in gigahertz, as well as the number of equispaced frequency steps. The second line instructs the code on the number of modes that the users wish to be saved into the output file. The group of three lines that follows describes the materials present in the computational domain: the first line reports the number of different dielectric materials involved, in this first example only one, the second describes the different element labels which characterize each material, separated by commas, and the third lists the corresponding permittivity. Finally, the label used to denote nodes at p.e.c. boundaries is given. There is no need to specify nodes at p.m.c. boundaries because the code WG uses the electric field as working unknown, for which p.m.c. boundaries represent homogeneous Neumann boundary conditions that need not be explicitly enforced since they are natural boundary conditions for the formulation used.

The code WG first reads the mesh and the electromagnetic file, then performs a preprocessing phase during which p.e.c. nodes and edges are identified and the portion of the matrices $[A]$ and $[B]$ that is frequency independent is computed and stored into an auxiliary FORTRAN array. Although this procedure is memory consuming, the CPU time saved when the analysis is requested for many frequency values is significant.

When the preprocessing phase is ended, the frequency cycle is entered: for each step in the sweep over the desired frequency range the frequency-independent matrices are copied into the working array, augmented by the frequency-dependent terms, and modified to enforce boundary conditions. There are actually two ways of imposing the homogeneous Dirichelet boundary conditions at nodes and edges lying at p.e.c. boundaries. The simplest is probably the following: if i is the index of a p.e.c. node or edge, then setting the corresponding field to zero is equivalent to set $a_{ij} = a_{ji} = 0$, and $b_{ij} = b_{ji} = 0$ for $j = 1, \ldots, N$ and $a_{ii} = b_{ii} = 1$. The simplicity of this approach is traded against memory and CPU time, since unnecessary big matrices are stored and solved for, and the presence in the solution of unphysical, but easily recognizable, unitary eigenvalues. The second approach, here implemented, is slightly more involved than the previous one, but more efficient in terms of memory and CPU requirements. It consists in actually reducing the dimensions of matrices $[A]$ and $[B]$ by eliminating all the rows and columns corresponding to both nodes and edges lying at p.e.c. boundaries.

The generalized eigenvalue problem assembled is then solved by resorting to the SGGEV subroutine of the LAPACK software library. This routine is suitable to deal with real matrices as those arising in the case of lossless material. If the reader is interested in including losses in the analysis, what needs to be done is changing the definitions of the FORTRAN arrays whose elements involve the permittivity value from "real" to "complex" and using the LAPACK routine CGGEV, which is the complex version of SGGEV.

It must be noted that while $[B]$ is not singular, $[A]$ is so, and this has two implications: not all freely available solvers can deal with the eigenvalue problem at hand, and there are, in the solution, nonphysical zero eigenvalues. The postprocessing phase to separate the physical from nonphysical solutions must be done carefully since numerical errors give rise to small, but not exactly zero eigenvalues. The last task of the code WG is saving the computed eigenvalues in output files. The pure real or imaginary eigenvalues, representing the propagation constant of modes that either propagate or are in cutoff, are saved in an user-defined output file, which is given, by the program, the extension "OUT." Each record of this file contains the frequency at which the computation has been carried out followed by as many values as propagating or evanescent modes have been found among the first M modes that the user saved. Each of these values represent a computed eigenvalue. If the relative mode propagates, that is the propagation constant γ is pure imaginary, then the value of its imaginary part, which is positive, is stored. If the mode is under cutoff then the opposite value of its real part, that is negative, is stored. This format allows the user to directly plot this output file to obtain a Brillouin diagram, as those presented later.

In the general case of lossless inhomogeneous waveguides there is also the possibility of complex modes, as we will see in the next section. If complex eigenvalues are found, all of them are saved into a separate file with the same name defined by the user, but extension "CPL." This file presents as many records as the total number of complex eigenvalues founds. Each record contains the frequency value for which the guiding structure has been analyzed, followed by the imaginary (β) and real (α) parts of the propagation constant, respectively. Since complex modes, due to the symmetry of the matrices $[A]$ and $[B]$, come in conjugate pairs, only one eigenvalue of each pair, that with a positive imaginary part, is stored.

Furthermore, if requested by the user, the code can also store the eigenvector, to allow for the field representation of the modes. These files are created for each real mode stored in the output file and for each frequency sample, hence besides being time consuming the process will also write a very large number of files.

Each mode actually produces two distinct files, one storing the longitudinal component coefficients, the other the transverse component coefficients. File names begin with the user-defined string to which is appended a some text according to the pattern `_ev_$_m#_f#.out` where `$` is either `z` or `t` and indicates the longitudinal or transverse component, while # are two integer numbers indicating the mode, as ordered by the program, and the frequency point.

In case of the presence of complex modes they are also saved, as many as the real modes are, in files with names like `_Cev_$_m#_f#.out`.

3.6 Some Examples

The empty rectangular waveguide of the previous section has all real eigenvalues: negative if the associated mode is propagating at the given frequency, positive if it is in cut-off. Figure 3.5 shows the dispersion diagram for the chosen ($a =$ 18.35 mm, $b = 9.175$ mm) rectangular waveguide in the range 5–30 GHz when the computational domain is comprised of the full waveguide cross-section. The modes shown, ordered for increasing cutoff frequency, are: TE_{10}, $TE_{20} \equiv TE_{01}$, $TE_{11} \equiv TM_{11}$, $TE_{21} \equiv TM_{21}$, TE_{30}, TE_{31}. The latter is equivalent to TM_{31}, but this mode is not reported in the figure.

Figure 3.5 shows that the dispersion curves of degenerate *TE* modes are pretty coincident, and this is reasonable since they are computed and approximated on the same domain, comprised of only the edge elements. On the other hand, when a *TE* mode is degenerate with a *TM* mode, the difference between the two dispersion curves is larger that in the previous case and this can be explained by noting that since *TM* modes have a nonzero E_z component, which is approximated in the nodal element domain, *TE* and *TM* modes have two different basis function sets and this is the primary cause of the larger gap in the computed results.

Generally speaking, the *TM* modes are more sensitive to numerical errors when using the electric field as working variables because their eigenvectors have three

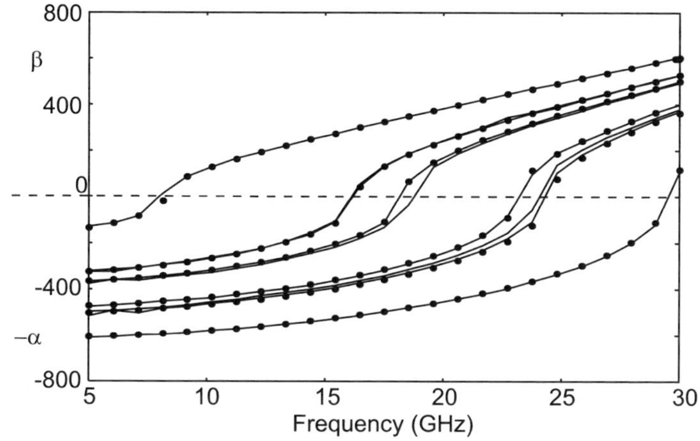

Figure 3.5 Dispersion diagram for the rectangular waveguide of Figure 3.4(a). Lines: FEM computation; bullets: analytic solution.

nonzero components in space and thus both nodal and vectorial basis functions must be used in the finite element approximation, while only edge elements are used in the approximation of *TE* modes eigenvectors. This is evident from the fact that the dispersion curves of *TM* modes are farther away from the theoretical results than the dispersion curves of *TE* modes. The situation is expected to be inverted when adopting the magnetic field as a working variable.

The mesh used for the analysis of the empty rectangular waveguide presents 91 nodes, 36 of which are on the p.e.c. boundary, and 234 edges, 36 of which are on the p.e.c. boundary, for a total of 253 unknowns. The two problems defined on half cross-sections have 49 nodes and 120 edges each; the p.e.c. closed half-waveguide has 24 nodes and 24 edges on the p.e.c. boundaries, for a total of 121 degrees of freedom, while the p.m.c. closed half-waveguide has 19 nodes and 18 edges on the p.e.c. boundaries, for a total of 132 degrees of freedom. It is thus evident that memory requirements for mesh storage are cut in a half, while matrices $[A]$ and $[B]$ have a size which is roughly one-quarter of those of the full waveguide problem leading to a saving of CPU time of approximately 85% for each frequency point. Of course if the whole modal spectra is desired, both the half cross-section problems must be analyzed, but the CPU time saving is still as high as 70%.

As a second much more complex and interesting example, the same shielded microstrip configuration introduced in Chapter 1 is analyzed again removing the quasistatic approximation previously adopted. By exploiting the full-wave formulation developed in this chapter, all the higher order modes of the structure can be found and, in particular, also the complex modes which may arise in lossless inhomogeneous structures [22, 23]. Please note that this time both the box and the strip must have the same label, 1, being both made of perfect electric conductor.

The mesh employed for the analysis is the same as that presented earlier (Figures 2.2 and 2.4), while the input electromagnetic data file is changed in the following Example 3.3.

Example 3.3

```
10.0, 20.0, 40          !Frequencies: Start, End, Steps
7                       !Modes
2                       !Materials
1, 2                    !Material Label(s)
1., 8.875               !Epsilon(s)
1                       !PEC Label
```

Figure 3.6 shows the real and imaginary part of the normalized (with respect to k_0) propagation constant $\tilde{\gamma}$ and the results compare well with those presented in [24]. The almost horizontal continuous line is relative to the fundamental mode whereas the dashed line represents the value of the fundamental mode normalized propagation constant computed using the quasistatic approximation of Chapter 1. This latter approximation leads to a fundamental mode that does not exhibit dispersion having a normalized phase propagation constant which is frequency independent and equal to $\tilde{\beta} = 2.9$. This is the usual behavior of a TEM mode. The rigorous FEM analysis based on the full-wave formulation shows that this is not the case: the fundamental mode exhibits a certain degree of dispersion and thus is not TEM. However, since the quasistatic approximation provides a fairly good simple model, it is a so called quasi-TEM mode.

It must be also noted that when examining only half structures introducing a magnetic wall at the symmetry plane, only even modes are detected. On the other hand, this kind of enclosure is preferable to that obtained by introducing an electric wall at the plane of symmetry because for a microstrip line the fundamental quasi-TEM mode and the first higher order modes are even.

Another very interesting phenomenon is the presence of a complex mode, that is a mode characterized by a complex value of the normalized propagation constant $\tilde{\gamma}$, even though the guiding structure has been assumed lossless. When losses are present, matrices $[A]$ and $[B]$ become complex and non-Hermitian, thus the

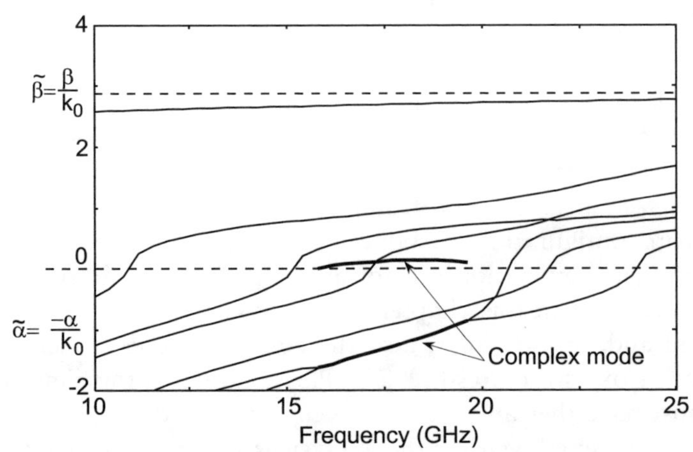

Figure 3.6 Normalized dispersion diagram for a boxed microstrip line. The presence of a couple of complex modes is evidenced with a heavier line.

eigenvalues may be complex. On the other hand, lossless structures givet rise to real symmetric finite element matrices $[A]$ and $[B]$, thus the generalized eigenvalue problem may have either real or complex conjugate pairs eigenvalues. This is actually the case for the lossless shielded microstrip line, in which the 6th and 7th modes degenerates into two conjugate complex modes in the range of frequency from 15.76 GHz to 19.61 GHz. The attenuation constant α of the modes is continuous, while a small phase propagation constant β arises in the above-mentioned frequency range and disappear at either higher or lower frequency where the two complex modes split again into two ordinary modes. This kind of complex modes is of course of a different nature from that which arise in a lossy waveguide. Complex values of the propagation constant in a lossy waveguide do not appear in complex conjugate pairs and the real part denotes a decaying in the amplitude of an ordinary mode due to the presence of losses.

The file with extension "CPL" created by the code WG when analyzing the shielded microstrip line contains many complex eigenvalues and each of them corresponds to a couple of ordinary modes which degenerate into a complex one. Tracking these mode changes is a hard task and very difficult to be accomplished automatically, both for the degeneration problem and for the intersections that the dispersion curves may exhibits. For instance, Figure 3.6 shows that the 3rd and 4th modes as well as the 5th and 6th modes, cross each other and this is a typical behavior of the dispersion curves of inhomogeneous guiding structures. The output file created by the analysis programs may thus require some user interventions before being used to create good plots. The user may have to recognize crossing points as well as to realign data when abrupt jumps occur in dispersion curves, due to the fact that either two real eigenvalues merge into a complex one or a complex mode splits into two real ones.

As an example, the files necessary to obtain Figure 3.6 have been included in the disc. File N1.DAT contains dispersion diagrams of the modes whose propagation constant always remains either real or purely imaginary, file N2A.DAT contains the lower frequency range (10–15.76 GHz) where eigenvalues 6 and 7 have their normal behavior, and file N2B.DAT contains the corresponding higher frequency range (19.61–25 GHz). File N3.DAT contains the complex modes (15.76–19.61 GHz) and, finally, the file GNUB is a file to be loaded in GNUPLOT that manages all four. For Matlab users this same task is accomplished by program `matb.m`.

The new Matlab postprocessing codes described in Chapter 2 and included in this second edition allow for a very flexible result presentation. In particular, as already noted, this second edition codes output field maps besides eigenvalues. A single frequency simulation at 17.5 GHz has been performed, requiring that the first two modes are saved. This leads to saving both the first two propagating modes at the selected frequency, and the first two complex modes, for a total of 8 output files, one for each component for each mode and frequency. The two propagating modes are the quasi-TEM fundamental mode and the quasi-TE_{10} mode which characterize the shielding box seen as a rectangular waveguide, loaded by the dielectric slab and the microstrip. Figure 3.7 clearly shows how the transverse component of the electric field is basically that of the static case (Chapter 1), whereas it is also evident how e_z is not zero and has a nonvanishing value at the dielectric-air interface, as expected.

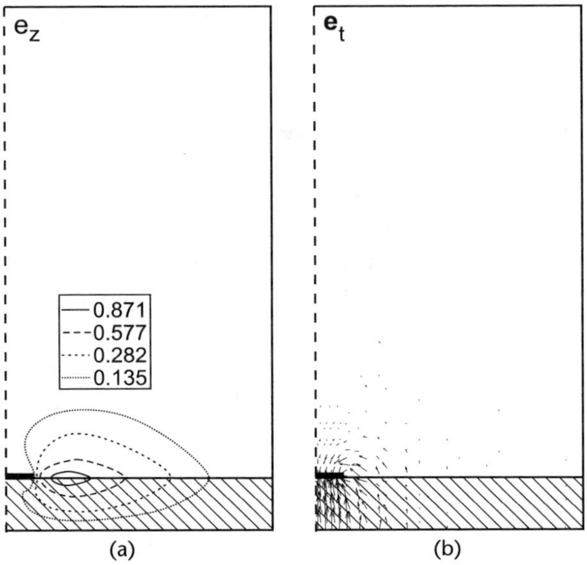

Figure 3.7 Quasi-TEM mode in a boxed microstrip. Longitudinal component (a) and transverse component (b) of the electric field.

It is of extreme interest also that the field distribution of the first complex mode, shown in Figure 3.8. It is apparent from the field maps that the complex mode is centered on the dielectric-air interface, and not on the strip, whose effect is present but not dominant on the transverse component of the field, and neglectable on the longitudinal one.

As the third and last example, the boxed unilateral finline depicted in Figure 3.9 is analyzed. This, with a substrate characterized by $\epsilon_r = 2.22$ is the configuration

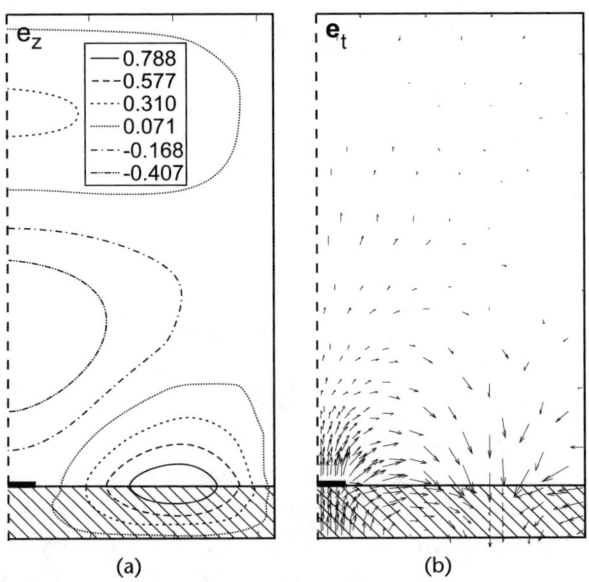

Figure 3.8 First complex mode in a boxed microstrip. Longitudinal component (a) and transverse component (b) of the electric field.

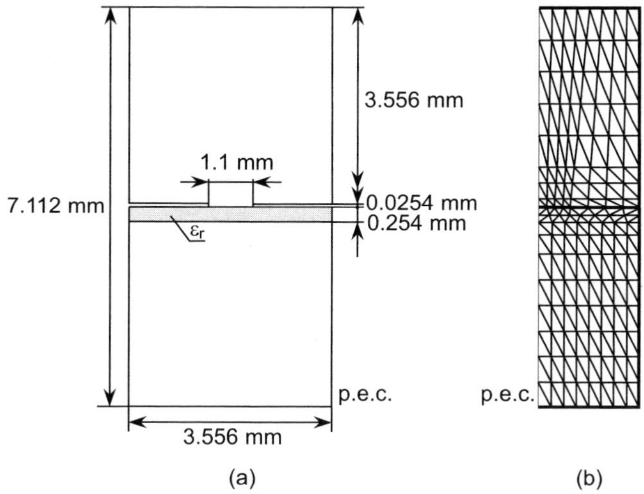

Figure 3.9 (a, b) Finline structure and finite element mesh adopted.

analyzed in a paper by Rozzi, Pierantoni and Farina [25] through a field representation in terms of the fields of the modes at cutoff.

To reduce the computational domain it is convenient to exploit the geometric symmetry, and to this end some considerations need to be done to determine how to close the half domain. First, the finline is a simply connected structure, thus no quasi-TEM mode may be present. Second, we must distinguish the modes which are actually supported by the finline, and the modes which indeed are those of a loaded rectangular waveguide. The former are characterized by a high electromagnetic energy density in the gap of the finline, while the latter have a field more smoothly distributed in the waveguide cross-section. It is easy to understand how lowest finline modes present the two metallic fins with opposite polarities, thus with the lines of the **E** field starting at one fin and ending at the other. This implies that,

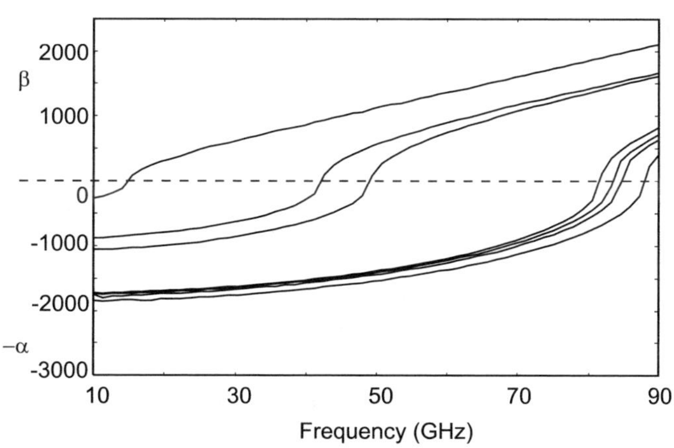

Figure 3.10 Dispersion diagram for the finline of Figure 3.9.

Table 3.1 Cutoff Frequencies for the First Seven Modes

Mode	*[25]*	*FEM*	*% Diff.*
1	15.561	14.879	4.4%
2	42.123	42.108	0.0%
3	49.699	48.961	1.5%
4	82.397	81.453	1.1%
5	84.029	83.503	0.6%
6	84.370	84.704	0.4%
7	91.281	87.951	3.7%

at the plane of symmetry, $E_x \neq 0$, while $E_y = 0$, being x and y the horizontal and vertical axis, respectively. Thus the first modes of the structure are odd in nature and they are taken into account by enclosing the half domain with a p.e.c. wall instead of that with a p.m.c. as in the previous example.

Figure 3.10 reports the code results in the range 10–90 GHz. It must be noted that the longest edge in the mesh is about 0.5 mm, and thus the shortest free-space wavelength which can be safely approximated is 5 mm, corresponding to 60 GHz. To estimate the accuracy of the method Table 3.1 reports the cutoff frequency values of the first 7 modes both accordingly to Rozzi et al. [25] and computed with the present code. It must be noted that since a frequency step of 1.333 GHz has been used for the present analysis, the FEM column in the table is filled with interpolated values.

3.7 Disc Content

3.7.1 FORTRAN

The software here described is contained in the subdirectory WG of the directory CODES. The program files are WG.F, WG_SUBS.F, and ELEEN.F that must be compiled all together and with the library described in Chapter 2. The WG code, as with all FORTRAN codes in this book, runs interactively, asking the user to type in the name of mesh and data files and along with other data. This of course can be avoided by using standard input redirection features of your operating system.

The three subdirectories WG, STRIP, and FIN, contains all the input and mesh files of the examples discussed in the previous section. A more complete description of each file is contained in the README file in the directory WG itself.

3.7.2 Matlab

For the Matlab version of the codes the structure is similar, `wg.m` being the main code, but all subroutines are contained in a file of their own, hence `eleen.m` computes matrices G while `Globm.m` and `DimMat.m` compute the frequency-independent and frequency-dependent part of the system matrices, the latter performing also eigenproblem solution. `WriteRes.m` and `VriteResFields.m` saves eigenvalues and eigenvectros. The `wg.m` code, as with all the Matlab codes in this book, runs noninteractively, but as a function, that is all the data required must be passed to `wg.m` when launched. The syntax is:

```
wg(fnamemesh,fnameaux,fnameout,fieldflag)
```

Where `fnamemesh` is a string containing the name of the file describing the mesh, `fnameaux` is a string containing the name of the file containing the electromagnetic data, and `fnameout` is a string with the basic name of the files which will contain the output. Finally `fieldflag` is an integer which, if equal to 1, forces the output of eigenvectors.

References

[1] Lee, J. F., "Finite Element Analysis of Lossy Dielectric Waveguides," *IEEE Transactions on Microwave Theory and Techniques*, Vol. 42, pp. 1025–1031, 1994.

[2] Silvester, P. P., "Finite-Element Solution of Homogeneous Waveguide Problems," *Alta Frequenza*, Vol. 38, pp. 313–317, 1969.

[3] Cendes, Z. J. and Silvester, P. P., "Numerical Solution of Dielectric Loaded Waveguides: I-Finite Element Analysis," *IEEE Transactions on Microwave Theory and Techniques*, Vol. MTT-18, pp. 1124–1131, 1970.

[4] Ahmed, S. and Daly, P., "Finite-Element Methods for Inhomogeneous Waveguides," *IEE Proceedings*, Vol. 116, pp. 1661–1664, 1969.

[5] Silvester, P. P. and Ferrari, R. L., *Finite Elements for Electrical Engineers*, Cambridge, UK: Cambridge University Press, 3rd edition, 1996.

[6] Konrad, A., "Vector Variational Formulation of Electromagnetic Fields in Anisotropic Media," *IEEE Trans. Microwave Theory Tech.*, Vol. 24, pp. 553–559, 1976.

[7] Rahman, B. M. A. and Davies, J. B., "Penalty Function Improvement of Waveguide Solution by Finite Elements," *IEEE Transactions on Microwave Theory and Techniques*, Vol. MTT-32, pp. 922–928, 1984.

[8] Fernandez, F. A. and Lu, Y., "A Variational Finite Element Formulation for Dielectric Waveguides in Terms of Transverse Magnetic Fields," *IEEE Transactions on Magnetics*, Vol. 27, pp. 3864–3867, 1991.

[9] Hayata, K., et al., "Vectorial Finite-Element Method Without Any Spurious Solutions for Dielectric Wave-Guiding Problems Using Transverse Magnetic-Field Component," *IEEE Transactions on Microwave Theory and Techniques*, Vol. MTT-34, pp. 1120–1124, 1986.

[10] Hayata, K., Eguchi, M., and Koshiba, M., "Finite Element Formulation for Guided-Wave Problems Using Transverse Electric Field Component," *IEEE Transactions on Microwave Theory and Techniques*, Vol. 37, pp. 256–258, 1989.

[11] Davies, J. B., "Finite Element Analysis of Waveguides and Cavities: A Review," *IEEE Transactions on Magnetics*, Vol. 29, pp. 1578–1583, 1993.

[12] Rahman, B. M. A., Fernandez, F. A., and Davies, J. B., "Review of Finite Element Methods for Microwave and Optical Waveguides," *Proceedings IEEE*, Vol. 79, pp. 1442–1448, 1991.

[13] Dillon, B. M. and Webb, J. P., "A Comparison of Formulations for the Vector Finite Element Analysis of Waveguides," *IEEE Transactions on Microwave Theory and Techniques*, Vol. 42, pp. 308–316, 1994.

[14] Barton, M. L. and Cendes, Z. J., "New Vector Finite Elements for Three-Dimensional Magnetic Field Computation," *Journal of Applied Physics*, Vol. 61, pp. 3919–3921, 1987.

[15] Lee, J. F., Sun, D. K., and Cendes, Z. J., "Full-Wave Analysis of Dielectric Waveguides Using Tangential Vector Finite Elements," *IEEE Transactions on Microwave Theory and Techniques*, Vol. MTT-39, pp. 1262–1271, 1991.

[16] Graglia, R. D., Wilton, D. R., and Peterson, A. F., "Higher Order Interpolatory Vector Bases for Computational Electromagnetics," *IEEE Transactions on Antennas and Propagation*, Vol. 45, pp. 329–332, 1997.

[17] Crowley, C. W., Silvester, P. P., and Hurwitz Jr., H., "Covariant Projection Elements for 3D Vector Field Problems," *IEEE Transactions on Magnetics*, Vol. 24, pp. 397–400, 1988.

[18] Peterson, A. F., "Vector Finite Element Formulation for Scattering from Two-Dimensional Heterogeneous Bodies," *IEEE Transactions on Antennas and Propagation*, Vol. 43, pp. 357–365, 1994.

[19] Cendes, Z. J., "Vector Finite Elements for Electromagnetic Field Computation," *IEEE Transactions on Magnetics*, Vol. 27, pp. 3958–3966, 1991.

[20] Sun, D., et al., "Spurious Modes in Finite-Element Methods," *IEEE Antennas Propagat. Magazine*, Vol. 37, pp. 12–24, 1995.

[21] Lu, Y., Zhu, S., and Fernandez, A. F., "The Efficient Solution of Large Sparse Nonsymmetric and Complex Eigensystems by Subspace Iteration," *IEEE Transactions on Magnetics*, Vol. 30, pp. 3582–358, 1994.

[22] Strube, J. and Arndt, F., "Rigorous Hybrid-Mode Analysis of the Transition From Rectangular Waveguide to Shielded Dielectric Image Guide," *IEEE Transactions on Microwave Theory and Technique*, Vol. MTT-33, pp. 391–401, 1985.

[23] Itoh, W.X. Huang T., "Complex Modes in Lossless Shielded Microstrip Lines," *IEEE Transactions on Microwave Theory and Technique*, Vol. 36, pp. 163–165, 1988.

[24] Valor, L. and Zapata, J., Efficient Finite Element Analysis of Waveguides with Lossy Inhomogeneous Anisotropic Materials Characterized by Arbitrary Permittivity and Permeability Tensors," *IEEE Transactions on Microwave Theory and Techniques*, Vol. 43, pp. 2452–2459, 1995.

[25] Rozzi, T., Pierantoni, L., and Farina, M., "Eigenvalue Approach to the Efficent Determination of Hybrid and Complex Spectrum of Inhomogeneous Closed Waveguide," *IEEE Transaction on Microwave Theory and Techniques*, Vol. 45, pp. 345–353, 1997.

CHAPTER 4

Microwave Guiding Structures: Devices and Circuits

The design of a high-performance microwave network requires an accurate characterization of each component, which can have a complex structure in terms of both geometry and material composition. In such case, the finite element method represents an excellent tool for the analysis and design since it easily allows taking into account all details of each device.

Characterization of microwave components requires a complete different approach than that discussed in the previous chapter for the characterization of a guiding structure, that is, for the computation of its dispersion diagram. The difference is twofold: first, even in the simplest case of closed waveguides a hybrid technique must be used to terminate the computational domain of analysis; and second, the problem to be solved is a deterministic one. As for the first difference, it must be noted that the effect of a discontinuity in a guiding structure may extend infinitely along the line itself, hence, in order to limit the analysis to a small region around the junction, a hybrid technique must be used to terminate the finite element mesh. The fact that the problem is of the deterministic type is a direct consequence of the final goal of the analysis: when analyzing a microwave component, the designer is usually interested in determining the behavior of the device when it is excited with a given incident field, most probably determined by computing the dispersion diagram and the fundamental mode of the guiding structures connected to the device under analysis. This leads to the formulation of a deterministic problem, whose solution provides the field distribution inside the component or junction when excited with a known field. A postprocessing phase is then usually necessary to recover the quantities of interest to the designer (scattering parameters, input impedance, power density, and so on).

Many different schemes has been proposed to solve this class of problems in the past years. Some of them combine the finite element method with integral equations [1, 2], other with modal expansions [3, 4], and others use an ingenious recursive scheme to iteratively enlarge the computational domain [5, 6].

The simple and accurate procedure to extract the scattering parameters of microwave components presented in this chapter belongs to the class that makes use of a modal expansion to represent the electromagnetic field into the guiding structures connected to the junctions. It will be described referring to the simple but yet important case of E- or H-plane rectangular waveguide junctions, but the formulation can be extended in a straightforward manner to the case of general junctions requiring a full three-dimensional treatment, as it will be done in Part II of this book. The same formulation can also be used when the modal expansion of the field in the guiding structures connected to the junction is not analytically known, but can be numerically computed solving the transverse problem as described in the previous chapter. The code described in Section 4.4 and included in the CD-ROM provided with this book implements the formulation outlined in the following sections and has been successfully employed for the analysis and design of E- and H- plane discontinuities and filters [4, 7].

4.1 The Finite Element—Modal Expansion Formulation: H-Plane Case

Rectangular waveguide junctions with a uniform cross-section along the direction perpendicular to the plane containing the waveguide axis (direction of propagation) and the electric or magnetic field of the fundamental TE_{10} mode are usually referred to as E-plane and H-plane junctions, respectively. This kind of devices is frequently used in microwave circuits as directional couplers, power dividers and combiners, phase shifters, filters, and stubs. Other components almost always present in waveguide circuits are E- and H-plane bends and corners which may have significant impact on the performance of sophisticated networks as those necessary for antenna beamforming.

Assuming that the junction is excited with the fundamental mode TE_{10} through one or more of its ports, the minimum number of field components necessary for a correct analysis of the junction is reported in Table 4.1: one component of the electromagnetic field suffices in case of homogeneous E- or H-plane junctions or inhomogeneous H-plane junctions, while two components are needed for the analysis of inhomogeneous E-plane junctions [8].

The finite element formulation of the problem will be carried out considering the case of an H-plane junction first, the case of E-plane homogeneous junctions being addressed in the next section. The top view of the geometry as well as the reference systems used for the analysis of a generic H-plane junction are sketched in Figure 4.1.

The region Ω where the finite element method is applied is delimited by the perfectly conducting wall Γ_0 and by N arbitrarily defined reference planes or ports Γ_k $(k = 1,\ldots,N)$ where the device is connected to rectangular waveguides. Assuming that one ore more ports are fed with the fundamental TE_{10} mode, since the problem geometry is uniform along the y direction, only the TE_{m0} modes are excited into the waveguides and the electric field into the region Ω has the E_y component only. The set of modes that must be used to represent the electromagnetic field into each waveguide section can be further reduced by exploiting symmetry, when it is present.

Table 4.1 Minimum Number of Field Components to be Used for the Analysis of E- and H-plane Junctions in Rectangular Waveguide

Junction Type	*Homogeneous*	*Inhomogeneous*
H-plane	E_y	E_y
E-plane	H_x	H_x, E_x

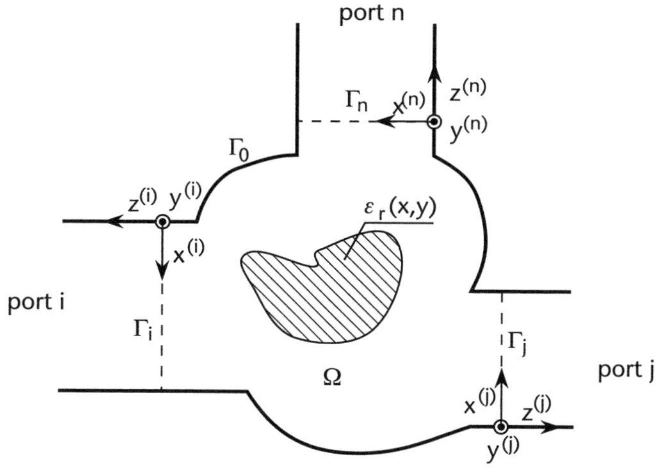

Figure 4.1 Generic rectangular waveguide multiport H-plane junction and reference systems adopted.

In general, the components of the electric and magnetic fields tangential to port k, $E^{(k)}_{y_{wg}}$, and $H^{(k)}_{x_{wg}}$, when the port (j) is fed with the TE_{10} mode can be expressed as

$$E^{(k)}_{y_{wg}}(x^{(k)}) = \delta_{kj} e_1^{(j)}(x^{(j)}) e^{j\beta_1^{(j)} z^{(j)}} + \sum_{m=1}^{\infty} B_m^{(k)} e_m^{(k)}(x^{(k)}) e^{-j\beta_m^{(k)} z^{(k)}} \tag{4.1}$$

$$j\omega\mu_0 H^{(k)}_{x_{wg}}(x^{(k)}) = \frac{\partial E^{(k)}_{y_{wg}}}{\partial z^{(k)}} \tag{4.2}$$

where $e_m^{(k)}, m = 1, 2, \ldots$ are the orthonormal modal functions of the TE_{m0} modes into the waveguide k:

$$e_m^{(k)}(x^{(k)}) = \frac{2}{\sqrt{a^{(k)} b}} \sqrt{\frac{k_0 Z_0}{\beta_m^{(k)}}} \sin\left(\frac{m\pi}{a^{(k)}} x^{(k)}\right) \tag{4.3}$$

In (4.1) to (4.3) k_0 and Z_0 are the free-space propagation constant and characteristic impedance, respectively; $\beta_m^{(k)}$ denotes the propagation constant into the waveguide k with dimensions $a^{(k)}$ and b:

$$\begin{cases} \beta_m^{(k)} = \sqrt{k_0^2 - k_{tm}^{(k)2}} & \text{for } k_0^2 \geq k_{tm}^{(k)2} \\ \beta_m^{(k)} = -j\sqrt{k_{tm}^{(k)2} - k_0^2} & \text{for } k_{tm}^{(k)2} > k_0^2 \end{cases} \tag{4.4}$$

and $k_{tm}^{(k)} = m\pi/a^{(k)}$ is the cutoff propagation constant of mode TE_{m0} into the waveguide k. In the preceding equations, all the waveguides connected to the junctions have been assumed empty.

Inside the arbitrarily shaped and inhomogeneous region Ω, the field cannot be expressed in terms of analytically known functions, thus the problem is solved seeking the solution of the scalar Helmholtz equation:

$$\nabla_t \cdot \left(\frac{1}{\mu_r} \nabla_t E_y \right) + k_0^2 \epsilon_r E_y = 0 \tag{4.5}$$

subject to the homogeneous Dirichlet boundary condition

$$E_y = 0 \tag{4.6}$$

at the metallic wall Γ_0 of the junction and the appropriate continuity conditions for the electric and magnetic fields at each port:

$$E_{y_{|\Gamma_k}} = E_{y_{wg}}^{(k)} \quad k = 1, \ldots, N \tag{4.7}$$

$$H_{x_{|\Gamma_k}} = H_{x_{wg}}^{(k)} \quad k = 1, \ldots, N \tag{4.8}$$

Recalling (4.2), the boundary condition (4.8) can be more conveniently expressed in terms of the outward normal derivative of the electric field at each port:

$$\frac{\partial E_y}{\partial n}_{|\Gamma_k} = \frac{\partial E_{y_{wg}}^{(k)}}{\partial z^{(k)}} \quad k = 1, \ldots, N \tag{4.9}$$

The boundary value problem described by the scalar Helmholtz equation (4.5) with boundary conditions (4.6), (4.7), and (4.9) may be recast in the weak form by using the weighted residual procedure. By choosing arbitrary differentiable ($C^{(0)}$) weighting functions W, $\bar{W}$, and $\bar{\bar{W}}$, (4.5), (4.6), (4.7), and (4.9) can be replaced by

$$\iint_\Omega W \nabla_t \cdot \left(\frac{1}{\mu_r} \nabla_t E_y \right) d\Omega + \iint_\Omega k_0^2 \epsilon_r W E_y d\Omega = 0 \tag{4.10}$$

$$\int_{\Gamma_0} \bar{W} E_y d\Gamma_0 = 0 \tag{4.11}$$

$$\int_{\Gamma_k} \bar{W} E_y d\Gamma_k = \int_{\Gamma_k} \bar{W} E_{y_{wg}}^{(k)} d\Gamma_k \quad k = 1, \ldots, N \tag{4.12}$$

$$\int_{\Gamma_k} \bar{\bar{W}} \frac{\partial E_y}{\partial n} d\Gamma_k = \int_{\Gamma_k} \bar{\bar{W}} \frac{\partial E_{y_{wg}}^{(k)}}{\partial z^{(k)}} d\Gamma_k \quad k = 1, \ldots, N \tag{4.13}$$

Without loss of generality, it is convenient selecting the weighting functions $\bar{W}$ and $\bar{\bar{W}}$ such that they are equal to the restriction of W at the contour $\partial\Omega$ of the region Ω. Furthermore, the trial function E_y is chosen among those satisfying the homogeneous Dirichlet condition (4.6) and W is restricted so that it vanishes at the Dirichlet boundary Γ_0. With these hypothesis, by applying Green's identity to (4.10) and introducing the boundary condition (4.13), leads to the weak form of the Helmholtz equation:

$$\iint_\Omega \nabla_t W \cdot \frac{1}{\mu_r} \nabla_t E_y d\Omega - k_0^2 \iint_\Omega \epsilon_r W E_y d\Omega - \sum_{k=1}^{N} \int_{\Gamma_k} W \frac{\partial E_{y_{wg}}^{(k)}}{\partial z^{(k)}} d\Gamma_k = 0 \tag{4.14}$$

which, together with (4.12) constitute the system of solving equations. The approximation to the exact solution is sought by a finite element discretization of such equations.

4.2 The Finite Element—Modal Expansion Formulation: E-Plane Case

The formulation of the electromagnetic problem in the case of homogeneous E-plane junctions (Figure 4.2) follows a path similar to that outlined in the previous section, but considering different components of the electromagnetic field [8, 9]. In particular, assuming the device fed with the dominant TE_{10} mode, due to the translational symmetry of the structures along the x axis (Figure 4.2) the electromagnetic field in each section of the E-plane junction has the same variation as that of the exciting field. Consequently, **E** and **H** inside the junction can be expressed as

$$\mathbf{H}(x,y,z) = \mathbf{H}_t(x,y,z) + \mathbf{H}_x(x,y,z) = \mathbf{h}_t(y,z)\cos\left(\frac{\pi}{a}x\right) + \hat{\mathbf{x}}h_x(y,z)\sin\left(\frac{\pi}{a}x\right) \quad (4.15)$$

$$\mathbf{E}(x,y,z) = \mathbf{E}_t(x,y,z) + \mathbf{E}_x(x,y,z) = \mathbf{e}_t(y,z)\cos\left(\frac{\pi}{a}x\right) + \hat{\mathbf{x}}e_x(y,z)\sin\left(\frac{\pi}{a}x\right) \quad (4.16)$$

Due to the symmetry of the homogeneous junction and type of excitation, the electric field scattered by the discontinuity does not have the e_x component, so h_x is the only component of the electromagnetic field in the x direction. This suggests to formulate the problem into the region Ω (Figure 4.2) in terms of h_x only, which satisfies the Helmholtz equation:

$$\nabla_t^2 h_x + k_t^2 h_x = 0 \quad (4.17)$$

with $k_t^2 = k_0^2 - \left(\frac{\pi}{a}\right)^2$. Besides, the above considerations suggest expressing the field in each waveguide connected to the junction in terms of LSE_{1n}^x $(n = 0, 1, ...)$ modes

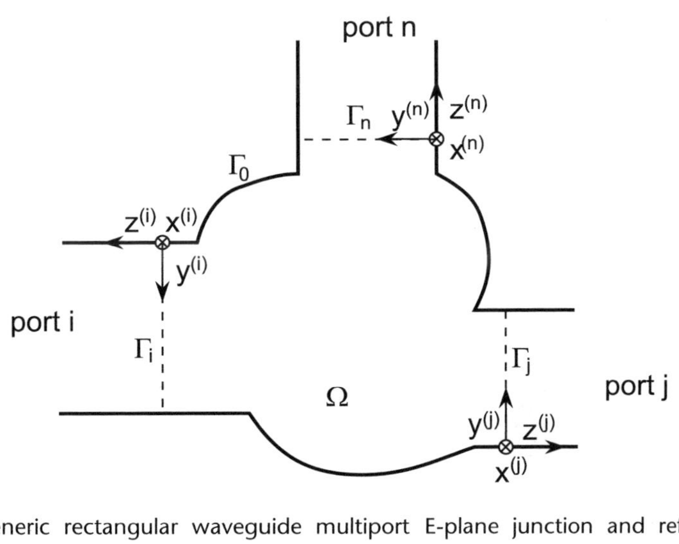

Figure 4.2 Generic rectangular waveguide multiport E-plane junction and reference systems adopted.

only. As a matter of fact, this set of modes has the $\mathbf{H}_x$ component of the magnetic field with the desired variation $\sin(\pi/ax)$ as in (4.15). Accordingly, the x component of the magnetic field into waveguide k, $h_{xwg}^{(k)}$, when the port (j) is fed with the TE_{10} mode, which is equivalent to the LSE_{10}^x, can be expressed as

$$h_{xwg}^{(k)}(y^{(k)}) = \delta_{kj} h_0^{(j)}(y^{(j)}) e^{j\beta_0^{(j)} z^{(j)}} - \sum_{n=0}^{\infty} B_n^{(k)} h_n^{(k)}(y^{(k)}) e^{-j\beta_n^{(k)} z^{(k)}} \tag{4.18}$$

where $h_n^{(k)}$ are the orthonormal modal functions of the LSE_{1n}^x modes into the waveguide k:

$$h_n^{(k)}(y^{(k)}) = \frac{G_n^{(k)}}{\sqrt{1+\delta_{0n}}} \cos\left(\frac{n\pi}{b^{(k)}} y^{(k)}\right) \tag{4.19}$$

with $\beta_n^{(k)}$ and G_n^k propagation constant and normalization factor, respectively, of the waveguide k with dimensions a and $b^{(k)}$:

$$\begin{cases} \beta_n^{(k)} = \sqrt{k_0^2 - \left(\frac{\pi}{a}\right)^2 - \left(\frac{n\pi}{b^{(k)}}\right)^2} & \text{for } k_0^2 \geq \left(\frac{\pi}{a}\right)^2 + \left(\frac{n\pi}{b^{(k)}}\right)^2 \\ \beta_n^{(k)} = -j\sqrt{\left(\frac{\pi}{a}\right)^2 + \left(\frac{n\pi}{b^{(k)}}\right)^2 - k_0^2} & \text{for } \left(\frac{\pi}{a}\right)^2 + \left(\frac{n\pi}{b^{(k)}}\right)^2 > k_0^2 \end{cases} \tag{4.20}$$

$$G_n^{(k)} = 2\sqrt{\frac{2Z_0}{ab^{(k)} k_0 \beta_n^{(k)} \left[k_0^2 - \left(\frac{\pi}{a}\right)^2\right]}} \tag{4.21}$$

Similarly to the case of H-plane junctions, the characterization of the device requires solving the Helmholtz equation (4.17) with boundary conditions:

$$\frac{\partial h_x}{\partial n} = 0 \tag{4.22}$$

at the metallic wall Γ_0 of the junction, and the continuity conditions for the tangential electric and magnetic fields at each port, that is for the components e_y, h_x, and h_y. It can be shown that these latter conditions are satisfied by enforcing at each port:

$$h_{x|_{\Gamma_k}} = h_{xwg}^{(k)} \quad k = 1, \ldots, N \tag{4.23}$$

$$\frac{\partial h_x}{\partial n}\bigg|_{\Gamma_k} = \frac{\partial h_{xwg}^{(k)}}{\partial z^{(k)}} \quad k = 1, \ldots, N \tag{4.24}$$

As a matter of fact, the transverse components of the electric and magnetic fields, $\mathbf{h}_t(y,z)$ and $\mathbf{e}_t(y,z)$, can be obtained from h_x by resorting to the transversalized (with respect to x) Maxwell's equations, that, in the case when $e_x = 0$, yield:

$$\mathbf{h}_t = \frac{\frac{\pi}{a}}{k_t^2} \nabla_t h_x \tag{4.25}$$

$$\mathbf{e}_t = j\frac{k_0 Z_0}{k_t^2}\hat{\mathbf{x}} \times \nabla_t h_x \tag{4.26}$$

From the above expressions it is easy to see that (4.23 and 4.24) actually also enforce the continuity of h_y and e_y. Following a weighted residual procedure similar to that outlined in the case of the H-plane junction, the final equations that must be discretized by means of finite elements to obtain the solution are:

$$\iint_\Omega \nabla_t W \cdot \nabla_t h_x d\Omega - k_t^2 \iint_\Omega W h_x d\Omega - \sum_{k=1}^{N} \int_{\Gamma_k} W \frac{\partial h_{xwg}^{(k)}}{\partial z^{(k)}} d\Gamma_k = 0 \tag{4.27}$$

$$\int_{\Gamma_k} \bar{W} h_x d\Gamma_k = \int_{\Gamma_k} \bar{W} h_{xwg}^{(k)} d\Gamma_k \quad k = 1, \ldots, N \tag{4.28}$$

The case of inhomogeneous E-plane junctions is more involved: the presence of dielectric discontinuities couples the h_x and e_x components of the electromagnetic field [8], so that it is necessary to solve simultaneously the scalar Helmholtz equations pertaining to both such components. The coupling between the two differential equations is provided by the boundary conditions at interfaces and can be introduced into the finite element formulation of the problem as outlined in [9, 10]. Moreover, the presence of the e_x component of the electric field renders necessary to express the electromagnetic field into each waveguide as the superposition of both longitudinal section electric and magnetic modes LSE_{1n}^x $(n = 0, 1, \ldots)$ and LSM_{1n}^x $(n = 1, 2, \ldots)$, characterized by $e_x = 0$ and $h_x = 0$, respectively, and continuity conditions must be enforced at each port on both h_x and e_x as well as their outward normal derivatives.

For sake of simplicity the code EHDEV (EH devices) described in the following section and included in the CD-ROM both as FORTRAN and Matlab code, is suitable to analyze rectangular waveguide discontinuities on either E- or H-plane, but it can be easily extended to deal with multiport junctions connected to waveguides with different dimensions. Indeed, the formulation outlined and the implementation described in the following section both refer to this more general case. The H-plane discontinuity or junction may include inhomogeneous dielectric material, whereas E-plane devices must be homogeneous.

4.3 Implementation

The solution of (4.12) and (4.14) in the case of H-plane junctions or (4.27) and (4.28) for E-plane junctions using finite elements in the framework of the weighted residual procedure outlined in the previous sections consists of the five steps:

1. Subdividing the region into finite elements;
2. Approximating the unknown function on each element;
3. Defining weighting functions and express the residue on each element;
4. Summing up contributions from all elements to obtain the residue on the whole domain;
5. Annihilating the residue and solve the linear system of equations obtained.

The implementation will be described in the following by referring to the case of H-plane junctions, but all arguments apply also to the E-plane case. As usual with finite elements, the solution on each element (e) is sought among the approximating functions $\bar{E}_y^{(e)}$ of the form:

$$\bar{E}_y^{(e)}(x,y) = \sum_{j=1}^{N^{(e)}} \bar{E}_{y_j}^{(e)} \alpha_j^{(e)}(x,y) \tag{4.29}$$

where $\bar{E}_{y_j}^{(e)}$ and $\alpha_j^{(e)}(x,y)$ $(j = 1,\ldots,N^{(e)})$ are the coefficients and the set of nodal-shape functions, respectively, each associated to one node of element (e). A set of weighting functions $W_i^{(e)}$ equal to the shape functions, $W_i^{(e)} = \alpha_i^{(e)}$ $(i = 1,\ldots,N^{(e)})$, is also introduced, and the residue $R_i^{(e)}$ relative to the ith weighting function on element (e) is expressed as:

$$\begin{aligned} &\frac{1}{\mu_r}\iint_{\Delta^{(e)}} \nabla_t\alpha_i^{(e)} \cdot \nabla_t\bar{E}_y^{(e)} d\Omega - k_0^2\epsilon_r \iint_{\Delta^{(e)}} \alpha_i^{(e)}\bar{E}_y^{(e)} d\Omega \\ &\quad - \sum_{k=1}^{N} \int_{\Gamma_k^{(e)}} \alpha_i^{(e)} \frac{\partial E_{y_{wg}}^{(k)}}{\partial z^{(k)}} d\Gamma_k \quad i = 1,\ldots,N^{(e)} \end{aligned} \tag{4.30}$$

or, in matrix form:

$$\frac{1}{\mu_r}[S^{(e)}]\cdot[\bar{E}_y^{(e)}] - k_0^2\epsilon_r[T^{(e)}]\cdot[\bar{E}_y^{(e)}] + \sum_{k=1}^{N}\left\{[C_k^{(e)}]\cdot[B_k] - [H_k^{(e)}]\right\} = [R^{(e)}] \tag{4.31}$$

In the above equation, $[S^{(e)}]$ and $[T^{(e)}]$ are the usual local matrices of scalar nodal element and $[\bar{E}_y^{(e)}]$, is the vector of nodal unknown coefficients of element (e). $[B_k]$ is a column vector whose j-th entry is the amplitude of the transmitted mode j at the port (k), and $[C_k^{(e)}]$ and $[H_k^{(e)}]$ come from the contour integrals at the $k = 1,\ldots, N$ ports.

In particular, when M modes are used to represent the field inside the waveguides connected to the device, $[C_k^{(e)}]$ is a matrix with as many rows as there are bases in the element ($N^{(e)} = 3$, for first-order triangle) and as many columns as there are modes on the port (M in our case). Its generic entry is

$$\begin{aligned} C_{k_{im}}^{(e)} &= j\beta_m^{(k)} \int_{\Gamma_k^{(e)}} \alpha_i^{(e)} e_m^{(k)} d\Gamma_k^{(e)} \\ &= j\beta_m^{(k)} \frac{2}{\sqrt{a^{(k)}b}} \sqrt{\frac{k_0 Z_0}{\beta_m^{(k)}}} \int_{\Gamma_k^{(e)}} \alpha_i^{(e)} \sin\left(\frac{m\pi}{a^{(k)}} x^{(k)}\right) d\Gamma_k^{(e)} \end{aligned} \tag{4.32}$$

and $[H_k^{(e)}]$ is a $(N^{(e)} \times 1)$ column vector with entries

$$\begin{aligned} H_{k_i}^{(e)} &= j\beta_1^{(l)} \delta_{kl} \int_{\Gamma_k^{(e)}} \alpha_i^{(e)} e_1^{(l)} d\Gamma_k^{(e)} \\ &= \delta_{kl} j\beta_1^{(l)} \frac{2}{\sqrt{a^{(l)}b}} \sqrt{\frac{k_0 Z_0}{\beta_1^{(l)}}} \int_{\Gamma_k^{(e)}} \alpha_i^{(e)} \sin\left(\frac{\pi}{a^{(l)}} x^{(l)}\right) d\Gamma_k^{(e)} \end{aligned} \tag{4.33}$$

where (l) is the port fed and δ_{kl} is the Kronecker's delta. The local residual (4.31) relative to each element can be assembled into a single linear system of equations. To this end, the matrices $[S^{(e)}]$ and $[T^{(e)}]$, with their coefficients $\frac{1}{\mu_r}$ and $-k_0^2\epsilon_r$, respectively, can be assembled into a global $(N_n \times N_n)$ matrix $[F]$, with N_n total number of nodes, by following the usual assembly strategy. As for the matrix $[C_k^{(e)}]$ and the vector $[H_k^{(e)}]$, they can be assembled into an $(N_n \times (N \times M))$ matrix $[C]$ and into an $(N_n \times 1)$ column vector $[H^{inc}]$, respectively. For both entities, the assembling strategy to recover the row index in the global matrix/vector is the same as that used for the local matrices $[S^{(e)}]$ and $[T^{(e)}]$. The assembling scheme for the column index j of matrix $[C]$ is readily determined by the formula $j = M \times (k-1) + m$, which means that the unknown amplitudes of transmitted modes at different ports are sequentially arranged in a unique column vector $[B]$ with dimension $((N \times M) \times 1)$. Annihilating the residue of the assembled system leads to the matrix equation:

$$[F] \cdot [\bar{E}] + [C] \cdot [B] = [H^{inc}] \tag{4.34}$$

in which the unknowns are the column vectors $[\bar{E}]$ and $[B]$. The former contains the coefficients of the finite element approximation of the electric field, the latter stores the amplitude of the transmitted field at the ports. The linear system of (4.34) cannot be solved since it has more unknowns $(N_n + N \times M)$ than equations (N_n). The missing equations must be built starting from the boundary condition (4.12) which has not been used yet. To exploit such equation, it is necessary to chose the weighting functions $\bar{W}$. It should be noted that $(N \times M)$ equations are needed, hence it is appropriate to introduce M weighting functions for each of the N ports of the junctions. A particularly convenient choice for the basis functions $\bar{W}_m^{(k)}$, $m = 1,\ldots,M$ to be used at port (k) is:

$$\bar{W}_m^{(k)} = \sin\left(\frac{m\pi}{a^{(k)}} x^{(k)}\right) \tag{4.35}$$

using these weighting functions the residue $R_m^{(k)}$ of the boundary condition (4.12) relative to the m-th weighting function at port (k) is expressed as:

$$\begin{aligned} &\int_{\Gamma_k} \sin\left(\frac{m\pi}{a^{(k)}} x^{(k)}\right) \bar{E}_y d\Gamma_k - \sqrt{\frac{a^{(k)}}{b}} \sqrt{\frac{k_0 Z_0}{\beta_m^{(k)}}} B_m^{(k)} \\ &\qquad - \delta_{kl} \sqrt{\frac{a^{(l)}}{b}} \sqrt{\frac{k_0 Z_0}{\beta_1^{(l)}}} = R_m^{(k)} \quad m = 1,\ldots,M \end{aligned} \tag{4.36}$$

where, again, (l) is the port fed and δ_{kl} is the Kronecker's delta. The above equation, relative to port (k), can be put in the matrix form:

$$[D_k] \cdot [\bar{E}_y^{(k)}] + [A_k] \cdot [B_k] - [E_k] = [R^{(k)}] \tag{4.37}$$

where the column vector $[\bar{E}_y^{(k)}]$ $(N^{(k)} \times 1)$ stores the finite element coefficients of the electric field associated to the $N^{(k)}$ nodes at port (k); $[D_k]$ is a rectangular matrix $(M \times N^{(k)})$ with generic entry $D_{k_{mj}}$

$$D_{k_{mj}} = \int_{\Gamma_k} \sin\left(\frac{m\pi}{a^{(k)}} x^{(k)}\right) \alpha_j^{(k)} d\Gamma_k \tag{4.38}$$

the matrix $[A_k]$ is diagonal with entries

$$A_{k_{mm}} = -\sqrt{\frac{a^{(k)}}{b}}\sqrt{\frac{k_0 Z_0}{\beta_m^{(k)}}} \quad m = 1, \ldots, M \tag{4.39}$$

and the column vector $[E_k]$ $(M \times 1)$ takes into account the incident field, if any, at port (k). It has only one nonzero entry, that are pertinent to the fundamental TE_{10} mode:

$$E_{k_1} = \delta_{kl}\sqrt{\frac{a^{(l)}}{b}}\sqrt{\frac{k_0 Z_0}{\beta_1^{(l)}}} \qquad E_{k_m} = 0 \qquad m = 2, \ldots, M \tag{4.40}$$

The continuity boundary conditions at each port, expressed by (4.37), can be assembled into a global system and the residue annihilated. Combining such system with that in (4.34), eventually yields the total system of equations to be solved, which has the structure:

$$\begin{bmatrix} [A] & [D] \\ [C] & [F] \end{bmatrix} \cdot \begin{bmatrix} [B] \\ [E] \end{bmatrix} = \begin{bmatrix} [E^{inc}] \\ [H^{inc}] \end{bmatrix} \tag{4.41}$$

where $[A]$ is a diagonal matrix, $[D]$ and $[C]$ are sparse rectangular matrices and $[F]$ is the sparse and symmetric finite element matrix.

The implementation of the analysis technique for homogeneous E-plane junctions is similar to that reported above and leads to a system of linear equations with the same structure as that in (4.41), even though with a doubled number of unknowns. The reader can easily derive the expression of the entries of the matrices $[C_k^{(e)}]$, $[A_k]$, and $[D_k]$, as well as the entries of the known vector for such case.

4.4 The Code EHDEV

The code EHDEV is the first code of this book to exploit a hybrid formulation, thus it deserves a thorough description, also because the solution of the block matrix (4.41) is performed through a custom-tailored block solver which will be employed also in other codes later on.

The main program reads the input data from two different data files: a mesh data file and an electromagnetic data file. The former is generated by the mesh generator described in Chapter 2, with the following rules:

- Dimensions are in millimeters.
- Any of the elements that the mesh generator can produce may be used, but second-order quadrilateral elements are strongly recommended for their excellent behavior in modeling curved boundaries. These elements have been used to produce the results reported in this chapter. Matlab mesh generators can handle only first-order triangles, and much finer meshes need to be used to attain similar results.
- Nodes on p.e.c. and p.m.c. contours must be identified by their own label: we will denote p.e.c. nodes with label 1 and p.m.c. nodes with label 2.
- Nodes on junction's ports deserve their own label too. The code here presented is for two-port devices, so two different labels must be defined. Conventionally, nodes on port 1 are assigned a label 11, and nodes on port 2 a label 12. Please note that nodes that are both on a port and on a metallic wall are assigned the label of the corresponding port, since the program automatically assumes that the first and last node of each port also lies at a metallic surface.

On the other hand, the electromagnetic data file has the format and contains the data described in Example 4.1.

Example 4.1

```
2               !# Number of Ports
18.35, 9.175    !# Waveguide dimensions a and b [mm]
11, 12          !# Port labels
1, 2            !# p.e.c - p.m.c. labels
10              !# Number of modes
H               !# Plane [E—H]
2               !# Number of materials
1. 4.           !# permittivity(ies)
1. 1.           !# permeability(ies)
1 2             !# Element label(s)
3               !# No of Frequency point(s)
10, 10.5, 11    !# Frequencies [GHz]
```

In Example 4.1, the first row of the file stores the number of ports of the junction, fixed to two for the current implementation, and the second row the dimension a and b of the waveguides connected to the discontinuity under analysis, which must be all equal. After that, the labels for the nodes lying at the ports are reported, conventionally 11 and 12, followed by the labels used to denote nodes at p.e.c. and p.m.c. materials, here put equal to 1 and 2. Then, the next record contains the number of modes to be used in the modal expansion at each port. This latter is a very important parameter and its influence on numerical results will be shown later on in this section. Generally speaking, the farther the reference planes from the discontinuity, the less the number of modes which must be included into the modal expansion to get accurate results, but the bigger the mesh. Establishing the ports close to the discontinuity provides a smaller mesh and hence a smaller number of nodal unknowns but this is traded against a greater number of modes needed to correctly terminate the computational domain. Lastly, a single character `E` or `H` states if the problem concerns an E- or H-plane device, respectively.

The subsequent group of four lines contains the number of different dielectric materials included in the computational domain, a list of the different permittivity and permeability values, and the labels which characterize each element belonging to a given material. This four lines must be present even if the junction is on the E-plane, even though in this latter case their content is ignored. The last record in the input electromagnetic data file stores the number of analysis to be performed and the frequency values, in GHz, to be used. The approach of defining every single frequency point might appear awkward, but being able to nonuniformly sample the band is valuable, especially when resonant behaviors occur, as it will be shown in the last example.

The flow of the program is then straightforward: first all quantities are scaled to the MKS system and all the frequency points are checked to ensure that the analysis is executed for frequency values above the cutoff of the fundamental mode TE_{10}. A warning message is also issued if for some frequency values the waveguide is overmoded, even though the code correctly handles this case.

The linear system of equations is then assembled and solved. Since the final coefficient matrix of (4.41), as well as the similar one for the E-plane case, has three out of four submatrices that are very sparse, it is convenient to separate each block employing different memorization schemes to limit memory requirements. In particular, matrix $[A]$ is diagonal thanks to the orthonormality of the modal expansion functions, and thus its construction is straightforward from (4.39). Matrix $[F]$ in (4.41) is stored using the standard band storage scheme presented in Chapters 1 and 2, and it is easily assembled starting from the elemental matrices $[S^{(e)}]$ and $[T^{(e)}]$ and following the rules outlined in the same chapters. As for the matrices $[C]$ and $[D]$, they have nonzero elements only into rows and columns, respectively, relative to nodes located at the ports Γ_k, and thus only those rows or columns need to be stored. However, for the solution of the linear system of (4.41) it is necessary to know the actual position of each of the nonzero rows/columns of such matrices, and for this purpose the nodes lying at ports are sought and stored in the FORTRAN vector NCYL in the same order as the corresponding rows/columns appear in the matrices $[C]$ and $[D]$. In this scheme, NCYL(I) represents the number, in the global numberings scheme, of the node associated to the ith nonzero rows (columns) of matrix $[C]$ ($[D]$). The Matlab code is somewhat simpler in this regard inasmuch as all matrices are considered sparse, which allows for the largest possible memory saving yet completely transparent to the user who does not need to keep track of nonzero rows or column as in FORTRAN.

The entries of the above-mentioned matrices can be easily computed in closed form by recognizing that the integrals appearing in (4.32) and (4.38) can be regarded as the imaginary part, possibly scaled, of one or more integrals of the kind:

$$\int_0^1 P_j^{(n)} e^{ax} \tag{4.42}$$

being $P_j^{(n)}$ $(j = 0, \ldots, n)$ the jth interpolation polynomial of order n on the interval $[0, 1]$ centered at the point $x_j = j/n$, and a a complex number. These integrals are analytically known and their values are provided by the library routine PJN-EXPA (FORTRAN) and `pjnexpa.m` (Matlab). This subroutine is quite general and it is exploited also by the code discussed in the following chapter. For this reason it is included in the library files, rather than within each program.

The function computes the integral (4.42) for polynomials up to order 4 by resorting to an appropriate linear combination of the analytical formula

$$\int x^n e^{ax} dx = \frac{1}{a^{n+1}} \left[\left(\sum_{i=1}^{n} \frac{n!}{i!} a^i x^i e^{ax} \right) - n! e^{ax} \right] \tag{4.43}$$

Since this contains the term $1/a^{n+1}$ it may be used only if $|a| > \epsilon$, with ϵ a small number. If this is not the case, the exponential in (4.42) is substituted by its Taylor expansion. In the subroutine the first four terms are considered, thus exploiting the approximation:

$$\begin{aligned} \int x^n e^{ax} dx &\simeq \int x^n \left[1 + ax + \frac{a^2}{2} x^2 + \frac{a^3}{6} x^3 \right] dx \\ &= \frac{1}{(n+1)!} x^{n+1} + \frac{a}{(n+2)!} x^{n+2} + \frac{a^2}{2(n+3)!} x^{n+3} \\ &+ \frac{a^3}{6(n+4)!} x^{n+4} \end{aligned} \tag{4.44}$$

These formulas are of course valid also for complex values of a.

The solution of the system (4.41) is computed by exploiting the block nature of the global coefficient matrix. The system is first recast into the two equations:

$$[A] \cdot [B] + [D] \cdot [\bar{E}] = [E^{inc}] \tag{4.45}$$

$$[C] \cdot [B] + [F] \cdot [\bar{E}] = [H^{inc}] \tag{4.46}$$

then, since the main goal of the analysis are the modal expansion coefficients at the ports, stored into the vector $[B]$, it is convenient using (4.46) to eliminate the unknown $[\bar{E}]$ from (4.45). By left multiplying (4.46) times $[F]^{-1}$ one obtains:

$$[\bar{E}] = [F]^{-1} \cdot [H^{inc}] - [F]^{-1} \cdot [C] \cdot [B] \tag{4.47}$$

that, when substituted into (4.45) leads to

$$\left([A] - [D] \cdot [F]^{-1} \cdot [C] \right) \cdot [B] = [E^{inc}] - [D] \cdot [F]^{-1} \cdot [H^{inc}] \tag{4.48}$$

The above linear system of equations has a full coefficient matrix but has dimensions much smaller than the original one (4.41). It is important to note that the matrix $[F]$ does not need to be completely inverted, but only those columns corresponding to nodes located at the ports need to be computed. As a matter of fact, only such columns are involved into the right multiplication of $[F]^{-1}$ times $[D]$ and can be efficiently computed by exploiting the banded nature of the matrix $[F]$. Once the linear system (4.48) has been solved, the field distribution within the junction, that is the unknown vector $[\bar{E}]$ of nodal values, can be computed, if needed, through (4.47).

Again, this method of solution is not unique to this program so the FORTRAN subroutine BLOCK_SOLVER and the Matlab function `BlockSolver.m` that implement it have been included into the FORTRAN and Matlab library tree, respectively.

For Matlab, very efficient iterative solvers are available expressively tailored for sparse matrices. Since sparse storage is used, these routines are exploited within `BlockSolver.m` to efficiently compute the needed columns of $[F]^{-1}$.

4.5 Some Examples

As a very first example, a rectangular waveguide segment is considered. The waveguide dimensions are a and b, with $a > b$. This indeed is nothing more than a fixed-value phase shifter and we expect, for the ideal case, $S_{11} = 0$, $S_{21} = e^{-j\beta l}$ being l the length of the waveguide segment and β the propagation constant of the guided wave, that is $\beta = \sqrt{k_0^2 - k_{th}^2}$, where $k_0 = 2\pi f/c$, with c the speed of light, and $k_{th} = \pi/a$ for the TE_{10} mode. The test has been carried out considering different grids over a 10 mm long rectangular waveguide with $a = 18.35$ mm, $b = 9.175$ mm. The mesh, comprised of second-order isoparametric elements, has been varied from a very coarse one (one single element), to a relatively fine mesh, exhibiting eight element sides on each port and six element sides along the waveguide walls. The analysis has been carried out in the frequency range of 10 to 15 GHz, which is the unimodal band of the rectangulr waveguide. Ten modes have been used to represent the field at each port.

Figure 4.3 reports the results for a 1×1, 2×2, and 4×3 element meshes. As it can be seen the modulus is already very well computed even with only 2 elements, since it is mainly linked to the number of elements on each port, while the phase is best computed with more elements, since its error is inversely proportional to the number of elements along the waveguide axis.

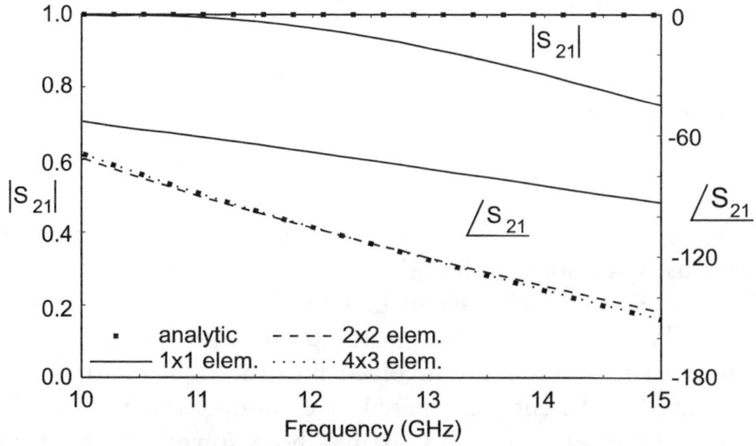

Figure 4.3 Scattering coefficients of a segment of rectangular waveguide for different meshes. Comparison with analytic results.

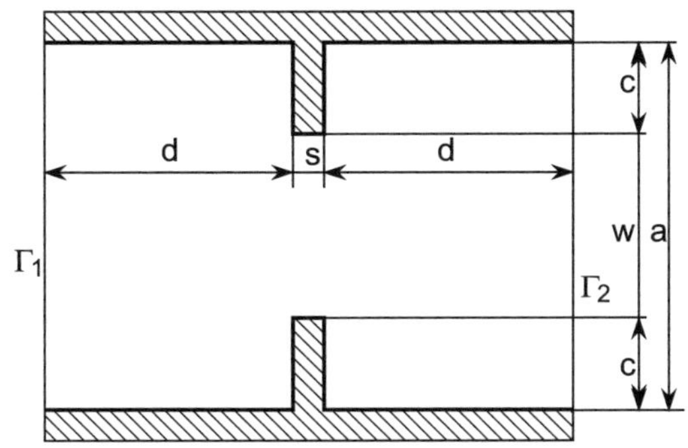

Figure 4.4 Iris geometry.

The disc contains many other finer meshes, whose results are not reported in the figure for sake of clarity since they are superimposed on each other.

As a second example, the H-plane iris in a rectangular waveguide of Figure 4.4 is analyzed. The waveguide dimensions are a and b, the iris is centered, has thickness s, width w and the reference planes or ports are symmetrically placed at a distance d from both sides of the iris.

Considering again the $a = 18.35$ mm, $b = 9.175$ mm rectangular waveguide, the iris dimensions are chosen as follow: $w = 9.175$ mm, $s = 0.5$ mm, $c = 4.587$ mm. The analysis is again in the $10 - 15$ GHz range, and even though no curved boundaries are present, the geometry is discretized by using second-order quadrilateral elements.

The question arises of how many modes are necessary to correctly match a given discontinuity at a given port Γ_i. This number may be assessed by analytically computing the attenuation of the various modes in cutoff from the discontinuity to the port. A rule of thumb is that only those modes which have an attenuation of less than 30 dB need to be considered, whereas all the others can be neglected. Using this rule, it is then possible to estimate the number M of modes to be included into the analysis.

To investigate the influence of higher order modes, the ports have been placed quite close to the iris, at a distance $d = 1$ mm. Figures 4.5 and 4.6 show the amplitude and phase, respectively, of S_{11} and S_{21} computed with $M = 1, 3, 5, 7, 11$. The solution is compared, in the same figures, with that provided by the mode matching method including 32 modes. As it can be seen, 3 modes are already enough to achieve an accurate solution, while with 7 or more the agreement with the reference is excellent.

The computations have been made with M odd since, due to symmetry, only odd modes are excited by the iris. Figure 4.7 shows the amplitude of the TE_{n0} scattered modes arising at Γ_1 and Γ_2 when they are placed 1 mm away from the iris, for an unitary incident TE_{10} mode. This analysis has been performed at a frequency of 10 GHz and shows that, indeed, even modes are always negligible, while only the first two odd modes have a significant amplitude. Considering a higher number of modes increases the computing time with little improvements in accuracy.

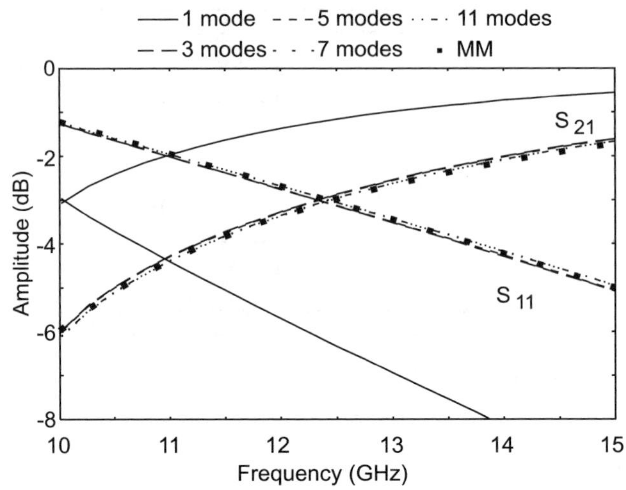

Figure 4.5 Iris analysis $|S_{11}|$ and $|S_{21}|$ for different number M of modes.

Furthermore, it is easy to see, and it is left to the reader as an exercise, that at larger distances d from the iris, even fewer modes are necessary to correctly match the FEM approximation to the modal expansion. Indeed, at a distance $d = 7$ mm from the iris it suffices matching only the fundamental mode. As a matter of fact, since for $d = 1$ mm the TE_{30} mode is the only relevant higher order mode, and its attenuation constant at 10 GHz is $\alpha = 469$ Np/m, such mode is subject to an attenuation of -24.4 dB passing from 1 mm to 7 mm, thus becoming negligible for most practical applications.

As a third example, to take full advantage of the formulation based on second-order quadrilateral isoparametric elements, the analysis of various waveguide bends is considered.

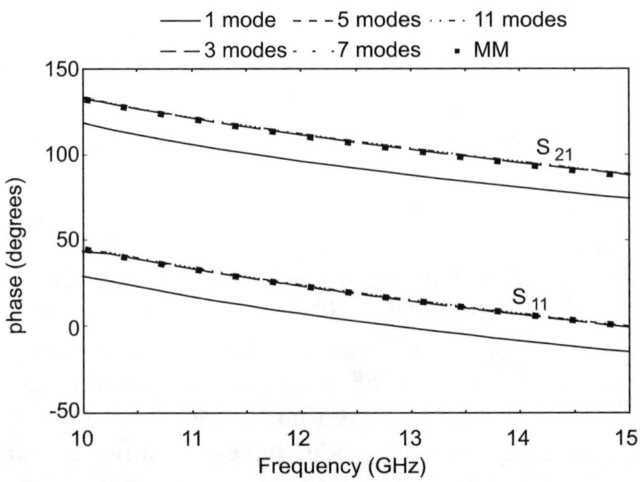

Figure 4.6 Iris analysis, Phase of S_{11} and of S_{21} for different number M of modes.

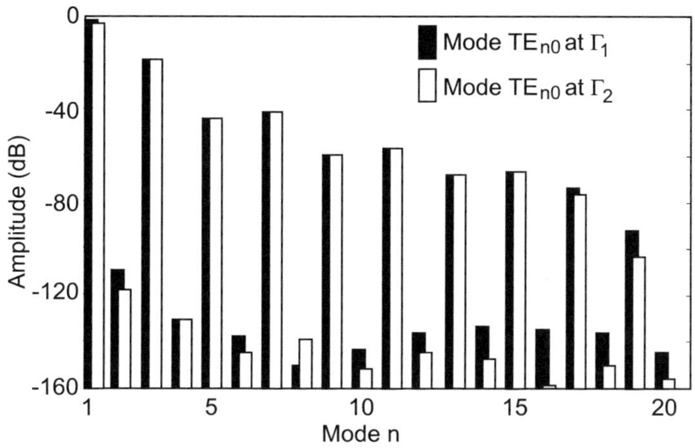

Figure 4.7 Amplitude of the TE_{n0} scattered modes arising at Γ_1 and Γ_2 for an unitary incident TE_{10} mode.

Four different cases are considered in the following (Figure 4.8):

(a) Square bends;
(b) Fully mitered bends;
(c) Partially mitered bends;
(d) Circular bends.

All the configurations analyzed are 90° bends, in the usual rectanguar waveguide, with internal radius $r_1(\phi) = 0$. Their geometry is evident from Figure 4.8, except for bend (c), which exhibits a degree of freedom in the length p of the straight portion of the exterior waveguide wall. In the present analysis $p = 3$ mm and the reference planes or ports have been placed at a distance equal to 2 mm from the inner 90° edge of the bend.

Figure 4.9 shows the reflection coefficient S_{11} for the different H-plane bends sketched in Figure 4.8 over the unimodal frequency band for the rectanguar wave-

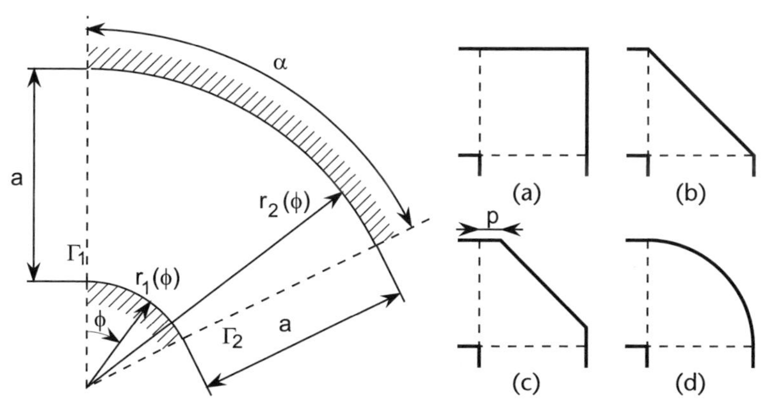

Figure 4.8 Geometry of a circular waveguide bend (left), and the four geometries analyzed in the text (right).

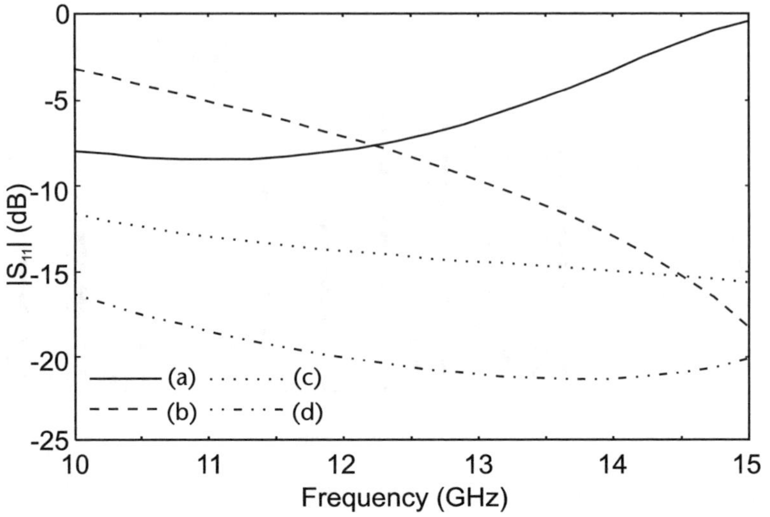

Figure 4.9 S_{11} for the H-plane bend configurations of Figure 4.8.

guide at hand. Bend (a) has the worst behavior while bend (d) has the best. On the other hand, bend (a) is easy to manufacture and thus has low cost, while the opposite applies to bend (d). Figure 4.10 reports the corresponding values of S_{21}.

The results show that bend (a) behaves quite well at low frequencies, but the mirror effect of the p.e.c. wall opposite to port 1 becomes more and more evident as frequency increases. Bend (b) has an opposite behavior, since the waveguide cross-section is smaller in the middle of the bend, thus giving high S_{11} at low frequencies, while the mirror-effect of the 45° wall enhances performances at high frequencies. Bend (c), having a geometry which has something in common with

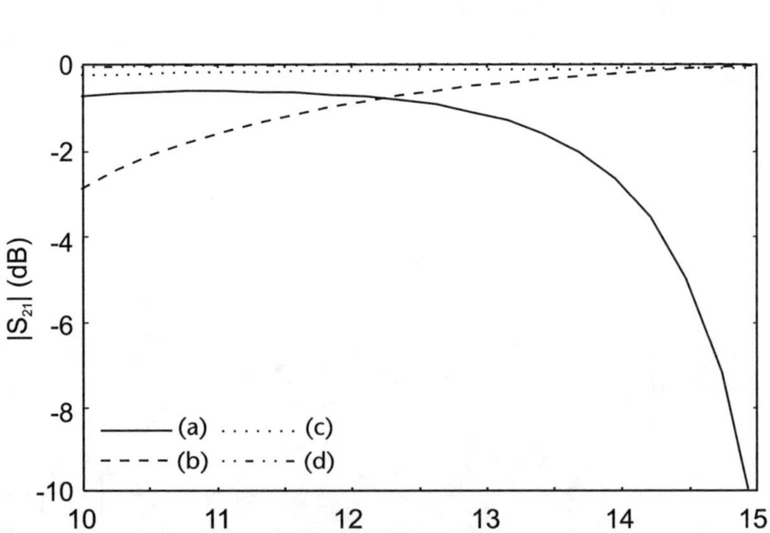

Figure 4.10 S_{21} for the H-plane bend configurations of Figure 4.8.

both the previous bends, exhibits a behavior which takes the best of both, providing a low S_{11} over a wide frequency band, while bend (d), being the one with less discontinuities, provides the best response of all.

The accompanying CD includes also the meshes and the `.aux` files for the same bends in E-plane. It is easy to understand how, by writing inner and outer boundaries as a function of ϕ as in Figure 4.8 and by parameterizing this function with some sort of spline it is possible to cast the EHDEV code in an optimizer to attain optimal performances [11].

As a final example, a two-post filter in WR90 is presented [12]. WR90 has $a = 22.86$ mm and $b = 10.16$ mm. Posts are metallic, have a radius $r = 1$ mm and are placed on the waveguide axis 20 mm apart, with ports placed 10 mm from the axis of the posts ($d = 10$ mm, Figure 4.11). The analysis is performed again with second-order quadrilateral elements on the unimodal WR90 band (from 7 to 13 GHz). Figure 4.12 shows the passband behavior of the filter. Please note that the filter is very narrow band, hence several frequency points are necessary. For the sake of speed the enclosed `.aux` file does an uneven discretization in frequency so as to have more points where S_{11} has the most abrupt changes. Finally Figure 4.13 shows the phase of the scattering coefficients.

4.6 Disc Content

4.6.1 FORTRAN

The software described in this chapter is stored in the subdirectory EHDEV of the directory CODES. The program files EHDEV.F and EHD_SUBS.F must be compiled all together and with the library described in Chapter 2 to generate the executable.

The three subdirectories PSHIFT, IRIS, BEND, and POSTS contain all the input and mesh files of the examples discussed in the previous section. A more complete description of each file is contained in the README file in the directory EHDEV itself.

The output file contains the generalized scattering matrix of the device in a format very easy to understand but not suitable for immediate plotting. Help

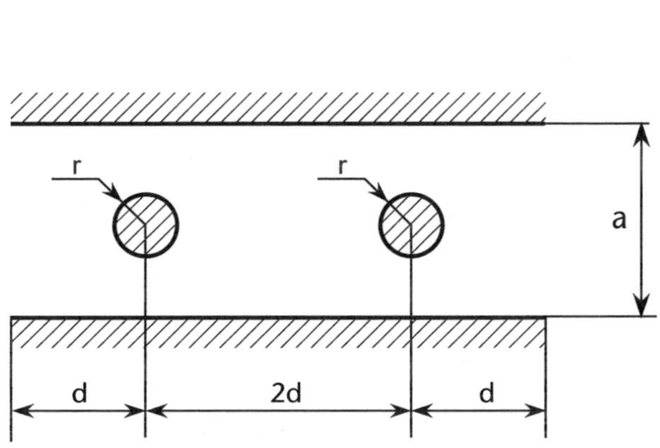

Figure 4.11 Geometry of the two-post WR90 filter.

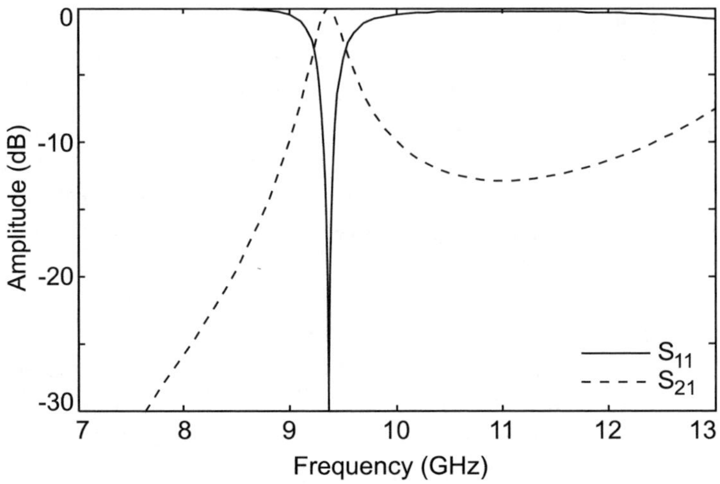

Figure 4.12 Amplitude of S_{11} and S_{21} for the filter in Figure 4.2.

plotting the result file, be it produced by the FORTRAN or the Matlab code, can be visualized using the Matlab function described here below.

4.6.2 MATLAB

For the Matlab version the structure is similar, except that, as usual, every subroutine is given as a separate `.m` file and the code is called like a function:

```
ehdev(fnamemesh,fnameaux,fnameout,fieldflag)
```

where the various parameters have the same meaning as in `wg.m` function described in Chapter 3.

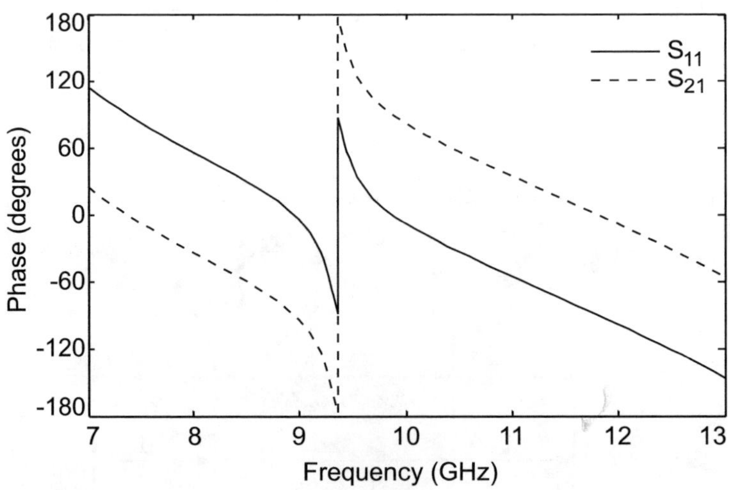

Figure 4.13 Phase of S_{11} and S_{21} for the filter in Figure 4.2.

Furthermore, the function GetGSM.m and SeeGSM.m included in the CD-ROM can be used to load and visualize, respectively, the output files generated by the codes (either FORTRAN or Matlab). For the first, its usage is:

```
[f,S] = GetGSM(gsmfile);
```

being gsmfile the user-defined output file name generated by EHDEV. The output variable f is a vector containing the frequency points, while S is a multidimensional array S(Np,Np,Nf,Nm) being Np the number of ports, Nf the number of frequency points and Nm the number of modes at each port.

While GetGSM.m is a useful library function to be embedded in user-written codes, to quickly have a video display or a printout of the results SeeGSM.m function can be used:

```
SeeGSM(gsmfile,pi,pj,mk,flag)
```

Here gsmfile is still the file name, while pi and pj are the indexes of the *S* parameter one desires to plot. mk is the mode index of the outgoing wave (the incident wave being always a TE_{10}) and flag being one of:

- ad: plots amplitude, dB;
- al: plots amplitude, linear;
- pr: plots phase, radiants;
- pd: plots phase, degrees.

Note that pi, pj, and mk can be arrays. In this case all possible combinations are generated and plotted. For example:

```
SeeGSM(gsmfile,[1 2],[1],[1 2],flag)
```

would plot 4 curves, relative to $S_{11}^{(1)}$ and $S_{21}^{(1)}$ where scattering mode is first one, and to $S_{11}^{(2)}$ and $S_{21}^{(2)}$ where scattering mode is second one.

References

[1] Kiyoshi, I. and Masanori, K., "Numerical Analysis of H-Plane Waveguide Junctions by Combination of Finite and Boundary Elements," *IEEE Transactions on Microwave Theory and Techniques*, Vol. 36, pp. 1343–1351, 1988.

[2] Wu, K. L., et al., "Waveguide Discontinuity Analysis with a Coupled Finite-Boundary Element Method," *IEEE Transactions on Microwave Theory and Techniques*, Vol. 37, pp. 993–998, 1989.

[3] Lee, J. F. and Cendes, Z. J., "The Transfinite Element Method for Modeling MMIC Devices," *IEEE Transactions on Microwave Theory and Techniques*, Vol. 36, pp. 1639–1649, 1988.

[4] Coccioli, R., et al., "Design of Matched Bends in Rectangular Waveguides by FEM," *International Journal of Microwave and Millimeter-wave Computer-Aided Engineering*, Vol. 6, pp. 391–398, 1996.

[5] Loo, S. L. and Silvester, P. P., "Boundary-Marching Method for Discontinuity Analysis in Waveguides of Arbitrary-Cross Section," *IEEE Transactions on Microwave Theory and Techniques*, Vol. 40, pp. 1889–1893, 1992.

[6] Loo, S. L. and Silvester, P. P., "Finite Element Analysis of Inductive Strips in Unilateral Finlines," *IEEE Transactions on Microwave Theory and Techniques*, Vol. 41, pp. 298–304, 1993.

[7] Coccioli, R., et al., "Design of Tolerance-Corrected Filters Employing Half-Cylinder Posts," *IEEE Transactions on Microwave Theory and Techniques*, Vol. 46(1), pp. 116–118, Jan. 1998.

[8] Kanellopoulos, V. N. and Webb, J. P., "A Complete E–Plane Analysis of Waveguide Junctions Using the Finite Element Method," *IEEE Transactions on Microwave Theory and Techniques*, Vol. 38, pp. 290–295, 1990.

[9] Coccioli, R., Pelosi, G., and Silvester, P. P., "Finite Element Package for Inhomogeneous H- and E-Plane Junctions," In *Software for Electrical Engineering Analysis and Design (ELECTROSOFT 96)*, S. Miniato (Pisa), Italy, May 25-28, 1996.

[10] Pelosi, G., Coccioli, R., and Graglia, R. D., "A Finite Element Analysis of Electromagnetic Scattering from a Moving Inhomogeneous Dielectric Cylinder of Arbitrary Cross-Section," *Journal of Physics D: Applied Physics*, Vol. 27, pp. 2013–2018, 1994.

[11] Coccioli, R., Pelosi, R., and Selleri, S., "Optimization of Bends in Rectangular Waveguide by a Finite Element-Genetic Algorithm Procedure," *Microwave and Optical Technology Letters*, Vol. 16(5), pp. 287–290, Dec. 1997.

[12] Fedi, G., et al., "FEM Trained Artificial Neural Networks for the Analysis and Design of Cylindrical Posts in Rectangular Waveguide," *Electromagnetics*, Vol. 22(4), pp. 323–330, 2002.

CHAPTER 5

Scattering and Antennas: Hybrid Methods

The use of finite elements for solving practical radial propagation problems has been extensively studied and many fairly complex structures have been analyzed, ranging from simple patch antennas to penetrable fixed or moving cylinders of arbitrary cross-section, to conformal antennas.

The main difficulty encountered in the numerical solution of this class of problems is posed by the unbounded geometric region where Maxwell's equations need to be solved. In one sense, the finite element method may be viewed as a numerical technique for evaluating Green's function of a bounded, complex structure with known boundary conditions. Consequently, its application to antenna and scattering problems has required introducing the Sommerfeld radiation condition into the problem formulation. The most popular methods for doing so convert the original open problem (i.e., defined on an infinite geometric region), into an equivalent interior problem (i.e., defined on an bounded geometric region), by introducing a fictitious bounding surface Γ. Such a surface is chosen so that the enclosed region Ω contains either the sources and surrounding complex objects, in case of antenna problems, or all the obstacles in case of scattering problems (Figure 5.1). Within the bounded region Ω, the solution can be easily determined by the finite element method if an appropriate boundary condition establishing a relationship between the tangential electric and magnetic fields at Γ can be set. The boundary condition may be of two different kinds:

- *Nonlocal*: the relationship valid at a point P on the boundary Γ (Figure 5.1) involves the values of the field on the whole boundary;
- *Local*: the relationship valid at the point P is expressed in terms of the field values at the neighboring points of P.

The so-called absorbing boundary conditions (ABC) are of local type and their usage has gained popularity because of their attractive features. The most important is that of maintaining the sparse nature of the coefficient matrix of the linear system of equations resulting from the application of FEM. Local ABCs can be subdivided into three kinds, depending on the procedure used to derive them: analytical, numerical, and based on the perfectly matched layer (PML) concept. These will be more extensively treated in the next chapter.

Nonlocal boundary conditions are obtained by representing the electromagnetic field in the region exterior to Γ in terms of either a modal expansion or an integral representation and using such expression to derive a relationship between

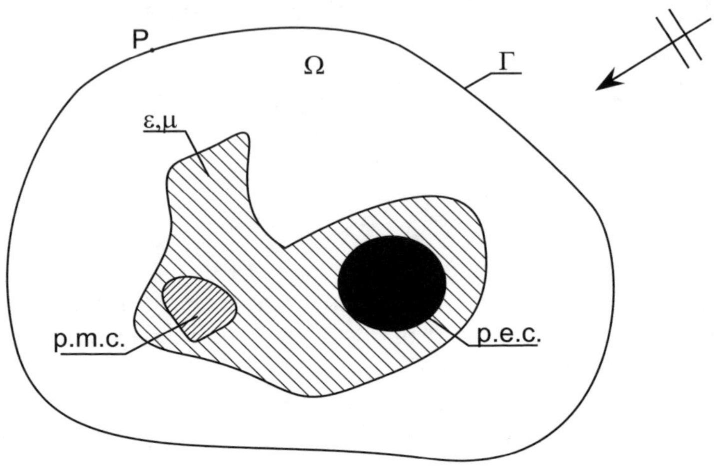

Figure 5.1 The finite element solution of open problems requires the introduction of a fictitious surface Γ, which bounds the spatial domain discretized with finite elements.

tangential electric and magnetic field components at Γ. Methods employing boundary conditions belonging to this group are often referred to as either hybrid methods, because they hybridize the FEM with other computational techniques, or exact mesh truncation schemes, because they derive the sought boundary conditions from an exact (in the limit) representation of the field in the exterior region. Examples of hybrid methods that derive nonlocal boundary conditions from an integral representation of the field are:

- Finite element-boundary integral method or FE-BI method [1–5];
- Finite element-MoM [6, 7];
- Field feedback formulation [8, 9];

while the following are examples of hybrid methods based on the usage of modal expansions:

- Unimoment method [10, 11];
- Finite element with extended boundary conditions [12];
- Transfinite element method [13, 14];
- Bymoment method [15–17].

Because hybrid techniques are based on the exact representation of the exterior field, they express the action at distance actually present in electromagnetic fields. This action is responsible for the nonlocal nature of the boundary conditions that they produce. The nonlocality of such boundary conditions spoils the sparsity of the finite element matrix, negatively affecting storage and CPU time requirements. On the other hand, due to the exact nature of the field representation in the exterior region, the fictitious surface Γ can be placed arbitrarily close to the scatterer, thus allowing a minimization of the geometrical domain where the finite element method is applied. The loss of sparsity may thus be compensated by at least some decrease in the matrix size.

An extensive review of hybrid finite element techniques applied to classic scattering and antenna problems is given by Silvester and Pelosi [18], who furnish many additional details.

In this chapter, the application of a hybrid FEM-modal expansion method to the analysis of the electromagnetic scattering from periodic structures with arbitrary cross-section will be shown. The formulation described allows analyzing both the TE_z ($E_z = 0$) and TM_z ($H_z = 0$) cases at normal incidence and is implemented using first- and second-order triangular or quadrilateral nodal elements by the code GRATING described at the end of the chapter.

5.1 Scattering by a Periodic Structure: Formulation

The problem of scattering from periodic structures is of importance for many applications in different fields such as optics, acoustics, and electromagnetics. A periodic model can be employed to analyze with sufficient accuracy many real problems as, for example, grating couplers [19], leaky-wave antennas [20], frequency selective surfaces [21], and artificially soft and hard surfaces [22–24]. A periodic model can used also to analyze the electromagnetic behavior of artificial photonic bandgap materials [25–27] and, together with a Monte Carlo–based simulation, may be employed to study the scattering of electromagnetic and acoustic waves by random rough surfaces [28]. Needless to say, currently studied metamaterials are often physically realized with periodic metallic inclusions in dielectrics [29–32].

FEM has been applied to the analysis of scattering from two-dimensional periodic surfaces uniform in one direction, say z, and periodic in the other, say x (Figure 5.2) both at normal and skew incidence [14, 33–35]. It has also been

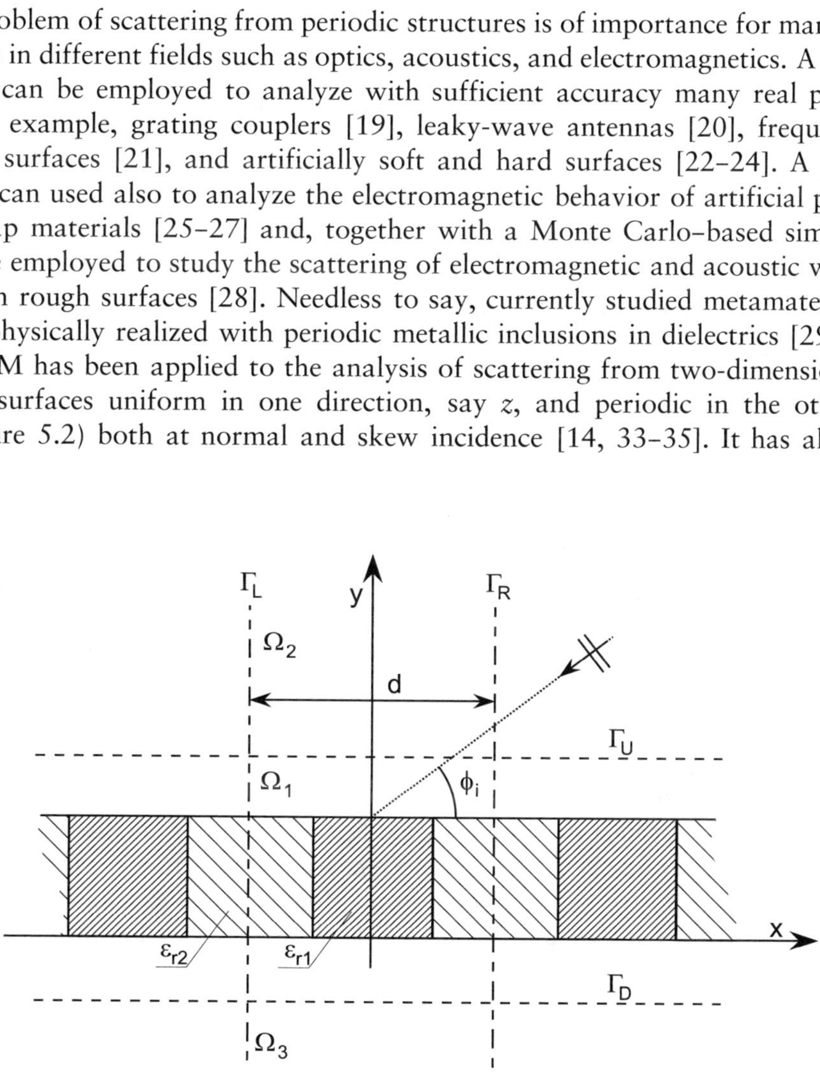

Figure 5.2 Geometry of the problem.

applied to the case of scattering from planar surfaces periodic along two directions [36, 37] and from artificial materials periodic in three directions [38].

The hybrid formulation outlined in this chapter is suitable for the analysis of planar two-dimensional periodic structure and hybridizes the finite element method with the Floquet's modal expansion to represent the scattered field [14]. The resulting hybrid method is flexible and accurate and allows analyzing periodic structures that can have arbitrary geometry and material composition of the unit cell.

The transverse section of a two-dimensional finite-thickness periodic structure is sketched in Figure 5.2. The incident field $u^i(x,y)$ is a plane wave independent of the z-coordinate with the electric field either parallel (TM_z, $H_z = 0$, $u^i = E_z$) or perpendicular (TE_z, $E_z = 0$, $u^i = H_z$) to the z-axis and impinging the structure at an angle ϕ_i:

$$u^i(x,y) = N_0^{e,h} e^{j(k\cos\phi_i x + k\sin\phi_i y)} \tag{5.1}$$

In the above equation k is the propagation constant of the medium above the periodic structure and $N_0^{e,h}$ is a normalizing constant.

Since the geometry of the scatterer is uniform along z and is a periodic function of x with period d, the total field $u^t = u^i + u^s$, sum of the incident (u^i) and the scattered (u^s) field, must fulfill the quasiperiodic condition

$$u^t(x+d,y) = u^t(x,y)e^{jkd\cos\phi_i} \tag{5.2}$$

In the above equations, as well as in the following analysis, the harmonic time dependence $e^{j\omega t}$ is assumed and suppressed. Equation (5.2) suggests that it is necessary to determine the total field u^t only in a basic cell of the periodic structure, for instance that with $-d/2 \leq x \leq d/2$, delimited by the fictitious boundaries Γ_L and Γ_R (Figure 5.2). This unit cell can be further partitioned into three different regions. Region Ω_1 spans the zone of complexity of the electromagnetic problem where the periodic structure is located and it is delimited by the fictitious boundaries Γ_U and Γ_D. Regions Ω_2 and Ω_3 consist of the homogeneous medium above and underneath the periodic structure, respectively. The wavenumber in these three regions will be denoted by $k_1(x,y) = k_0\sqrt{\epsilon_r(x,y)\mu_r(x,y)}$ and $k_2 = k_3 = k$, respectively.

At this point it is clear that the fictitious boundary $\Gamma = \partial\Omega_1$, made up of segments of the four lines Γ_U, Γ_D, Γ_R, and Γ_L, plays the same role as the fictitious surface Γ introduced in the previous section: it is used to subdivide the infinitely extending geometric domain of the scattering problem into an interior problem, spanning the bounded region Ω_1, and an exterior one, spanning regions Ω_2 and Ω_3. The interior problem is solved by using the finite element method, while the field in regions Ω_2 and Ω_3 is expressed in terms of Floquet modes. This representation is exploited to derive boundary conditions for the interior problem, in a way similar to that followed in Chapter 4 to deal with waveguide discontinuities.

As mentioned above, the scattered field u^s in the regions Ω_2 and Ω_3 is expanded in terms of Floquet modes as:

$$u^{s^{U,D}}(x,y) = \sum_{m=-\infty}^{\infty} B_m^{U,D}\psi_m(x)e^{-j\gamma_m|y-y^{U,D}|} \tag{5.3}$$

where the superscripts U and D denote the field in the upper (Ω_2) and the lower (Ω_3) region, respectively, and the coefficients $B_m^{U,D}$ are the unknown coefficients of the Floquet expansion. It is worth noting that, without loss of generality, the reference planes for the scattered field in the homogeneous regions Ω_2 and Ω_3 have been taken coincident with the planes $y = y^U$ and $y = y^D$, respectively, that is with the two boundaries Γ_U and Γ_D. In (5.3) the functions $\psi_m(x)$ give the transverse variation of the field:

$$\psi_m(x) = N_m^{e,h} e^{-jk_m x}, \quad m = 0, \pm 1, \pm 2, \ldots \tag{5.4}$$

with

$$k_m = \frac{2m\pi}{d} - k\cos\phi_i \tag{5.5}$$

and $N_m^{e,h}$ are normalization constants. The superscripts e and h denote the TM_z and TE_z case, respectively. The value of such constants is chosen so that the flux of the scattered power associated with the mth harmonic through a surface perpendicular to the y axis, of unitary length along the z direction and of width equal to one period d, is unitary. It is easy to show that this condition on the scattered power is realized when

$$N_m^e = \sqrt{\frac{2}{d}}\sqrt{\frac{k\zeta}{\gamma_m}} \tag{5.6}$$

$$N_m^h = \sqrt{\frac{2}{d}}\sqrt{\frac{k}{\gamma_m \zeta}} \tag{5.7}$$

In the above equations as well as in (5.3), γ_m is the propagation constant of the mth Floquet harmonic along the y direction, given by

$$\gamma_m = \sqrt{k^2 - k_m^2} \quad \text{for } k^2 \geq k_m^2 \tag{5.8}$$

$$\gamma_m = -j\sqrt{k_m^2 - k^2} \quad \text{for } k^2 < k_m^2 \tag{5.9}$$

Having expressed the scattered field in regions Ω_2 and Ω_3, the formulation of the scattering problem is very similar to that outlined for the analysis of waveguide discontinuities. Inside the electromagnetically complex region Ω_1 the field cannot be represented by an analytical expression, thus the problem is solved seeking the solution of the scalar Helmholtz equation:

$$\nabla_t \cdot \left(\frac{1}{p}\nabla_t u^t\right) + k_0^2 q u^t = 0 \tag{5.10}$$

with

$$u^t = E_z \quad p = p(x,y) = \mu_r \quad q = q(x,y) = \epsilon_r \quad TM_z \text{ case} \tag{5.11}$$

$$u^t = H_z \quad p = p(x,y) = \epsilon_r \quad q = q(x,y) = \mu_r \quad TE_z \text{ case} \tag{5.12}$$

Appropriate boundary conditions must be enforced to solve the partial differential (5.10) in region Ω_1. In particular, the continuity of the tangential component of the electric and magnetic field must be imposed at Γ_U and Γ_D. It is easily seen that both conditions are enforced when setting:

$$u^t(x,y)|_{\Gamma_U,\Gamma_D} = u^i(x,y)|_{\Gamma_U,\Gamma_D} + u^s(x,y)|_{\Gamma_U,\Gamma_D} \tag{5.13}$$

$$\frac{\partial u^t(x,y)}{\partial y}|_{\Gamma_U,\Gamma_D} = \frac{\partial u^i(x,y)}{\partial y}|_{\Gamma_U,\Gamma_D} + \frac{\partial u^s(x,y)}{\partial y}|_{\Gamma_U,\Gamma_D} \tag{5.14}$$

The periodic boundary condition (5.2) for the total electric and magnetic fields must be also enforced:

$$u^t(x,y)|_{\Gamma_R} = u^t(x,y)|_{\Gamma_L} e^{jkd\cos\phi_i} \tag{5.15}$$

$$\frac{1}{p}\frac{\partial u^t(x,y)}{\partial x}|_{\Gamma_R} = \frac{1}{p}\frac{\partial u^t(x,y)}{\partial x}|_{\Gamma_L} e^{jkd\cos\phi_i} \tag{5.16}$$

The above boundary conditions are valid in the assumption that the fictitious surfaces Γ_U and Γ_D are placed in homogeneous regions. On the contrary, Γ_R and Γ_L may coincide with material interfaces inside the periodic structure. Following the weighted residual procedure, the boundary value problem at hand may be recast into the weak form:

$$\iint_{\Omega_1}\left(\nabla_t u^t \cdot \frac{1}{p}\nabla_t W - k_0^2 q u^t W\right) dS - \int_{\Gamma}\frac{1}{p}\frac{\partial u^t}{\partial n} W d\gamma = 0 \tag{5.17}$$

$$\int_{\Gamma_U,\Gamma_D} \bar{W} u^t dx = \int_{\Gamma_U,\Gamma_D} \bar{W} u^i dx + \int_{\Gamma_U,\Gamma_D} \bar{W} u^s dx \tag{5.18}$$

$$\int_{\Gamma_U,\Gamma_D} \bar{\bar{W}}\frac{\partial u^t}{\partial n} dx = \int_{\Gamma_U,\Gamma_D} \bar{\bar{W}}\frac{\partial u^i}{\partial n} dx + \int_{\Gamma_U,\Gamma_D} \bar{\bar{W}}\frac{\partial u^s}{\partial n} dx \tag{5.19}$$

$$\int_{\Gamma_R} \bar{\bar{\bar{W}}} u^t dy = \int_{\Gamma_L} \bar{\bar{\bar{W}}} u^t e^{jkd\cos\phi_i} dy \tag{5.20}$$

$$\int_{\Gamma_R}\frac{1}{p}\bar{\bar{W}}\frac{\partial u^t}{\partial n} dy = -\int_{\Gamma_L}\frac{1}{p}\bar{\bar{W}}\frac{\partial u^t}{\partial n} e^{jkd\cos\phi_i} dy \tag{5.21}$$

where, $\Gamma = \Gamma_U + \Gamma_L + \Gamma_D + \Gamma_R$, W, $\bar{W}$, $\bar{\bar{W}}$, and $\bar{\bar{\bar{W}}}$ are suitable weighting functions (W must be differentiable), and $\partial/\partial n$ is the outward normal derivative evaluated at the boundary Γ. The minus sign in (5.21) takes into account the fact that the outward unit normal is opposite at Γ_R and Γ_L. As in the case of waveguide devices, the weighting function $\bar{\bar{W}}$ can be taken, without loss of generality, equal to the restriction of W to the contour Γ of the region Ω_1. If perfectly conducting material is present into the unit cell, the homogeneous Dirichlet (TM_z case, $u = E_z$) or Neumann (TE_z case, $u = H_z$) boundary conditions at its surface can be enforced in the usual way. The contour integral appearing in (5.17) can of course be rewritten as

$$\begin{aligned}\int_{\Gamma} \frac{1}{p}\frac{\partial u^t}{\partial n} W d\gamma &= \int_{\Gamma_U} \frac{1}{p}\frac{\partial u^t}{\partial n} W d\gamma + \int_{\Gamma_L} \frac{1}{p}\frac{\partial u^t}{\partial n} W d\gamma \\ &+ \int_{\Gamma_D} \frac{1}{p}\frac{\partial u^t}{\partial n} W d\gamma + \int_{\Gamma_R} \frac{1}{p}\frac{\partial u^t}{\partial n} W d\gamma\end{aligned} \tag{5.22}$$

and the terms relative to the upper and lower contours Γ_U and Γ_D can be used for introducing the boundary condition (5.19) into the weak form of the Helmholtz equation (5.17). It is important to note that the normal derivative of the incident and the scattered field can be analytically evaluated from (5.1) and (5.3). Also the other two boundary integral terms in (5.22), defined along Γ_R and Γ_L, can be exploited for enforcing boundary conditions. To this end, let's consider a weighting function W_l centered at the node l on the left boundary Γ_L. The weak form of the Helmholtz equation weighted with W_l is

$$\begin{aligned}\iint_{\Omega_l} \left(\nabla_t u^t \cdot \frac{1}{p}\nabla_t W_l - k_0^2 q u^t W_l\right) dS &- \int_{\Gamma_U} \frac{1}{p}\frac{\partial u^s}{\partial n} W_l d\gamma \\ &- \int_{\Gamma_L} \frac{1}{p}\frac{\partial u^t}{\partial n} W_l d\gamma \\ &- \int_{\Gamma_D} \frac{1}{p}\frac{\partial u^s}{\partial n} W_l d\gamma \\ &= \int_{\Gamma_U} \frac{1}{p}\frac{\partial u^i}{\partial n} W_l d\gamma\end{aligned} \tag{5.23}$$

where Ω_l is the region spanned by the weighting function W_l. In the above equation the continuity condition (5.19) has been also introduced, assuming that the incident field is coming from the region Ω_2 above the periodic structure. It should also be noted that the boundary integral on Γ_R is missing. This is because the weighting function used in the weighted residual procedure are usually taken equal to the basis functions used to expand the unknown field (Galerkin formulation) and these latter do not span all the computational domain in finite element methods. Consequently, W_l vanishes at Γ_R.

In order to simplify the formulation of the problem, the finite element mesh is built with the same number of nodes along the boundaries Γ_L and Γ_R, and equally located. That is, for each node at Γ_R, there is a corresponding node at Γ_L having the same y-coordinate. In this hypothesis, let W_r denote the weighting function centered in the node r on the right boundary Γ_R having the same y-coordinate of the node l (Figure 5.3). The weak form of the Helmholtz equation associated to W_r is

$$\begin{aligned}\iint_{\Omega_r} \left(\nabla_t u^t \cdot \frac{1}{p}\nabla_t W_r - k_0^2 q u^t W_r\right) dS &- \int_{\Gamma_U} \frac{1}{p}\frac{\partial u^t}{\partial n} W_r d\gamma \\ &- \int_{\Gamma_D} \frac{1}{p}\frac{\partial u^t}{\partial n} W_r d\gamma - \int_{\Gamma_R} \frac{1}{p}\frac{\partial u^t}{\partial n} W_r d\gamma = 0\end{aligned} \tag{5.24}$$

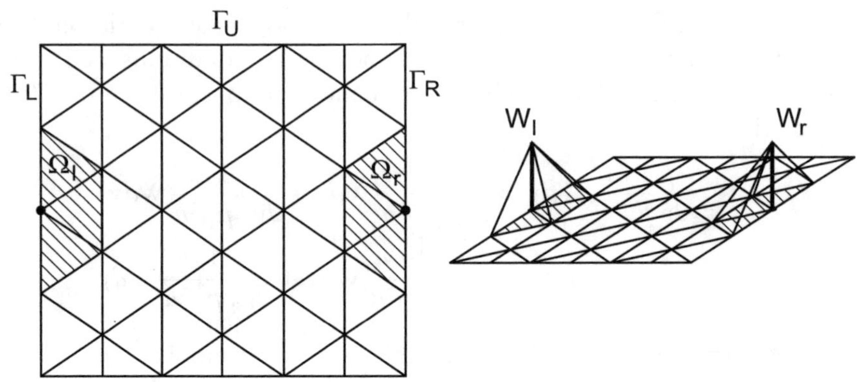

Figure 5.3 Weighting functions centered at nodes lying at the contours Γ_L and Γ_R.

where Ω_r is the region spanned by W_r. With the mesh built as described above, the restrictions $\bar{\bar{W}}_r$ of W_r along Γ_R and $\bar{\bar{W}}_l$ of W_l along Γ_L, are equal, even if Ω_r and Ω_l are not. Then, combining (5.24) and (5.21) yields

$$
\begin{aligned}
-\int_{\Gamma_L} \frac{1}{p}\frac{\partial u^t}{\partial n} W_l d\gamma = e^{-jkd\cos\phi_i} \Bigg\{ & \iint_{\Omega_r} \left(\frac{1}{p} \nabla_t u^t \cdot \nabla_t W_r - k_0^2 q u^t W_r \right) dS \\
& - \int_{\Gamma_U} \frac{1}{p}\frac{\partial u^t}{\partial n} W_r d\gamma - \int_{\Gamma_D} \frac{1}{p}\frac{\partial u^t}{\partial n} W_r d\gamma \Bigg\}
\end{aligned}
\tag{5.25}
$$

which can be introduced into the weak form of the Helmholtz equation (5.23) to give

$$
\begin{aligned}
& \iint_{\Omega_l} \left(\nabla_t u^t \cdot \frac{1}{p} \nabla_t W_l - k_0^2 q u^t W_l \right) dS \\
& - \int_{\Gamma_U} \frac{1}{p}\frac{\partial u^s}{\partial n} W_l d\gamma - \int_{\Gamma_D} \frac{1}{p}\frac{\partial u^s}{\partial n} W_l d\gamma \\
& + e^{-jkd\cos\phi_i} \Bigg\{ \iint_{\Omega_r} \left(\nabla_t u^t \cdot \frac{1}{p} \nabla_t W_r - k_0^2 q u^t W_r \right) dS \\
& - \int_{\Gamma_U} \frac{1}{p}\frac{\partial u^s}{\partial n} W_r d\gamma - \int_{\Gamma_D} \frac{1}{p}\frac{\partial u^s}{\partial n} W_r d\gamma \Bigg\} \\
& = \int_{\Gamma_U} \frac{1}{p}\frac{\partial u^i}{\partial n} W_l d\gamma + e^{-jkd\cos\phi_i} \left\{ \int_{\Gamma_U} \frac{1}{p}\frac{\partial u^i}{\partial n} W_r d\gamma \right\}
\end{aligned}
\tag{5.26}
$$

The above weak form of the Helmholtz equation includes the continuity conditions (5.19) and the periodicity condition (5.21) on the normal derivative of the field.

The periodic boundary condition on the field itself (5.15) or its weak form (5.20) is not introduced directly into the weak form of the Helmholtz equation. However, it can be easily enforced on a node-by-node basis for instance by choosing weighting functions $\bar{\bar{\bar{W}}}$ in (5.20) which are Dirac's delta functions centered at nodes lying at Γ_R and Γ_L. The N_R equations so obtained, where N_R is the number of nodes at Γ_R, replace the N_R equations obtained by weighting the Helmholtz equation with the functions W_r. As a matter of fact, these latter equations have already been used to enforce the periodic boundary condition on the derivative of the field at Γ_L as shown in (5.27), and must not be used again to set up the final system of equations.

Finally, the continuity of the longitudinal field component at Γ_U and Γ_D must be imposed (5.18). This can be conveniently done by choosing a set of weighting functions $\bar{W}_m = \left[\psi_m(x)/N_m^{e,h}\right]^*$, where $*$ denotes the complex conjugate. With this choice, the continuity conditions for the field at Γ_U and Γ_D is expressed by the integral equations:

$$\int_{\Gamma_U,\Gamma_D} u^t e^{jk_m x} dx = \int_{\Gamma_U,\Gamma_D} u^i e^{jk_m x} dx + \int_{\Gamma_U,\Gamma_D} u^s e^{jk_m x} dx \tag{5.27}$$

with $m = 0, \pm 1, \pm 2, \ldots, \pm M$ where M is the order of the highest Floquet harmonic used for representing the scattered field into regions Ω_2 and Ω_3. Using (5.1) and (5.3) to express the incident and scattered fields, respectively, assuming the incident field coming from the region Ω_2, and taking also into account the orthogonality of the functions $e^{jk_m x}$, the above equations give

$$\int_{\Gamma_U} u^t e^{jk_m x} dx = d\delta_{0m} N_0^{e,h} e^{jk\sin\phi_i y_U} + dN_m^{e,h} B_m^U \tag{5.28}$$

$$\int_{\Gamma_D} u^t e^{jk_m x} dx = dN_m^{e,h} B_m^D \tag{5.29}$$

where δ_{ij} is the Kronecker's delta.

To summarize the formulation adopted, let us assume that the mesh built has N nodes, N_R of which are located at Γ_R, and that the scattered field is represented with $2M + 1$ harmonics in each of the homogeneous regions Ω_2 and Ω_3. Then, the scattering problem has a total of $N + 2(2M + 1)$ unknowns. The linear system of equations built by following the formulation here outlined is made up of:

- $2(2M + 1)$ equations derived from (5.28) and (5.29) with $m = 0, \pm 1, \pm 2, \ldots, \pm M$, expressing the continuity of the tangential (z-directed) component of the field at Γ_U and Γ_D.
- $N - N_R$ equations derived from the weak form of the Helmholtz equation (5.23) with weighting functions associated to all nodes of the mesh, but those on the boundary Γ_R. These equations embed the continuity condition (5.19) at Γ_U and Γ_D as well as the periodicity condition (5.21) at Γ_R and Γ_L, as shown by (5.27) for the weighting function W_l.
- N_R equations obtained by enforcing the periodicity condition (5.15) on a node by node basis for each of the nodes located at Γ_R.

5.2 Numerical Implementation

To set up the finite element solution, region Ω_1 is divided into N_e finite elements and in each of them the approximation $\bar{u}^{t(e)}$ of the exact solution is expressed via a linear combination of nodal basis functions. For sake of clarity, let us assume that triangular first-order elements are used, so that over the element (e)

$$\bar{u}^{t(e)}(x,y) = \sum_{j=1}^{3} \alpha_j^{(e)}(x,y)\bar{u}_j^{t(e)} \tag{5.30}$$

where $\alpha_j^{(e)}(x,y)$, $j = 1,2,3$, are the interpolating first-order shape functions defined on element (e) and $\bar{u}_j^{t(e)}$, $j = 1,2,3$, are the unknown values of the total field at the jth local node of the same element. By introducing this expression into (5.28) and (5.29) that enforce the field continuity at Γ_U and Γ_D, and annihilating the residue relative to each weighting function $e^{jk_m x}$, with $m = 0, \pm1, \pm2, \ldots, \pm M$, a linear system of equations is obtained. This can be assembled into a single matrix equation similar to (4.37) set up in Chapter 4 to enforce field continuity at the ports of a waveguide device:

$$[A] \cdot [B] + [D] \cdot [\bar{U}] = \left[U^{inc}\right] \tag{5.31}$$

In (5.31), $[B]$ is a $2(2M+1)$ column vector containing the unknown amplitudes of the Floquet harmonics of the field scattered in regions Ω_2 and Ω_3, $[\bar{U}]$ is a $(N \times 1)$ column vector storing the N unknown values of the field at the nodes of the finite element mesh in region Ω_1, and $\left[U^{inc}\right]$ is the forcing term, which accounts for the incident field. If the entries of the vector $[B]$ are ordered so that all the coefficients B_m^U with $m = -M, \ldots, M$ occupy the first $2M+1$ positions, followed by the $(2M+1)$ coefficients B_m^D with $m = -M, \ldots, M$, and assuming that the incident plane wave propagates from region Ω_2 toward region Ω_3, then the only nonzero entry of vector $\left[U^{inc}\right]$ is:

$$U_{M+1}^{inc} = -dN_0^{e,h} e^{jk\cos\phi_i y_U} \tag{5.32}$$

provided that the same order is taken for the weighting functions in (5.28) and (5.29). Under the same assumption, the matrix $[A]$ is diagonal, with elements

$$A_{M+1+m,M+1+m} = dN_m^{e,h} \quad \text{for } m = -M, \ldots, M \tag{5.33}$$

$$A_{j,j} = A_{j-(2M+1),j-(2M+1)} \quad \text{for } j = 2M+2, \ldots, 2(2M+1) \tag{5.34}$$

Matrix $[D]$ in (5.31) is a sparse $2(2M+1) \times N$ matrix with nonzero entries only in those columns corresponding to nodes at Γ_U, for what concern the first $2M+1$ rows, and at Γ_D as for rows from $2M+2$ to $2(2M+1)$. In particular, if vectors j_U $[j_D]$ is thought of as a linear array containing the labels, in the global numbering scheme, of the N_U $[N_D]$ nodes at Γ_U $[\Gamma_D]$, then the nonzero entries of matrix $[D]$ are given by:

$$D_{M+1+m,j_U(j)} = -\sum_{(e)} \int_{\Gamma_U^{(e)}} \alpha_j^{(e)} e^{jk_m x} d\gamma$$
$$\text{for} \quad m = -M, \ldots, M; \quad j = 1, \ldots, N_U \tag{5.35}$$

$$D_{3M+2+m,j_D(j)} = -\sum_{(e)} \int_{\Gamma_D^{(e)}} \alpha_j^{(e)} e^{jk_m x} d\gamma$$
$$\text{for} \quad m = -M, \ldots, M; \quad j = 1, \ldots, N_D \tag{5.36}$$

where the summation is extended to the elements having one edge along Γ_U $[\Gamma_D]$ and with a node on that edge with label $j_U(j)$ $[j_D(j)]$ in the global numbering scheme.

The second set of linear equations to be solved together with (5.31) is derived by introducing the finite element approximation (5.30) of the sought solution into the weak form (5.17) of the Helmholtz equation. Choosing weighting functions W equal to the shape functions $\alpha_i^{(e)}$ (Galerkin formulation) and taking into account also the boundary conditions (5.19), (5.20), and (5.21) as outlined in the previous section, after annihilating the residue on each element and carrying out the usual assembly, one gets the matrix equation:

$$[C] \cdot [B] + [F] \cdot [\bar{U}] = \left[U^{inc'}\right] \tag{5.37}$$

that is similar to (4.34) relative to the waveguide discontinuity problem. Here [F] is the $(N \times N)$ finite element coefficient matrix accounting for propagation into the complex region Ω_1, [C] is a sparse $N \times 2(2M+1)$ matrix deriving from the continuity condition of the derivative of the field at Γ_U and Γ_D, while $\left[U^{inc'}\right]$ is the forcing vector associated to the normal derivative of the incident field. This latter vector has nonzero elements only at rows associated to nodes at Γ_U:

$$U^{inc'}_{j_{U(j)}} = jk \sin \phi_i e^{jk \sin \phi_i y_U} \sum_{(e)} \int_{\Gamma_U^{(e)}} \alpha_j^{(e)} e^{jk \cos \phi_i x} d\gamma$$
$$\text{for} \quad m = -M, \ldots, M; \quad j = 1, \ldots, N_U. \tag{5.38}$$

[C] has a structure similar to $[D]$, with nonzero entries only on those rows corresponding to nodes at Γ_U, for what concern the first $2M+1$ columns, and at Γ_D as for columns from $2M+2$ to $2(2M+1)$. Its nonzero entries can be expressed as

$$C_{j_U(j),M+1+m} = j\gamma_m N_m^{e,h} \sum_{(e)} \int_{\Gamma_U^{(e)}} \alpha_j^{(e)} e^{-jk_m x} d\gamma$$
$$\text{for} \quad m = -M, \ldots, M; \quad j = 1, \ldots, N_U \tag{5.39}$$

$$C_{j_D(j),3M+2+m} = j\gamma_m N_m^{e,h} \sum_{(e)} \int_{\Gamma_D^{(e)}} \alpha_j^{(e)} e^{-jk_m x} d\gamma$$
$$\text{for} \quad m = -M, \ldots, M; \quad j = 1, \ldots, N_D \tag{5.40}$$

When assembling matrix (5.37), special care must be given to local matrices involving shape functions centered at nodes located at boundaries Γ_R and Γ_L. As a matter of fact, (5.27) shows that rows of the finite element matrix $[F]$ associated to nodes at Γ_R must be multiplied times the phase shift $e^{-jkd\cos\phi_i}$ and added to the row corresponding to the equivalent node at Γ_L, (i.e., with the same y-coordinate). Similar treatment must be given to the rows of both matrix $[C]$ and vector $\left[U^{inc'}\right]$ associated to the weighting function centered at the node at the upper-right corner of region Ω_1, at the intersection between Γ_U and Γ_R.

Moreover, enforcing the periodicity condition (5.15) on a node-by-node basis, implies substituting the finite element equations associated to weighting functions centered at nodes located at Γ_R with the condition (5.15) itself. Consequently, the finite element matrix F has a structure slightly changed with respect to the usual one. In particular, the changes introduced spoil its symmetry even though they do not negatively affect sparsity. A possible remedy, at the cost of a more involved assembly strategy, is provided by the formulation developed in [38, 39] where the periodic part of the field is used as the unknown of the problem.

Equations (5.31) and (5.37) constitute the global system of equations to be solved and can be assembled into a block matrix equation similar to that built in the case of waveguide discontinuity characterization:

$$\begin{bmatrix} [A] & [D] \\ [C] & [F] \end{bmatrix} \begin{bmatrix} [B] \\ [\bar{U}] \end{bmatrix} = \begin{bmatrix} \left[U^{inc}\right] \\ \left[U^{inc'}\right] \end{bmatrix} \tag{5.41}$$

which can be solved using the same strategy.

5.3 The Code GRATING

The code GRATING implements the formulation described in the previous sections and follows the standards used throughout this book. It needs two input data files: one describing the geometry, and the other specifying all the data necessary for the electromagnetic analysis. The mesh data file can be built using the mesh generator described in Chapter 2 and, to enforce periodicity, it must be such that for each node on Γ_L there is a correspondent node on Γ_R, with the same y-coordinate.

Since the code must be able to distinguish the different kinds of nodes, a careful node labeling is necessary: nodes at p.e.c. or p.m.c. must have their own labels, different from those assigned to nodes at Γ_U, Γ_D, Γ_L and Γ_R. In addition, particular labels must be used to denote nodes belonging to more than one of the above-mentioned groups.

For instance, a given label *fl* must be assigned to nodes located at Γ_U. Nodes at Γ_L and Γ_R need different labels, say l and r, respectively, while nodes at a p.e.c. or p.m.c. surface should be characterized by a label *pec* and *pmc*, respectively. Then, nodes belonging to more than one of this special boundaries must be denoted by a label of their own as in Figure 5.4.

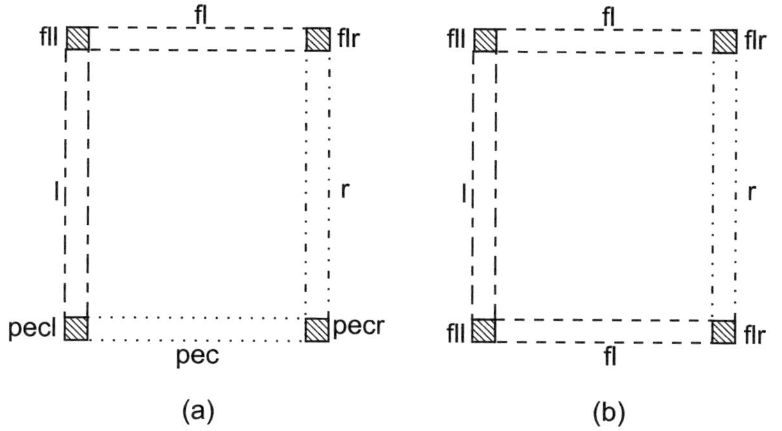

Figure 5.4 General node labeling scheme for the unit cell of a periodic structure with: (a) the p.e.c. substrate and (b) the opened lower boundary.

After having read the mesh data file, the code GRATING reads data needed for the electromagnetic analysis from the electromagnetic data file, which has the format described in Example 5.1.

Example 5.1

```
15, 5, -15        ! Floquet Labels [fll fl flr]
10,    -10        ! Left and Right Labels [l r]
11, 1, -11        ! PEC Labels [pecl pec pecr]
12, 2, -12        ! PMC Labels [pmcl pmc pmcr]
2                 ! Floquet Harmonics Above [MU]
2                 ! Floquet Harmonics Below [MD]
2                 ! No. of different dielectrics
1., 4.            ! Relative Epsilon(s)
1., 1.            ! Relative Mu(s)
1, 2              ! Element Label(s)
70.0              ! Incidence Angle
1.,1.,1           ! Frequency stepping [Start, End, No. of Steps]
E                 ! Polarization [E or H]
0                 ! Compute and save interior field [1=yes, 0=no]
```

In the above sample file, the full set of node labels, the number M of Floquet harmonics is used to represent the scattered field, as well as the number and characteristic physical constants of the materials comprising the periodic structure can be recognized. The last part of the file contains data on the incident angle, the frequency sweep, and the field polarization to be used for the analysis. The user can also set a flag to choose whether the field in region Ω_1 is computed and stored or not. In the latter case only the coefficients of the Floquet harmonics of the scattered field are evaluated.

A word is deserved on the implementation of the frequency sweep. The mesh dimensions are considered as given in terms of the free-space wavelength λ_0 of the reference frequency, so that when building the mesh it is convenient choosing a value equal to one for the period d of the cell, and scaling accordingly the dimension along y. The frequency sweep is then simulated by rescaling the whole mesh to the desired value of the ratio d/λ_0, between the two values indicated in the 12th record of the input electromagnetic data file.

When all the necessary input data have been read, the code GRATING performs the following operations:

- Builds a vector containing all the nodes on Γ_U and Γ_L;
- Finds the nodes on Γ_L and Γ_R to enforce periodicity;
- Centers the mesh on the y-axis.

Then, the actual electromagnetic analysis takes place. For each frequency value, the coefficient matrix and right-hand side vector of the linear system of (5.41) must be built as outlined in the previous section, and the resulting matrix equation can be solved using the block solver described in the previous chapter. The code creates the following output files:

- A set of files `SCATT.###.OUT` and `TRANS.###.OUT`, one for each frequency point, containing the coefficients of the Floquet modal expansion of the scattered field in region Ω_2 and Ω_3, respectively.
- If the flag in the last record of the input electromagnetic data file is set to 1, then the nodal values of the FEM expansion of the field in region Ω_1 (E_z for TM_z case, H_z for TE_z case), are stored in the file `FIELD.###.OUT`. This file can be postprocessed by using the postprocessing utility described in Chapter 2, to obtain the field map into region Ω_1 as shown for instance in Figure 5.12.
- A pair of summary files `REFLET.OUT` and `TRANSMIT.OUT` containing the data relative to the reflected and, eventually, transmitted power for each frequency point.

While in FORTRAN these name files are the same in every run, in the Matlab code it is possible to specify a word to prepend to all the above-mentioned file names to distinguish results from separate runs.

5.4 Some Examples

As a first example, a problem whose solution is analytically known is chosen. The code GRATING is quite general and can be employed to analyze structures discretized with any of the four kinds of nodal elements described in the book. Our first example will also provide a comparison between results obtained using first-order triangular elements and second-order quadrilateral elements, to prove the superiority of the latter, and of second-order elements in general, in terms of accuracy. The comparison is made analyzing a very simple structure: a dielectrically coated perfectly conducting plane illuminated by a TM_z polarized plane wave. The coating is $\lambda_0/2$ thick and the fictitious surface Γ_U is placed λ_0 above the dielectric interface, being λ_0 the free-space wavelength. The example can be used to estimate the absolute error of the numerical solution, due both to discretization and round-off errors.

By choosing an incidence angle $\phi_i = 70°$, and a dielectric material with $\epsilon_{r_1} = 4, \mu_{r_1} = 1$, the propagation angle ϕ_1 into the dielectric slab can be derived from the Snell law and is equal to 80.15 degrees. The analytical solution can be obtained from the transmission-line analogy using the equivalent circuit shown in Figure 5.5. Such circuit is comprised of a semi-infinite branch of a transmission line with characteristic impedance $Z_{y_0} = Z_0/\sin\phi_i = 401.2\Omega$, terminated with a short-circuited transmission line with characteristic impedance $Z_{y_1} = Z_0/(\sqrt{\epsilon_{r_1}}\sin\phi_1) = 191.3\Omega$ and length $d_1 = 0.985\lambda_{y_1}$. The code GRATING computes the reflection coefficient at a distance $3\lambda_0/2$ from the conducting plane (section BB' of Figure 5.5), that is at a distance $d_0 = 0.940\lambda_{y_0}$ from the section AA'. The reflection coefficient at the section BB' can be readily evaluated using the transmission lines formulas:

$$Z_{AA'} = jZ_{y_1}\tan(2\pi d_1/\lambda_{y_1}) = -j17.76\Omega \tag{5.42}$$

$$\Gamma_{AA'} = \frac{Z_{AA'} - Z_{y_0}}{Z_{AA'} + Z_{y_0}} = e^{j3.230} \tag{5.43}$$

$$\Gamma_{BB'} = \Gamma_{AA'}e^{-2j2\pi d_0/\lambda_{y_0}} = e^{j3.230}e^{-j11.809} = e^{-j8.579} \tag{5.44}$$

Hence, the phase of the reflection coefficient at section BB' is $-131.5°$. Figure 5.6 shows the phase of $\Gamma_{BB'}$ computed with different FEM meshes using triangular first-order elements or quadrilateral second-order elements. To have a fair comparison, the number of nodes on Γ_U has been chosen as the parameter. For example, the coarsest mesh has 5 nodes on Γ_U, thus exhibiting 4 first-order triangular elements, but only two second-order quadrilateral, the discretization on the vertical side is chosen proportionally. In general, given a second-order quadrilateral mesh with NE elements and NN nodes, the corresponding first-order triangular has $8 * NE$ elements and $NN + NE$ nodes. If $NE << NN$ the comparison can be said to be between meshes with almost the same number of unknowns. It is evident from Figure 5.6 that the second-order quadrilateral elements provide more

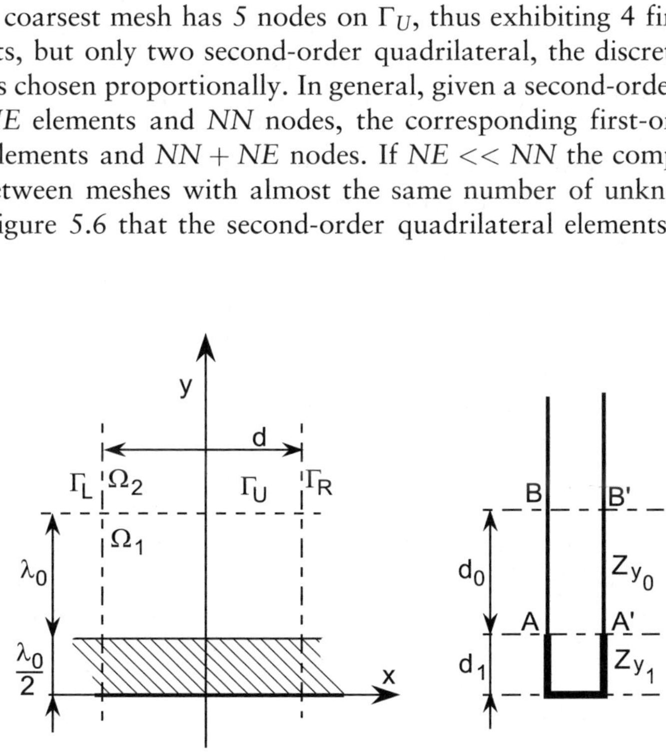

Figure 5.5 Coated plane and its equivalent circuit.

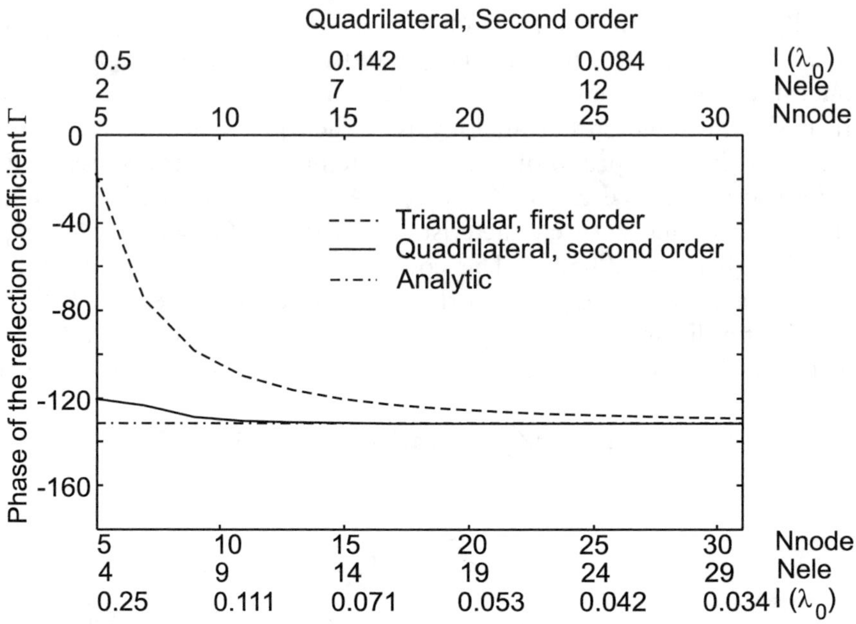

Figure 5.6 Phase of the reflection coefficient $\Gamma_{BB'}$ for different meshes and different elements.

accurate results at a comparable discretization, and this is particularly true for coarse meshes.

For instance, with 11 nodes on Γ_U the first-order triangular mesh has $NE = 400, NN = 231$, the computed phase of the reflection coefficient has an error of 16.5%, and provides the field distribution shown in Figure 5.7. On the other hand, the "equivalent" second-order quadrilateral mesh has $NE = 50, NN = 181$ and, by using 21% less unknowns, provides a phase of the reflection coefficient affected by an error of only 1%, and the field distribution depicted in Figure 5.8.

The same structure can be analyzed without the p.e.c. surface at the bottom. In such case the finite element mesh must be enlarged to also span part of the free-space region underneath the dielectric slab as shown in Figure 5.9. The same figure reports also the amplitude of the transmission coefficient for values of the incident angle ϕ_i ranging from 10 degrees to 90 degrees. The incident plane wave has TM_z polarization. It can be observed that the dielectric screen is perfectly transparent for those frequency values at which the length of the equivalent transmission line of the slab is a multiple of half a wavelength, in the dielectric. Being the dielectric half a free-space wavelength thick at the reference frequency, and being $\epsilon_r = 4$, these maximums of transmitted power occur for $d_1/\lambda_0 = 0.5$, $n = 1, 2, \ldots$ at $\phi = 90°$. As ϕ decreases then, the total transmission happens for higher frequency values. It must be noted that there is no total reflection since the dielectric slab has $\epsilon_1 > \epsilon_0$.

These calculations have been carried out using an extremely fine first-order triangular mesh, stored in the enclosed CD-ROM. It is left to the reader to check the behavior of such a structure when a second-order mesh is employed.

The next two examples are relative to practical problems for which the analytical solution is not available. The first one is the periodic grating sketched in Figure 5.10 comprised of three finite thickness metallic strips embedded into

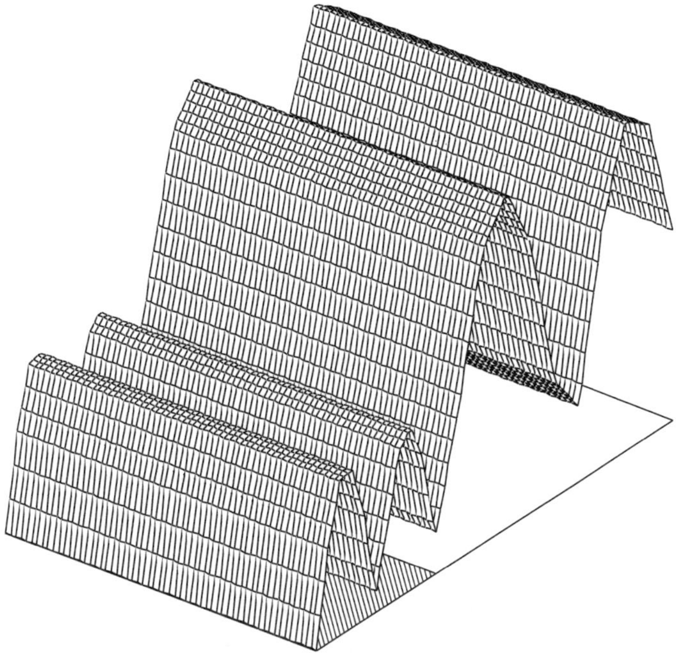

Figure 5.7 Magnitude of the field in region Ω_1, computed by using first-order triangular elements: $NE = 400, NN = 231$.

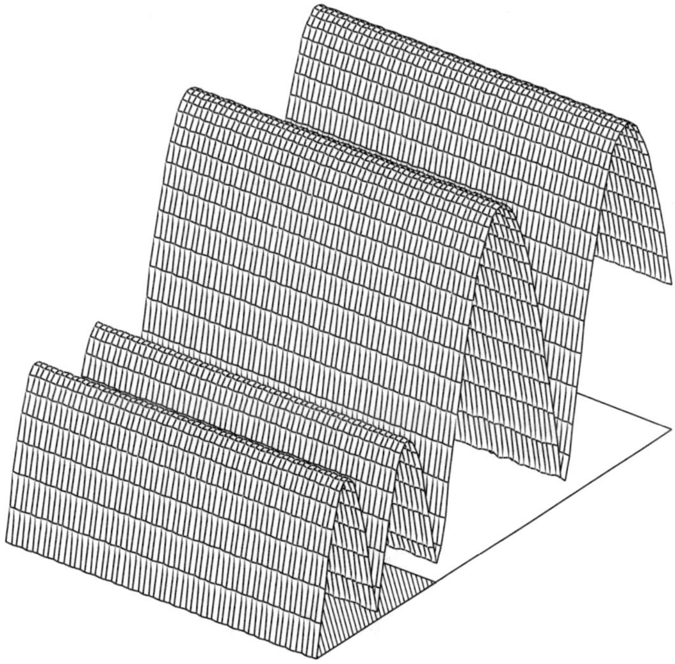

Figure 5.8 Magnitude of the field in region Ω_1, computed by using second-order quadrilateral elements: $NE = 50, NN = 181$.

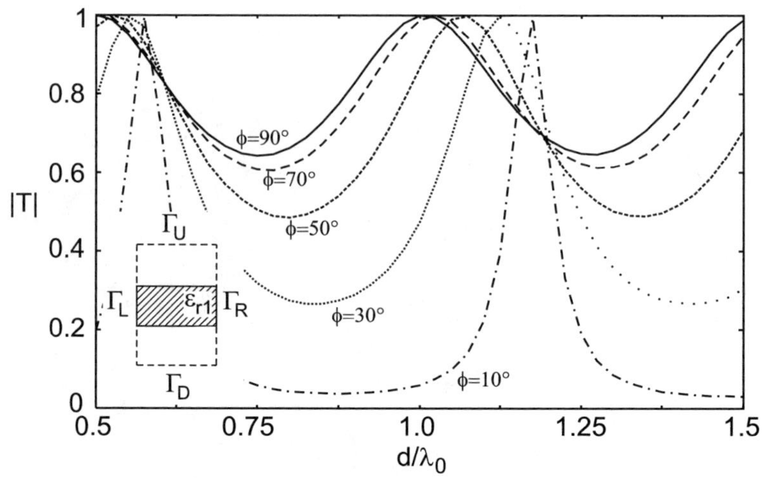

Figure 5.9 Amplitude of the transmission coefficient through a dielectric screen ($\epsilon_r = 4$) for different values of the incidence angle: TM_z polarization.

a grounded dielectric slab. The discretization of such geometry requires a fairly complex subdivision into blocks to be meshed by our mesh generator. The input file for the mesh generator is included in the program disc, but can also be easily reconstructed or modified to analyze different configurations. In this latter case, it is important to keep in mind the node labeling scheme shown in Figure 5.4.

The structure is analyzed at a single frequency, for which the period of the structure is equal to $0.7\lambda_0$ and the unit cell is discretized by using again a triangular first-order mesh (Figure 5.11). It must be noted that such a fine mesh, so fragmented into blocks, may give rise to a FEM coefficient matrix with a very wide

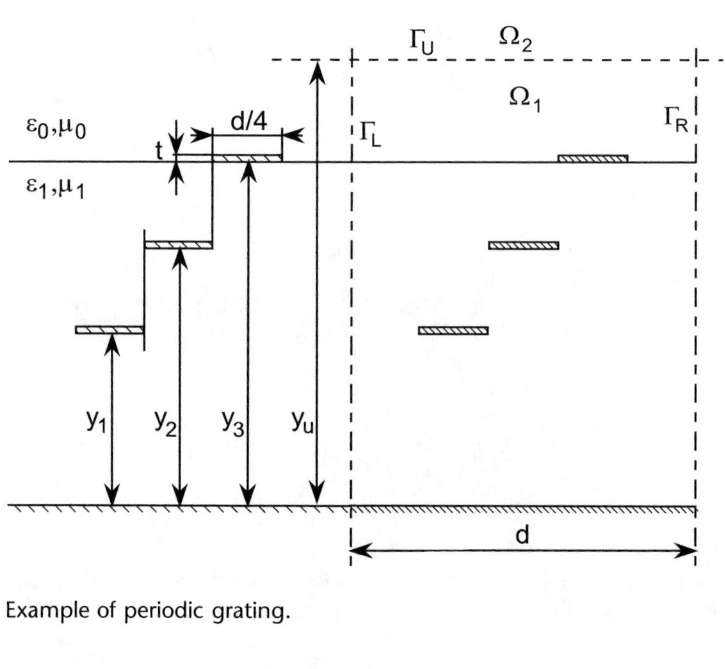

Figure 5.10 Example of periodic grating.

Figure 5.11 Final mesh for the periodic grating under exam.

band, hence the usage of the numbering optimization tools described in Chapter 2, which can also take into account the periodic nature of the structure, is strongly recommended.

For this example, the field inside the structure has been chosen as output and its contour map in region Ω_1 is displayed in Figure 5.12.

The last structure analyzed is a two-dimensional, entirely dielectric periodic structure comprised of a triangular lattice of dielectric rods. This and similar structures are often referred to as photonic crystals [40] or electromagnetic band-gap (EBG) materials [41] and have the property that they can forbid electromagnetic wave propagation for certain frequency values. The word "photonic" in their name derives from the fact that this kind of material has been originally investigated at optical frequency, while the word "crystal" suggests that such structures may act on electromagnetic waves in the same way as natural crystals act on electron waves.

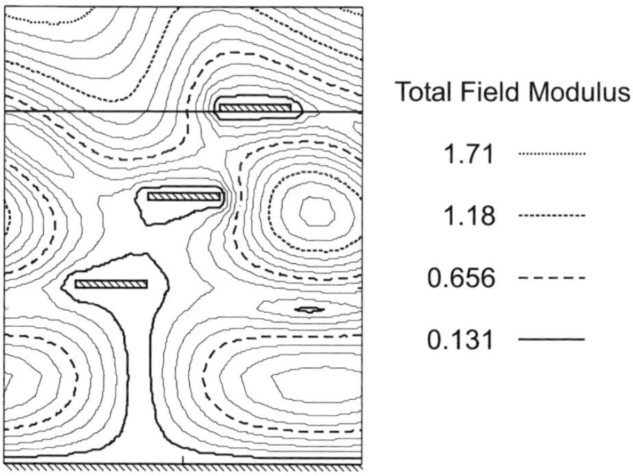

Figure 5.12 Contour map of the modulus of the total field.

The EBG material considered is sketched in Figure 5.13, and the goal of our analysis is the computation of the transmission coefficient of a crystal made of several layers of dielectric rods. The characteristics of such a structure depend on many parameters, as the lattice constant d, the ratio r/d between the radius r of the rods and d, as well as the value of the dielectric permittivity of the rods. For the particular value of the ratio r/d chosen for this analysis ($r/d = 0.2$), the crystal presents good stop-band behavior when the number of layers n of dielectric rods is quite high, say $n > 15$. The unit cell to be analyzed with the code GRATING then would extend for one crystal lattice constant d along the x direction and would span all the layers comprising the structure along the y direction (Figure 5.13). Such a cell would be very large in terms of FEM unknowns. The codes here provided are well able to analyze crystals with an arbitrary number of layers of rods, the upper limit being only the available computer resources. However, the number of layers chosen for the analysis has been intentionally reduced to a maximum of $n = 4$, to keep system requirements at the level of an ordinary PC. If the reader likes to investigate further and has the computer resources, then all that it is necessary to do, in the FORTRAN code, is increase the parameters of the program that define the maximum number of nodes, elements, and others. A similar modification may be needed in the mesh generator as well as in the numbering scheme optimization tools. Alternatively the code could be ported to FORTRAN90, taking advantage of the more flexible memory allocation/deallocation commands available. Of course Matlab codes, thanks to better memory management, do not have any hard limits to the mesh dimension but rather are limited only by the available memory.

The computed transmission coefficient of the crystal sketched in Figure 5.13 is plotted in Figure 5.14 versus the normalized lattice constant d/λ_0 for different values of the dielectric rods permittivity, and different numbers of layers of rods.

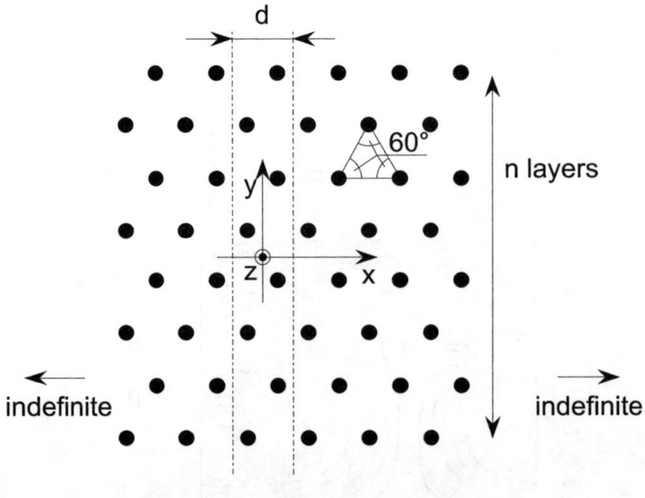

Figure 5.13 Structure of a photonic crystal made by $n = 8$ layers of dielectric rods with axis parallel to z. Each layer is infinitely extending along the x direction. Successive layers are arranged to form a triangular lattice in the $x - y$ plane.

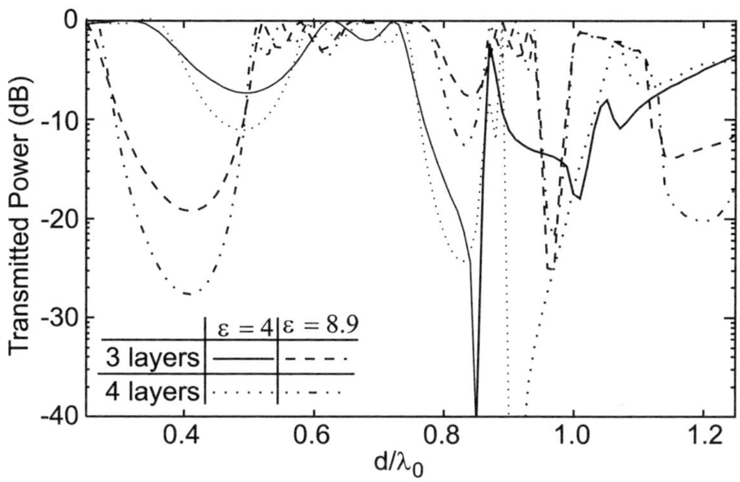

Figure 5.14 Transmitted power versus normalized lattice constant for the photonic crystal sketched in Figure 5.13.

It can be noted that a *forbidden gap* arises for values of the normalized lattice constant around 0.5 and 0.4 when the rod permittivity is $\epsilon_r = 4$, and $\epsilon_r = 8.9$, respectively.

5.5 Disc Content

5.5.1 FORTRAN

The software described here is contained in the subdirectory GRATING of the directory CODES in the enclosed CD-ROM. The program files are GRATING.F and G_SUBS.F. They must be compiled and linked with the library described in Chapter 2.

The four subdirectories CPLANE, WINDOW, GRATING and PHOCRY, contain all the input and mesh files of the examples discussed in the previous section. A more complete description of each file is given in the README file in the directory GRATING itself.

5.5.2 Matlab

Matlab subdirectory CODES contains the `grating.m` function, which must be called according to

```
grating(fnamemesh,fnameaux,fnameout,fieldflag)
```

where the first three inputs are mesh, auxiliary, and word to prepend to all output file names, respectively, while the last flag, if set to 1, makes the code compute, and save, fieldmaps.

The Matlab version contains the same subdirectories for the examples as there are in the FORTRAN version.

References

[1] Silvester, P. and Hsieh, M. S., "Finite-Element Solution of 2-Dimensional Exterior-Field Problems," *IEE Proceedings*, Vol. 118, pp. 1743–1747, 1971.

[2] McDonald, B. H. and Wexler, A., "Finite-Element Solution of Unbounded Field Problems," *IEEE Transactions on Microwave Theory and Techniques*, Vol. MTT-20, pp. 841–847, 1972.

[3] Jin, J. M. and Liepa, V. V., "Application of Hybrid Finite Element Method to Electromagnetic Scattering from Coated Cylinders," *IEEE Transactions on Antennas and Propagation*, Vol. 36, pp. 50–54, 1988.

[4] Jin, J. M. and Liepa, V. V., "A Note on Hybrid Finite Element Method for Solving Scattering Problems," *IEEE Transactions on Antennas and Propagation*, Vol. 36, pp. 1486–1490, 1988.

[5] Jin, J. M. and Volakis, J. L., "Electromagnetic Scattering by and Transmission Through a "Three-Dimensional Slot in a Thick Conducting Plane," *IEEE Transactions on Antennas and Propagation*, Vol. 39, pp. 543–550, 1991.

[6] Yuan, X., "Three-Dimensional Electromagnetic Scattering from Inhomogeneous Objects by the Hybrid Moment and Finite Element Method," *IEEE Transactions on Microwave Theory and Techniques*, Vol. 38, pp. 1053–1058, 1990.

[7] Yuan, X., Lynch, D. R., and Strohbehn, J. W., "Coupling of Finite Element and Moment Methods for Electromagnetics Scattering from Inhomogeneous Objects," *IEEE Transactions on Antennas and Propagation*, Vol. 38, pp. 386–393, 1990.

[8] Morgan, M. A. and Welch, B. E., "The Field Feedback Formulation for Electromagnetic Scattering Computations," *IEEE Transactions on Antennas and Propagation*, Vol. AP-34, pp. 1377–1382, 1986.

[9] Morgan, M. A. and Welch, T. B., "Field Feedback Computation of Scattering by Two-Dimensional Penetrable Objects," *IEEE Transactions on Antennas and Propagation*, Vol. 40, pp. 445–450, 1992.

[10] Mei, K. K., "Unimoment Method of Solving Antenna and Scattering Problems," *IEEE Transactions on Antennas and Propagation*, Vol. AP-22, pp. 760–766, 1974.

[11] Mei, K. K., "Unimoment Method for Electromagnetic Wave Scattering," *Journal of Electromagnetic Waves and Applications*, Vol. 1, pp. 201–222, 1987.

[12] Morgan, M. A., et al., "Finite Element-Boudary Integral Formulation for Electromagnetic Scattering," *Wave Motion*, Vol. 6, pp. 91–103, 1984.

[13] Lee, J. F. and Cendes, Z. J., "The Transfinite Element Method for Modeling MMIC Devices," *IEEE Transactions on Microwave Theory and Techniques*, Vol. 36, pp. 1639–1649, 1988.

[14] Pelosi, G., Freni, A., and Coccioli, R., "A Hybrid Technique for Analyzing the Scattering from Periodic Structures," *IEE Proceedings, Part H*, Vol. 140, pp. 65–70, 1993.

[15] Cangellaris, A. C. and Lee, R., "The Bymoment Method for Two-Dimensional Electromagnetic Scattering," *IEEE Transactions on Antennas and Propagation*, Vol. 38, pp. 1429–1437, 1990.

[16] Cangellaris, A. C. and Lee, R., "Finite Element Analysis of Electromagnetic Scattering From Inhomogeneous Cylinders at Oblique Incidence," *IEEE Transactions on Antennas and Propagation*, Vol. 39, pp. 645–650, 1991.

[17] Pekel, U. and Lee, R., "A Three-Dimensional Application of the Bymoment Method for Electromagnetic Scattering," *Journal of Electromagnetic Waves and Applications*, Vol. 9, pp. 973–991, 1995.

[18] Silvester, P. P. and Pelosi, G., *Finite Elements for Wave Electromagnetics*, New York: IEEE Press, 1994.

[19] Tamir, T. and Peng, S. T., "Analysis and Design of Grating Couplers," *Appl. Phys.*, Vol. 14, pp. 235–254, 1977.

[20] Guglielmi, M. and Boccalone, G., "A Novel Theory for Dielectric-Inset Waveguide Leaky-Wave Antennas," *IEEE Trans. Antennas Propagat.*, Vol. AP-39, pp. 497–504, 1991.

[21] Mittra, R. J., Chan, C. H., and Cwik, T., "Techniques for Analyzing Frequency Selective Surfaces—A Review," *Proceedings of the IEEE*, Vol. 76, pp. 1593–1615, 1988.

[22] Kildal, P. S., "Definition of Artificially Soft and Hard Surfaces for Electromagnetic Waves," *Electron. Lett.*, Vol. 24, pp. 168–170, 1988.

[23] Kildal, P. S., "Artificially Soft and Hard Surfaces in Electromagnetics," *IEEE Trans. on Antennas Propagat.*, Vol. AP-38, pp. 1537–1544, 1990.

[24] Anders Aas, J., Plane-Wave Reflection Properties of Two Artificially Hard Surfaces," *IEEE Trans. Antennas Propagat.*, Vol. AP-39, pp. 651–656, 1991.

[25] Pendry, J. B., "Photonic Band Structures," *Journal of Modern Optics*, Vol. 41, pp. 209–229, 1994.

[26] Yang, H. Y. D., "Finite Difference Analysis of Two-Dimensional Photonic Crystals," *IEEE Transactions on Microwave Theory and Techniques*, Vol. 44, pp. 2688–2695, 1997.

[27] Coccioli, R., Pelosi, G., and Itoh, T., "Finite Element Analysis of Out-of-Plane Propagation in Two-Dimensional Photonic Crystals," *IEEE AP-S International Meeting*, Montreal, July 13-18, 1997.

[28] Lou, S. H., et al., "Monte Carlo Simulations of Scattering of Waves by a Random Rough Surface with the Finite Element Method and the Finite Difference Method," *Microwave and Optical Technology Letters*, Vol. 3, pp. 150–154, 1990.

[29] Ishimaru, A., et al., "Generalized Constitutive Relations for Metamaterials Based on the Quasi-Static Lorentz Theory," *IEEE Transactions on Antennas and Propagation*, Vol. 51(10), pp. 2550–2557, Oct. 2003.

[30] Grbic, A. and Eleftheriades, G. V., "Dispersion Analysis of a Microstrip-Based Negative Refractive Index Periodic Structure," *IEEE Microwave and Wireless Components Letters*, Vol. 13(4), pp. 155–157, April 2003.

[31] Markley, L. and Eleftheriades, G. V., "A Negative-Refractive-Index Metamaterial for Incident Plane Waves of Arbitrary Polarization," *IEEE Antennas and Wireless Propagation Letters*, Vol. 6, pp. 28–32, 2007.

[32] Mosallaei, H. and Sarabandi, K., "Design and Modeling of Patch Antenna Printed on Magneto-Dielectric Embedded-Circuit Metasubstrate," *IEEE Transactions on Antennas and Propagation*, Vol. 55(1), pp. 45–52, Jan. 2007.

[33] Gedney, S. D. , Lee, J. F., and Mittra, R., "A Combined FEM/MoM Approach to Analyze the Plane Wave Diffraction by Arbitrary Gratings," *IEEE Transactions on Microwave Theory and Techniques*, Vol. 40, pp. 363–370, 1992.

[34] Calamia, M., et al., "Finite Element Analysis of Electromagnetic Scattering from Periodic Structures at Oblique Incidence," *Electronics Letters*, Vol. 29, pp. 847–848, 1993.

[35] Cecchini, R., Coccioli, R., and Pelosi, G., "PERIODIC^3: A Software Package for the Analysis of Artificially Anisotropic Surfaces," *IEEE Antennas Propagat. Magazine*, Vol. 37, pp. 83–86, 1995.

[36] Mias, C., Webb, J. P., and Ferrari, R. L., "Finite Element Analysis of Electromagnetic Plane Wave Scattering from an Infinite Doubly Periodic Structure," *COMPEL—The International Journal for Computation and Mathematics in Electrical and Electronic Engineering*, Vol. 13, Supplement A, pp. 393–398, 1994.

[37] Lucas, E. W. and Fontana, T. P., "A Three-Dimensional Hybrid Finite Element/Boundary Element Method for the Unified Radiation and Scattering Analysis of General Infinite Periodic Array," *IEEE Transactions on Antennas and Propagation*, Vol. 43, pp. 145–154, 1995.

[38] Ferrari, R., "Spatially Periodic Structures," in *Finite Element Software for Microwave Engineering*, Itoh, T., Pelosi, G., and Silvester, P. P., (eds.), New York: John Wiley & Sons, 1996.

[39] Fernandez, F. A. and Lu, Y., "A Variational Finite Element Formulation for Dielectric Waveguides in Terms of Transverse Magnetic Fields," *IEEE Transactions on Magnetics*, Vol. 27, pp. 3864–3867, 1991.

[40] Joannopoulos, J. D., Meade, R. D., and Winn, J. N., *Photonic Crystals: Molding the Flow of Light*, Princeton, NJ: Princeton University Press, 1995.

[41] Engheta, N. and Ziolkowski, R. W., *Metamaterials: Physics and Engineering Explorations*, Piscataway, NJ: Wiley-IEEE Press, 2006.

CHAPTER 6

Scattering and Antennas: Absorbing Boundary Conditions

As mentioned in the previous chapter, in addition to using the finite element method in the framework of a hybrid technique, scattering and antenna problems may be solved by employing absorbing boundary conditions (ABC) to terminate the computational domain at a fictitious surface Γ (Figure 6.1). The name ABC alludes to the requirement that all outward-traveling waves must be completely absorbed at the fictitious surface Γ as if it were not present at all, and no reflection into the computational domain should occur.

Theoretical development of ABC began in the applied mathematics field at the end of the 1970s, when Engquist and Majda [1] developed absorbing boundary conditions for the numerical solution of the scalar wave equation in time domain using the theory of pseudodifferential operators. Their theory was developed in the simple case of a plane wave normally impinging a plane interface and, although very simple, has been applied to various scattering and radiation problems in the framework of both finite elements [2] and finite difference-time domain methods [3]. A few years later, Bayliss, Turkel, et al. [4, 5] proposed an alternative approach to develop boundary conditions able to absorb outward-traveling waves at a circular fictitious boundary. This second kind of ABC is more flexible than the previous one and the analytical technique to derive them, pointed out by the authors, has been subsequently used also to developed ABC for the vector wave equation [6].

Since their first appearance in the applied mathematics field, the usage of absorbing boundary conditions spread slowly in electromagnetics, where it gained popularity and has become widely used to solve antenna and scattering problems. Researchers in electromagnetics proposed also two other kinds of ABC: numerically derived ABC [7–9] and those based on the perfectly matched layer (PML) concept [10].

The most advantageous feature of ABC, shared by all kinds of them, is that they are of local type, in the sense that at some point P on the fictitious surface Γ they are expressed in terms of the electromagnetic field at P and at its neighboring points. This implies that their usage does not spoil the sparsity of the coefficient matrix of the linear system of equations produced by FEM. Needless to say, this characteristic makes them very attractive for numerical computations.

Two kinds of absorbing boundary conditions are described in this chapter and implemented in the enclosed codes: the Bayliss and Turkel type, and those based on the PML concept. The next section gives a brief description of the analytical derivation conceived by Bayliss and Turkel, while their usage in the framework of the FEM solution of a scattering problem is shown in Section 6.2, and their implementation is detailed in Section 6.3. Section 6.4 outlines the theory behind the perfectly matched anisotropic absorber (PMA), which is the kind of PML used to terminate the computational domain in this book. Its usage in the framework of an FEM solution of an antenna problem is shown in Section 6.5, and its implementation is described in Section 6.6. The structure of the two codes presented in this chapter is reported next. Section 6.7 describes the code CYL designed to solve the problem of scattering from cylinders with arbitrary cross-section, while Section 6.8 includes application examples. The second code, described in Section 6.9 and named OWG, computes radiation from an open-ended waveguide. Some examples showing its capabilities are presented in Section 6.10.

6.1 Analytic ABC

The derivation of analytic absorbing boundary conditions starts from the asymptotic expansion of the outward-traveling electromagnetic field. Considering a cylindrical coordinate system (ρ, ϕ), it has been shown that the outgoing solution of the two-dimensional Helmholtz equation can be represented by the convergent expansion [11]:

$$u(\rho, \phi) = H_0^{(2)}(k\rho) \sum_{i=0}^{\infty} \frac{f_i(\phi)}{\rho^i} + H_1^{(2)}(k\rho) \sum_{i=0}^{\infty} \frac{g_i(\phi)}{\rho^i} \tag{6.1}$$

where $H_0^{(2)}$ and $H_1^{(2)}$ are the zeroth and 1st order Hankel functions of the second kind, respectively, and $f_i(\phi)$ and $g_i(\phi)$ are unknown expansion functions of the ϕ coordinate alone. To derive the above expansion, the time dependence $e^{j\omega t}$ has been assumed. Introducing the asymptotic approximation of the Hankel functions [12], the outward-traveling solution can be expressed as:

$$u(\rho, \phi) = \sqrt{\frac{2}{\pi k\rho}} e^{-(jk\rho - \pi/2)} \sum_{i=0}^{\infty} \frac{F_i(\phi)}{\rho^i} \tag{6.2}$$

Bayliss and Turkel [4] started from this expression to develop the sequence of differential operators of increasing order

$$B_0 = 1 \tag{6.3}$$

$$B_1 = \frac{\partial}{\partial \rho} + jk + \frac{1}{2\rho} \tag{6.4}$$

$$B_m = \left(\frac{\partial}{\partial \rho} + jk + \frac{4m - 3}{2\rho} \right) B_{m-1} \tag{6.5}$$

which, when applied to the outgoing electromagnetic field at a circular contour Γ, give rise to absorbing boundary conditions of increasing accuracy for the solution of the problem in the region enclosed by Γ (Figure 6.1). As a matter of fact, for any function u_i of the form

$$u_i = e^{-jk\rho} \frac{F_i(\phi)}{\rho^{i+1/2}} \tag{6.6}$$

it is easy to see that $B_m u_i = 0$, for $1 \leq i \leq m$ and for any function u having an expansion of the form (6.2)

$$B_m u = O\left[\frac{1}{\rho^{2m+\frac{1}{2}}}\right] \tag{6.7}$$

This means that in numerical computations in which the exterior region is truncated at a circular fictitious surface Γ with radius $\rho = \rho_1$, enforcing the boundary condition

$$B_m u|_{\rho=\rho_1} = 0 \tag{6.8}$$

actually matches the numerical solution u to the first m terms of the series expansion (6.2). In other words, the computed field inside Γ is forced to have the same asymptotic behavior as the exact analytic solution.

In the following, the second-order absorbing boundary condition $B_2 u = 0$ will be used to solve antenna and scattering problems. From the recursive relation in (6.5), the second-order operator B_2 may be expanded as

$$B_2 = \frac{\partial^2}{\partial\rho^2} + \left(2jk + \frac{3}{\rho}\right)\frac{\partial}{\partial\rho} + \frac{3jk}{\rho} + \frac{5}{4\rho^2} - k^2 \tag{6.9}$$

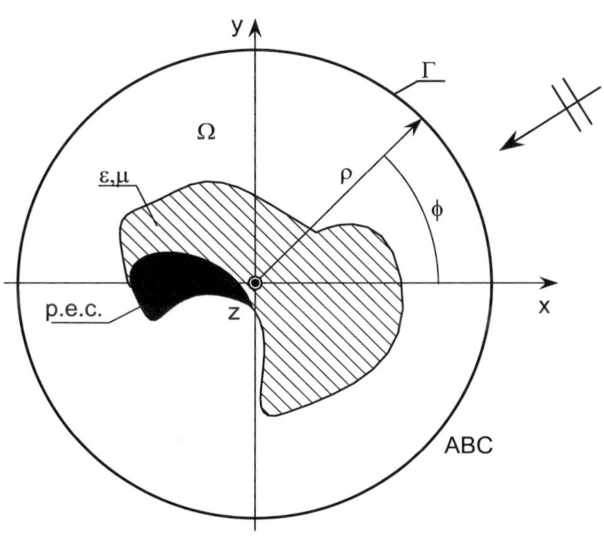

Figure 6.1 Inhomogeneous cylinder with arbitrary cross-section enclosed by a fictitious circular boundary Γ.

The presence of the second-order derivative with respect to ρ makes the implementation cumbersome because it would require using the basis function at least twice differentiable with respect to ρ to expand the field u. This requirement on the basis function can be avoided by resorting to the Helmholtz equation in cylindrical coordinate, from which is possible to get:

$$\frac{\partial^2}{\partial \rho^2} = -\frac{1}{\rho^2}\frac{\partial^2}{\partial \phi^2} - \frac{1}{\rho}\frac{\partial}{\partial \rho} - k^2 \tag{6.10}$$

Substituting (6.10) into (6.9) yields:

$$B_2 = \frac{\partial}{\partial \rho} + a(\rho) - b(\rho)\frac{1}{\rho^2}\frac{\partial^2}{\partial \phi^2} \tag{6.11}$$

where

$$a(\rho) = jk + \frac{1}{2\rho} - \frac{b(\rho)}{4\rho^2}; \qquad b(\rho) = \frac{1}{2\left(jk + \frac{1}{\rho}\right)} \tag{6.12}$$

Equation (6.11) is a much more convenient expression because the second-order derivative in ϕ can be reduced to a first-order derivative by integration by parts along the contour Γ, as will be shown later on.

6.2 Scattering Problems: Formulation with Analytic ABC

To show how the ABC analytically derived by Bayliss and Turkel can be used in the framework of a finite element solution to a scattering problem, let's consider the case of an inhomogeneous dielectric cylinder illuminated by a plane wave (Figure 6.1). The cylinder may have an arbitrary cross-section and might even be not simply connected. The application of FEM to solve this problem requires introducing a fictitious surface Γ to terminate the computational domain. In order to apply analytical ABC at Γ, such contour should be taken circular with its center as close as possible to the center of the cylinder or to the center of the cluster of cylinders in case of a multiple-connected cross-section. The radius ρ_1 of Γ should be chosen so that the asymptotic expansion in (6.2) is a good approximation of the exact solution at Γ. Apparently, such value is unknown, but common practice and experience suggest to take it so that Γ is at a distance of at least 0.3λ from the scatterer. In the region Ω delimited by the fictitious boundary Γ the problem is formulated as a boundary value problem and solved via FEM. The longitudinal component of the total electromagnetic field inside Ω fulfills the Helmholtz equation

$$\nabla_t \cdot \left(\frac{1}{p}\nabla_t u^t\right) + k_0^2 q u^t = 0 \tag{6.13}$$

with

$$\begin{cases} u^t = E_z^t \quad p = p(x,y) = \mu_r(x,y) \quad q = q(x,y) = \epsilon_r(x,y) \quad TM_z \text{ case} \\ u^t = H_z^t \quad p = p(x,y) = \epsilon_r(x,y) \quad q = q(x,y) = \mu_r(x,y) \quad TE_z \text{ case} \end{cases} \tag{6.14}$$

In general, there might be also a homogeneous Dirichlet boundary Γ_D (either a p.e.c surface in case of TM_z polarization or a p.m.c surface in case of TE_z polarization), at which $u^t = 0$, and a homogeneous Neumann boundary Γ_N (either a p.m.c surface in case of TM_z polarization or a p.e.c surface in case of TE_z polarization), at which $\partial u^t/\partial n = 0$. Finally, the second-order absorbing operator B_2 of Bayliss and Turkel may be employed to write the boundary condition for the total field at the circular boundary Γ as:

$$\begin{aligned}\frac{\partial u^t}{\partial n}|_{\rho=\rho_1} &= \frac{\partial u^t}{\partial \rho}|_{\rho=\rho_1} = \left(\frac{\partial u^i}{\partial \rho} + \frac{\partial u^s}{\partial \rho}\right)|_{\rho=\rho_1} \\ &= \frac{\partial u^i}{\partial \rho}|_{\rho=\rho_1} - a(\rho_1)u^s + b(\rho_1)\frac{1}{\rho_1^2}\frac{\partial^2 u^s}{\partial \phi^2}|_{\rho=\rho_1}\end{aligned} \tag{6.15}$$

Following the usual weighted residual procedure, the boundary value problem at hand may be recast in the weak form:

$$\iint_\Omega W\nabla_t \cdot \frac{1}{p}\nabla_t u^t d\Omega + \iint_\Omega k_0^2 q W u^t d\Omega + \int_{\Gamma_D} W_D u^t d\gamma + \int_{\Gamma_N} W_N \frac{\partial u^t}{\partial n} d\gamma = 0 \tag{6.16}$$

$$\begin{aligned}\oint_\Gamma \bar{W}\frac{\partial u^t}{\partial n}d\gamma &= \oint_\Gamma \bar{W}\frac{\partial u^i}{\partial \rho}d\gamma + \oint_\Gamma \bar{W}\frac{\partial u^s}{\partial \rho}d\gamma \\ &= \oint_\Gamma \bar{W}\frac{\partial u^i}{\partial \rho}d\gamma - \oint_\Gamma \bar{W}\left[a(\rho_1)u^s - \frac{b(\rho_1)}{\rho_1^2}\frac{\partial^2 u^s}{\partial \phi^2}\right]d\gamma\end{aligned} \tag{6.17}$$

where W, W_D, W_N, and $\bar{W}$ are arbitrary weighting functions. Without loss of generality, the function u^t may be chosen to satisfy the Dirichlet boundary condition $u^t = 0$ at Γ_D, and the weighting functions W, W_D, W_N and $\bar{W}$ so that $W = 0$ at Γ_D, $W_N = -W$ at Γ_N, and $\bar{W} = W$ at Γ. In this hypothesis, after applying Green's theorem to the first integral in the left-hand side of (6.16), and introducing (6.17) into (6.16), the boundary value problem reduces to,

$$\begin{aligned}&\iint_\Omega \frac{1}{p}\nabla_t W \cdot \nabla_t u^t d\Omega - \iint_\Omega k_0^2 q W u^t d\Omega - \oint_\Gamma W\frac{\partial u^i}{\partial \rho}d\gamma \\ &\qquad + \oint_\Gamma W\left[a(\rho_1)u^s - \frac{b(\rho_1)}{\rho_1^2}\frac{\partial^2 u^s}{\partial \phi^2}\right]d\gamma = 0\end{aligned} \tag{6.18}$$

To derive the above equation, we have used the fact that $\partial\Omega = \Gamma_D \cup \Gamma_N \cup \Gamma$. Recalling also the relation $u^t = u^i + u^s$, and integrating by parts to eliminate the second derivative in ϕ, yields the final equation:

$$\begin{aligned}&\iint_\Omega \frac{1}{p}\nabla_t W \cdot \nabla_t u^s d\Omega - \iint_\Omega k_0^2 q W u^s d\Omega + \oint_\Gamma \left[W a(\rho_1)u^s + b(\rho_1)\frac{\partial W}{\partial \gamma}\frac{\partial u^s}{\partial \gamma}\right]d\gamma = \\ &\qquad - \iint_\Omega \frac{1}{p}\nabla_t W \cdot \nabla_t u^i d\Omega + \iint_\Omega k_0^2 q W u^i d\Omega + \oint_\Gamma W\frac{\partial u^i}{\partial \rho}d\gamma\end{aligned} \tag{6.19}$$

The approximation to the exact solution is then sought by a finite element discretization of the above equation. In solving (6.19), one must remember to explicitly enforce the essential Dirichlet boundary condition $u^t = 0$ at Γ_D, which can be expressed in terms of the scattered field as: $u^s = -u^i$ at Γ_D.

Of course analytic ABC can be used also to solve antenna problems. In such case the working unknown is the total field u^t that, due to the nature of the problem, is outward-traveling. Then, the Bayliss and Turkel ABC can be directly applied to u^t in this case and the weak form of the boundary value problem is easily derived.

6.3 Analytic ABC: Implementation

The reader should by now be familiar with the application of FEM in the framework of a weighted residual procedure to solve integrodifferential equations such as (6.19). The solution process requires subdividing the region Ω into N_e finite elements and, in each of them, the exact scattered field is approximated via a linear combination $\bar{u}^{s(e)}$ of nodal basis functions. Let us assume that triangular first-order elements are used, so that over the element (e)

$$\bar{u}^{s(e)}(x, y) = \sum_{j=1}^{3} \alpha_j^{(e)}(x, y)\bar{u}_j^{s(e)} \tag{6.20}$$

where $\alpha_j^{(e)}(x, y), j = 1, 2, 3$, are the interpolating first-order shape functions defined on element (e) and $\bar{u}_j^{s(e)}$ is the unknown value of the scattered field at the jth local node of the same element. As usual, on each triangular element of the mesh is introduced also a set of weighting functions that, for convenience, are chosen identical to the set of shape functions $\alpha_j^{(e)}(x, y)$. By introducing such weighting functions together with the approximation (6.20) of the scattered field into (6.19) and annihilating the residue related to each weighting function, one gets the matrix equation expressing the contribution from a single triangle to the global linear system of equations

$$\begin{aligned} \frac{1}{p}[S^{(e)}][E^{s(e)}] - k_0^2 q[T^{(e)}][E^{s(e)}] + a(\rho_1)L^{(k)}[A^{(k)}][E^{s(e)}] \\ + \frac{b(\rho_1)}{L^{(k)}}[B^{(k)}][E^{s(e)}] = [F^{(e)}] \end{aligned} \tag{6.21}$$

In (6.21), $[S^{(e)}]$ and $[T^{(e)}]$ are the usual local matrices described in Chapter 2, while the other two terms in the left-hand side, involving the matrices $[A^{(k)}]$ and $[B^{(k)}]$, come from the analytic absorbing boundary conditions and are present only when the element (e) has the edge k, with length L_k, lying on the fictitious boundary Γ. The entries of the matrices $[A^{(k)}]$ and $[B^{(k)}]$ are constant and can be computed as:

$$A_{ij}^{(k)} = \int_{\Gamma_k} \alpha_i \alpha_j d\gamma \tag{6.22}$$

$$B_{ij}^{(k)} = \int_{\Gamma_k} \frac{\partial \alpha_i}{\partial \gamma} \frac{\partial \alpha_j}{\partial \gamma} d\gamma \tag{6.23}$$

The values of the entries of the matrices $[A^{(k)}]$ and $[B^{(k)}]$ to enforce second-order Bayliss and Turkel ABC using elements of polynomial orders one through four have been published in [13]. For the case of first-order elements of our interest, they are:

$$[A^{(1)}] = \frac{1}{6}\begin{bmatrix} 2 & 1 & 0 \\ 1 & 2 & 0 \\ 0 & 0 & 0 \end{bmatrix} \quad [A^{(2)}] = \frac{1}{6}\begin{bmatrix} 0 & 0 & 0 \\ 0 & 2 & 1 \\ 0 & 1 & 2 \end{bmatrix} \quad [A^{(3)}] = \frac{1}{6}\begin{bmatrix} 2 & 0 & 1 \\ 0 & 0 & 0 \\ 1 & 0 & 2 \end{bmatrix} \tag{6.24}$$

$$[B^{(1)}] = \begin{bmatrix} 1 & -1 & 0 \\ -1 & 1 & 0 \\ 0 & 0 & 0 \end{bmatrix} \quad [B^{(2)}] = \begin{bmatrix} 0 & 0 & 0 \\ 0 & 1 & -1 \\ 0 & -1 & 1 \end{bmatrix} \quad [B^{(3)}] = \begin{bmatrix} 1 & 0 & -1 \\ 0 & 0 & 0 \\ -1 & 0 & 1 \end{bmatrix} \tag{6.25}$$

Finally, in (6.21), $[F^{(e)}]$ is the 3×1 column vector

$$[F^{(e)}] = -\frac{1}{p}[S^{(e)}][E^{i(e)}] + k_0^2 q[T^{(e)}][E^{i(e)}] + [E^{\prime i(e)}] \tag{6.26}$$

representing equivalent volumetric and superficial sources. It is worth noting that the volumetric equivalent sources, represented by the first two terms in the right-hand side of the above equation, give a nonzero contribution when p and q are not both equal to one, that is when the element (e) belongs to a dielectric or magnetic material. The last term in the right-hand side of the previous equation takes into account surface current at the fictitious contour Γ which may be thought of as coming from the application of the equivalent theorem at the boundary Γ itself [14]. The above expression for the vector $[F^{(e)}]$ has been derived by linearly approximating the plane wave incident field on the generic element (e) using the same shape functions as those used to expand the unknown scattered field:

$$E^i(x,y) = \sum_{j=1}^{3} E_j^i \alpha_j(x,y) \tag{6.27}$$

In this way, it is easy to express the first and second integrals in the right-hand side of (6.19) via the usual elemental matrices. The vector $[E^{\prime i(e)}]$ takes into account contributions coming from the line integral in the right-hand side of (6.19) and it is present only in those elements having a side along the contour Γ. Its entries can be computed in a number of different ways. Let's assume that element (e) has the local edge k, which connects the local vertexes k and $k+1$ (modulo 3), lying along Γ. Then, using first-order triangular elements, the element vector $[E^{\prime i(e)}]$ has the nonzero entries:

$$E'^{i(e)}_k = jk\rho_1 \int_{\phi_k}^{\phi_{k+1}} W_k \cos(\phi - \phi^i) e^{jk\rho_1 \cos(\phi - \phi^i)} d\phi \tag{6.28}$$

$$E'^{i(e)}_{k+1} = jk\rho_1 \int_{\phi_k}^{\phi_{k+1}} W_{k+1} \cos(\phi - \phi^i) e^{jk\rho_1 \cos(\phi - \phi^i)} d\phi \tag{6.29}$$

where W_k is the weighting function centered at the local node k, which is located along Γ at the angle ϕ_k. If the mesh is sufficiently fine so that the edge k of the triangle may be considered equal to the arc of Γ spanning the angles from $\phi = \phi_k$ to $\phi = \phi_{k+1}$, the entries of $[E'^{i(e)}]$ may be numerically computed using:

$$E'^{i(e)}_{k} = jk\rho_1\Delta\phi^{(e)} \int_0^1 (1-t)\cos(\phi_k - \phi^i + t\Delta\phi^{(e)})e^{jk\rho_1\cos(\phi_k-\phi^i+t\Delta\phi^{(e)})}dt \tag{6.30}$$

$$E'^{i(e)}_{k+1} = jk\rho_1\Delta\phi^{(e)} \int_0^1 t\cos(\phi_k - \phi^i + t\Delta\phi^{(e)})e^{jk\rho_1\cos(\phi_k-\phi^i+t\Delta\phi^{(e)})}dt \tag{6.31}$$

with $\Delta\phi^{(e)} = \phi_{k+1} - \phi_k$.

The assembling of the local system of (6.21) into the global one can be accomplished by following the usual strategy described in Chapter 2, that is, by taking into account the relation between the local and global numbering schemes. Apparently, the global system of equations obtained is sparse since the analytical ABC are modeled through the sparse matrices $[A^{(k)}]$ and $[B^{(k)}]$ that do not spoil the sparsity of the FEM matrix. The final system of equations may be solved simply by using one of the subroutines of the LAPACK software library designed to deal with banded matrices. It must be noted however, that the coefficients $a(\rho_1)$ and $b(\rho_1)$ in (6.21) are complex. Consequently, complex algebra must be used to solve the problem and this increases the memory requirements with respect to that necessary to solve an electromagnetic problem defined on the same computational domain, considered as closed, and involving the same lossless materials. For what concerns Matlab implementation, sparse matrices are used to fully benefit from the system-sparse nature. Matlab solver recognizes the complex nature of the system and chooses automatically its best internal solver.

6.4 The Concept of Perfectly Matched Anisotropic Absorber

Analytic absorbing boundary conditions are quite simple to implement and have the convenient characteristic of retaining the sparsity of the FEM matrix. However, their efficiency is negatively affected by the fact that they must be applied to a circular fictitious boundary. This limitation is particularly disadvantageous when the scatterer has a slender shape and, as a consequence, the computational domain enclosed by the fictitious circular contour Γ encompasses a large area of free-space where the application of the computationally intensive FEM is not actually needed. This disadvantage can be overcome by terminating the FEM grid with the so-called perfectly matched anisotropic absorber, an artificial material whose plane interface theoretically features perfect absorption of arbitrarily polarized incident plane waves at any frequency and for any incidence angle. The characteristic constants of such absorber have been pointed out by Sacks et al. [15], and their derivation is briefly surveyed here for completeness.

The general form of the frequency domain curl Maxwell's equations in a medium characterized by complex diagonal tensors is

$$\nabla \times \mathbf{E} = -j\omega[\mu]\mathbf{H} - [\sigma_M]\mathbf{H} = -j\omega[\bar{\mu}]\mathbf{H} \tag{6.32}$$

$$\nabla \times \mathbf{H} = j\omega[\epsilon]\mathbf{E} + [\sigma_E]\mathbf{E} = j\omega[\bar{\epsilon}]\mathbf{E} \tag{6.33}$$

where

$$[\bar{\mu}] = \mu_0 \begin{bmatrix} \mu_r^x + \dfrac{\sigma_M^x}{j\omega\mu_0} & 0 & 0 \\ 0 & \mu_r^y + \dfrac{\sigma_M^y}{j\omega\mu_0} & 0 \\ 0 & 0 & \mu_r^z + \dfrac{\sigma_M^z}{j\omega\mu_0} \end{bmatrix} \tag{6.34}$$

$$[\bar{\epsilon}] = \epsilon_0 \begin{bmatrix} \epsilon_r^x + \dfrac{\sigma_E^x}{j\omega\epsilon_0} & 0 & 0 \\ 0 & \epsilon_r^y + \dfrac{\sigma_E^y}{j\omega\epsilon_0} & 0 \\ 0 & 0 & \epsilon_r^z + \dfrac{\sigma_E^z}{j\omega\epsilon_0} \end{bmatrix} \tag{6.35}$$

To match the intrinsic impedance of the anisotropic medium to free space, the condition

$$\frac{[\bar{\epsilon}]}{\epsilon_0} = \frac{[\bar{\mu}]}{\mu_0} \tag{6.36}$$

must hold. Consequently, the tensors $[\bar{\epsilon}]$ and $[\bar{\mu}]$ can be written as

$$[\bar{\mu}] = \mu_0[\Lambda] = \mu_0 \begin{bmatrix} a & 0 & 0 \\ 0 & b & 0 \\ 0 & 0 & c \end{bmatrix} \qquad [\bar{\epsilon}] = \epsilon_0[\Lambda] = \epsilon_0 \begin{bmatrix} a & 0 & 0 \\ 0 & b & 0 \\ 0 & 0 & c \end{bmatrix} \tag{6.37}$$

In [15], the authors show that the dispersion relation for such a media is the ellipsoid

$$\frac{k_x^2}{bc} + \frac{k_y^2}{ac} + \frac{k_z^2}{ab} = k_0^2 \tag{6.38}$$

whose solutions are of the form

$$k_x = k_0\sqrt{bc}\,\sin\,\theta\,\cos\,\phi \tag{6.39}$$

$$k_y = k_0\sqrt{ac}\,\sin\,\theta\,\sin\,\phi \tag{6.40}$$

$$k_z = k_0\sqrt{ab}\,\cos\,\theta \tag{6.41}$$

It is clear then that the Cartesian components of the propagation vector can be controlled by an appropriate choice of the parameters a, b, and c. To understand how this property can be exploited to design a reflectionless interface, let's consider the planar interface depicted in Figure 6.2 between free-space and the anisotropic absorber described above. The interface coincides with the plane $x = 0$ and a plane wave coming from free-space impinges at an angle ϕ_i. The dispersion relation into the anisotropic media reduces then to:

$$k_x = k_0\sqrt{bc}\cos\phi \tag{6.42}$$

$$k_y = k_0\sqrt{ac}\sin\phi \tag{6.43}$$

$$k_z = 0 \tag{6.44}$$

The arbitrarily polarized plane wave impinging the interface with the absorber may be decomposed into a linear combination of TE_z and TM_z plane waves. By treating separately each polarization, it is an easy exercise to derive the generalization of Snell's law

$$\sin\phi_i = \sin\phi_r \tag{6.45}$$

$$\sqrt{ac}\sin\phi_t = \sin\phi_i \tag{6.46}$$

and the expressions of the respective reflection coefficients:

$$R^{TE} = \frac{\cos\phi_i - \sqrt{\frac{c}{b}}\cos\phi_t}{\cos\phi_i + \sqrt{\frac{c}{b}}\cos\phi_t} \tag{6.47}$$

$$R^{TM} = \frac{\sqrt{\frac{c}{b}}\cos\phi_t - \cos\phi_i}{\cos\phi_i + \sqrt{\frac{c}{b}}\cos\phi_t} \tag{6.48}$$

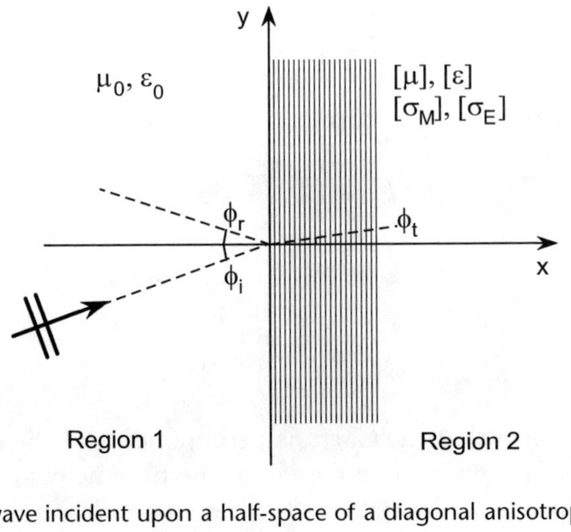

Figure 6.2 A plane wave incident upon a half-space of a diagonal anisotropic medium.

By inspection of the above relations, it is seen that imposing the requirement $\sqrt{ac} = 1$ simplifies the phase matching condition, and by also requiring $b = c$, the interface becomes reflectionless for any frequency, incidence angle, and polarization. In other words, if the tensors characterizing the material in region 2 of Figure 6.2 are:

$$[\bar{\mu}] = \mu_0[\Lambda] = \mu_0 \begin{bmatrix} a & 0 & 0 \\ 0 & \frac{1}{a} & 0 \\ 0 & 0 & \frac{1}{a} \end{bmatrix} \qquad [\bar{\epsilon}] = \epsilon_0[\Lambda] = \epsilon_0 \begin{bmatrix} a & 0 & 0 \\ 0 & \frac{1}{a} & 0 \\ 0 & 0 & \frac{1}{a} \end{bmatrix} \tag{6.49}$$

the plane interface at $x = 0$ is a perfectly matched interface to free space. Moreover, if $a = \alpha - j\beta$, the transmitted waves will be damped in the anisotropic absorber with an attenuation given by

$$\mathbf{E}_t(x, z) = \mathbf{E}_0 e^{-k_0 \beta \cos\phi_t x} e^{-jk_0(\sin\phi_t y + \alpha \cos\phi_t x)} \tag{6.50}$$

in the TM_z case and

$$\mathbf{H}_t(x, z) = \mathbf{H}_0 e^{-k_0 \beta \cos\phi_t x} e^{-jk_0(\sin\phi_t y + \alpha \cos\phi_t x)} \tag{6.51}$$

in the TE_z case.

It is worth noting that producing a reflectionless interface between free space and PMA along a plane $y = constant$ is as easy as using the material tensors:

$$[\bar{\mu}] = \mu_0[\Lambda] = \mu_0 \begin{bmatrix} \frac{1}{b} & 0 & 0 \\ 0 & b & 0 \\ 0 & 0 & \frac{1}{b} \end{bmatrix} \qquad [\bar{\epsilon}] = \epsilon_0[\Lambda] = \epsilon_0 \begin{bmatrix} \frac{1}{b} & 0 & 0 \\ 0 & b & 0 \\ 0 & 0 & \frac{1}{b} \end{bmatrix} \tag{6.52}$$

When this kind of material is used to absorb outward-traveling waves in FEM computations, two issues arise. The first is related to the fact that the perfectly matched absorber cannot extend to infinity but must be truncated to some finite thickness; the second regards the fact that the interface cannot be an infinitely extending plane, but the absorber must be "wrapped" around the computational domain to form a box, thus creating edges along the interface free-space-PMA. As for the first point, the PMA is usually taken sufficiently thick and backed by a perfectly conducting plane. The reflected field created by the p.e.c termination travels back toward the computational domain, but if the PMA is thick enough the reflected wave is strongly damped and does not affect significantly the results. Formulas for assessing the reflection coefficient of a finite thickness layer of PMA backed with a p.e.c. can be found in the literature [15]. The second issue is somewhat more delicate and must be carefully handled during the implementation phase to realize a good absorber. The common approach is to keep reflection as low as possible from the PMA by choosing the PMA properties at an edge region so that it would be perfectly matched to the adjacent side regions if the edge were of infinite extension. Developing the related analysis, one arrives at the interesting conclusion that the tensor $[\Lambda]$ for the edge region should be chosen equal to the matrix product of the tensors of the two adjacent sides (Figure 6.3). For instance, the

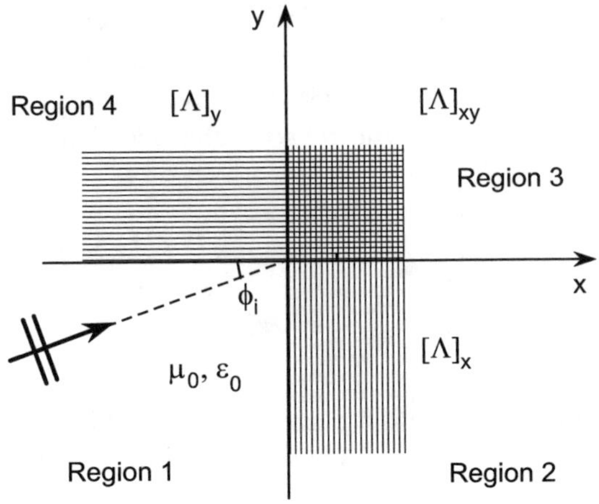

Figure 6.3 Modeling of edge regions in PMAs.

tensor $[\Lambda^{xy}]$ characterizing the PMA at an edge between plane interfaces free-space PMA at $x = constant$ and $y = constant$ is chosen as: $[\Lambda^{xy}] = [\Lambda^{x}] \cdot [\Lambda^{y}]$, where $[\Lambda^{x}]$ and $[\Lambda^{y}]$ are the tensors characterizing the PMA with interface $x = constant$ and $y = constant$, respectively.

6.5 Antenna Problems: Formulation Using PMA

The finite element formulation to be used when the perfectly matched anisotropic absorber is employed to terminate the computational domain will be now shown in the framework of the analysis of a simple antenna problem. Let's consider the classic configuration of a parallel plate waveguide fed with the first TE_1 mode and opened on a ground plane. In this case, the geometric region Ω of computations extends partly inside the waveguide and partly in free space (Figure 6.4). Its border $\partial\Omega$ is comprised of the electromagnetic input port of the antenna (Γ_1), the perfectly conducting surface of the waveguide walls and the ground plane (Γ_0), and the fictitious surface Γ_A, which must be able to absorb outward-traveling waves. If analytic ABC are used at Γ_A, then this latter part of the border should be chosen as a circular arc with its center located in the middle of the waveguide aperture. It is worth noting that in antenna problems the total field is outward-traveling, then when using analytic ABC in this context it is not necessary to formulate the problem in terms of the scattered field as done in Section 6.2.

On the other hand, when the PMA concept is used, Γ_A can be conveniently chosen as either a half-square or half-rectangle with its center still located in the middle of the antenna aperture (Figure 6.4). Using this second approach, Γ_A is considered as a perfectly conducting surface coated with a perfectly matched anisotropic absorber of thickness t. The problem can then be formulated as a boundary value problem in the usual way: the Helmholtz equation in the longitudinal component E_z of the electric field must be solved in the region Ω, with homogeneous Dirichlet boundary conditions at the perfectly conducting walls Γ_0 and Γ_A, and appropriate

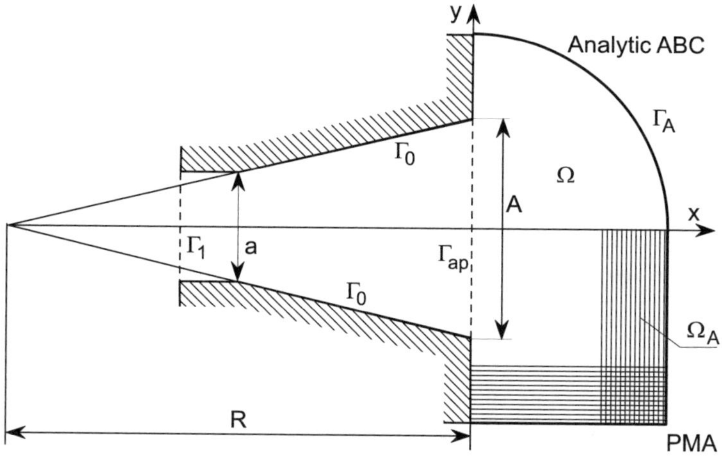

Figure 6.4 Computational region for the analysis of a parallel plate waveguide opened on a ground plane. The upper half of the figure shows the domain for the analytic ABC formulation, the lower half for the PMA formulation.

continuity conditions for the tangential field at the electromagnetic port Γ_1. These latter can be enforced using a technique similar to that described in Chapter 4 for the analysis of rectangular waveguide devices, so it will not be repeated here. On the contrary, this section focuses on the Helmholtz equation in the region Ω_A occupied by the anisotropic material.

From the curl Maxwell's equations in the anisotropic material, is possible to obtain the double curl equation:

$$\nabla \times \left([\Lambda]^{-1} \cdot \nabla \times \mathbf{F} \right) - k_0^2 [\Lambda] \mathbf{F} = 0 \tag{6.53}$$

where $\mathbf{F}$ may be either $\mathbf{E}$ or $\mathbf{H}$. For the sake of simplicity, in the following we will assume $\mathbf{F} = \mathbf{E}$. Introducing a differentiable vector-weighting function $\mathbf{W}$, applying the projection procedure to the above equation, and using Green's theorem yields:

$$\begin{aligned} &\iint_{\Omega_A} \nabla \times \mathbf{W} \cdot [\Lambda]^{-1} \cdot \nabla \times \mathbf{E} d\Omega - \iint_{\Omega_A} k_0^2 \mathbf{W} \cdot [\Lambda] \cdot \mathbf{E} d\Omega \\ &\qquad + jk_0 Z_0 \oint_{\partial\Omega_A} \mathbf{H} \times \hat{\mathbf{n}} \cdot \mathbf{W} d\gamma = 0 \end{aligned} \tag{6.54}$$

The line integral extends to the contour of the region Ω_A and can be omitted in the implementation because its value is zero. As a matter of fact, the contribution coming from Γ_A vanishes because Γ_A is a Dirichlet boundary at which both the tangential electric field and the weighting function are set to zero. The other part of the line integral extends to the interface between the PMA and free space and cancels out with a similar line integral coming from the weak form of the double curl equation in the free-space region $\Omega - \Omega_A$. The reason why they cancel out is because all the quantities in the integral are equal but the outward normal, which is opposite. The full vectorial equation (6.54) can be reduced to a scalar one for the particular problem at hand, in which $\mathbf{E} = E_z\hat{\mathbf{z}}$. In particular, by choosing $\mathbf{W} = W\hat{\mathbf{z}}$, and substituting into (6.54), gives the scalar weak form of the Helmholtz equation into the PMA:

$$\iint_{\Omega_A} a\frac{\partial W}{\partial y}\frac{\partial E_z}{\partial y}d\Omega + \iint_{\Omega_A} b\frac{\partial W}{\partial x}\frac{\partial E_z}{\partial x}d\Omega - \iint_{\Omega_A} k_0^2 c W E_z d\Omega = 0 \tag{6.55}$$

The above equation has been obtained by considering a diagonal tensor $[\Lambda]$ of the generic form,

$$[\Lambda] = \begin{bmatrix} a & 0 & 0 \\ 0 & \mathrm{b} & 0 \\ 0 & 0 & c \end{bmatrix} \tag{6.56}$$

and thus is valid in both the horizontal and vertical layer of the PMA. The equation valid in each layer, and in the corner region between two adjacent perpendicular layers, is obtained by appropriately choosing the value of the three parameters a, b, and c following the rules outlined in the previous section.

To summarize, the equations included in the correct formulation of the boundary value problem sketched in Figure 6.4 are: the weak form of the Helmholtz equation in the free-space region $\Omega - \Omega_A$; the weak form (6.55) of the Helmholtz equation in the perfectly matched anisotropic absorber region Ω_A; the equations coming from the imposition of continuity of electromagnetic field at port 1, that are similar to those presented in Chapter 4 for the analysis of waveguide devices. The essential homogeneous Dirichlet boundary condition at the surface Γ_0 must be also explicitly enforced.

The formulation of scattering problems using PMA and the scattered field as working unknown is a bit tricky and deserves a few words. The computational domain Ω can be subdivided in two nonoverlapping regions: Ω_A where the PMA is located, and $\Omega - \Omega_A$ where the real physical problem is defined. The latter region is completely "wrapped" in by the PMA region. The weak form of the Helmholtz equation in it can be derived following a procedure similar to that described in Section 6.2 and has the form:

$$\begin{aligned} &\iint_{\Omega-\Omega_A} \frac{1}{p}\nabla_t W \cdot \nabla_t u^s d\Omega - \iint_{\Omega-\Omega_A} k_0^2 q W u^s d\Omega = \\ &- \iint_{\Omega-\Omega_A} \frac{1}{p}\nabla_t W \cdot \nabla_t u^i d\Omega + \iint_{\Omega-\Omega_A} k_0^2 q W u^i d\Omega \end{aligned} \tag{6.57}$$

This is essentially (6.19) without the line integrals, that disappear because the region Ω is delimited by the outer interface with the PMA region Ω_A and possibly by the Dirichlet or Neumann contour representing the surface of perfectly electric or magnetic scatterers. On the other hand, the equation to be solved in the PMA region can be expressed as:

$$\begin{aligned} &\iint_{\Omega_A} a\frac{\partial W}{\partial y}\frac{\partial u^s}{\partial y}d\Omega + \iint_{\Omega_A} b\frac{\partial W}{\partial x}\frac{\partial u^s}{\partial x}d\Omega - \iint_{\Omega_A} k_0^2 c W u^s d\Omega = \\ &- \iint_{\Omega_A} \nabla_t W \cdot \nabla_t u^i d\Omega + \iint_{\Omega_A} k_0^2 W u^i d\Omega \end{aligned} \tag{6.58}$$

Also this equation is very similar to (6.19) but with the line integrals missing. The reason why they are identically zero here is because Ω_A is delimited by the interface with the interior region $\Omega - \Omega_A$ and by the homogeneous Dirichlet boundary Γ_A that terminate the PMA layer. It is worth noting that coefficients p and q in the right-hand side of (6.58) have been set to 1. As a matter of fact, the PMA layer must be present only for the scattered field that must be absorbed, but not for the the incident field.

6.6 PMA Implementation

The implementation of the algorithm outlined in the previous section is quite straightforward and most of the elements needed have already been discussed in the previous chapters. The construction of the FEM coefficient matrix for the solution of the Helmholtz equation in homogeneous and inhomogeneous regions was treated in Chapters 3, 4, and 5. The construction and assembling of the matrices needed to match the FEM solution to a modal expansion of the field have been discussed in Chapters 4 and 5. Thus we only need to outline here the construction and assembling of the FEM matrix for the solution of the Helmholtz equation into the anisotropic material used to realize the perfectly matched absorber. The weak form (6.55) of this latter equation can be discretized introducing in it the FEM approximation of the field and choosing weighting functions equal to the finite element shape functions. If first-order elements are used, the three equations related to the three weighting functions centered at the vertexes of a given element (e) can be written in the matrix form:

$$b[S_x^{(e)}][E^{s(e)}] + a[S_y^{(e)}][E^{s(e)}] - k_0^2 c[T^{(e)}][E^{s(e)}] = 0 \tag{6.59}$$

where $[E^{s(e)}]$ is the 3×1 column vector of the field nodal values. The entries of the local matrices $[S_x^{(e)}]$ and $[S_y^{(e)}]$ are given by

$$S_{x\,ij}^{(e)} = \iint_{\Omega^{(e)}} \frac{\partial \alpha_i^{(e)}}{\partial x} \frac{\partial \alpha_j^{(e)}}{\partial x} dS \tag{6.60}$$

$$S_{y\,ij}^{(e)} = \iint_{\Omega^{(e)}} \frac{\partial \alpha_i^{(e)}}{\partial y} \frac{\partial \alpha_j^{(e)}}{\partial y} dS \tag{6.61}$$

Since in the PMA layer the standard element matrices defined in Chapter 2 cannot be used, a new subroutine ELEA.F, and a corresponding `EleA.m` has been coded to treat the anisotropic material. This subroutine builds the two element matrices $[S_x^{(e)}]$ and $[S_y^{(e)}]$. It can be shown that, being $A^{(e)}$ the area of element (e):

$$[S_x^{(e)}] = \frac{1}{4A^{(e)}} \begin{bmatrix} (x_3 - x_2)^2 & (x_3 - x_2)(x_1 - x_3) & (x_3 - x_2)(x_2 - x_1) \\ (x_3 - x_2)(x_1 - x_3) & (x_1 - x_3)^2 & (x_1 - x_3)(x_2 - x_1) \\ (x_3 - x_2)(x_2 - x_1) & (x_2 - x_1)(x_1 - x_3) & (x_2 - x_1)^2 \end{bmatrix} \tag{6.62}$$

$$[S_y^{(e)}] = \frac{1}{4A^{(e)}} \begin{bmatrix} (y_3 - y_2)^2 & (y_3 - y_2)(y_1 - y_3) & (y_3 - y_2)(y_2 - y_1) \\ (y_3 - y_2)(y_1 - y_3) & (y_1 - y_3)^2 & (y_1 - y_3)(y_2 - y_1) \\ (y_3 - y_2)(y_2 - y_1) & (y_2 - y_1)(y_1 - y_3) & (y_2 - y_1)^2 \end{bmatrix} \tag{6.63}$$

with x_i and y_i Cartesian coordinates of the vertex i. The matrix $[T^{(e)}]$ in (6.59) is the usual one. The assembling strategy of the local matrix (6.59) into the global system of equation follows the usual rules. The only difference is now that the local matrices $[S_x^{(e)}]$ and $[S_y^{(e)}]$ must be multiplied times the coefficients a and b characterizing the PMA absorber before being assembled into the global FEM matrix.

6.7 The Code CYL

Two different codes are presented in this chapter. The first, CYL, is designed to deal with scattering from an inhomogeneous cylinder of arbitrary cross-section, possibly not simply connected, while the second, OWG, solves the problem of radiation from an open-ended waveguide computing also the return loss of the radiating element. This section is devoted to the description of the first of these two codes.

Similarly to the software presented in previous chapters, the code CYL requires two input files: the first describes the geometry and its discretization in finite elements, while the second contains data needed to specify the electromagnetic analysis to be performed. The geometric data file must be built complying with the following requirements:

- Dimensions are normalized to the wavelength λ at the frequency of interest.
- First-order triangular elements should be used.
- Nodes lying at p.e.c. and p.m.c. contours must be identified by their own label: we will denote p.e.c. nodes with label 1 and p.m.c. nodes with label 2.
- If analytic ABC are used, then the nodes at the circular fictitious contour Γ used to terminate the computational domain should be denoted by their own label: in the following examples by label 10. Furthermore, the center of the circular contour Γ must be placed at the origin $(0, 0)$ of the coordinate system.
- If, on the other hand, the user choses to employ a PMA layer to terminate the computational domain, then the elements, not the nodes, belonging to the horizontal and vertical layers of the absorber should have their own label. The same applies for the four corners. In our examples, elements belonging to the horizontal and vertical layers have been assigned label values 5 and 50, respectively, whereas a label value of 55 denotes corner elements. When the PMA technique is used to truncate the computational domain, the nodes at the outer boundary Γ should be labeled as p.e.c. nodes.

It is worth stressing that the geometric mesh must be built differently for the case of analytical ABC or PMA. When analytical ABC is used, the outer boundary Γ must be circular, preferably with its center located at the center of radiation of the scattered waves. This is an easy task if a circular cylinder is analyzed, but a much more difficult one if the scatterer has an odd-shaped cross-section. In this latter case, the usage of the PMA layer may be more convenient because it requires only the construction of a simple rectangular enclosure with some layers of elements to simulate the lossy anisotropic material. It may also provide more accurate results because the PMA performs much better in absorbing waves with arbitrary direction of propagation.

After having read the geometric data file, the code CYL opens and reads the electromagnetic data file, which has the format and contains the data described in Example 6.1 below:

Example 6.1

```
0.          !Angle Phi of plane wave incidence [degrees]
2           !ABC TYPE 1= Analytic 2nd order, 2=PMA
10          !ABC Node Label
5,50,55     !PMA element labels
(1.,-1.)    !PMA coefficient a
1           !PEC Node Label
2           !PMC Node Label
1           !Number of Dielectric Materials
(1.,0.)     !Epsilonr(s)
(1.,0.)     !Mur(s)
1           !Label(s)
E           !Incident Field Polarization: E = TM; H = TE
```

The first record of the file stores the incidence direction of the impinging plane wave, in degrees. This is given as the angle between the negative direction of the x axis and the vector $\mathbf{k}$ of propagation of the plane wave itself. The second record contains the type of ABC chosen to terminate the computational domain: 1 for the analytic ABC, 2 for PMA. Needless to say, the mesh must be created accordingly. Regardless of the ABC chosen, the next three records define:

1. The labels assigned to the nodes at the contour Γ where the analytic ABC must be applied;
2. The labels assigned to the elements belonging to the PMA layers, distinguishing among horizontal layer, vertical layer, and corners;
3. The value of the parameter a which defines the perfectly matched anisotropic absorber.

The subsequent lines define labels assigned to nodes at p.e.c. and p.m.c., the number of dielectric materials present in the computational domain and the values of their physical constants, as well as the value of the labels given to elements belonging to each material. The number of dielectric materials should not include the perfectly matched absorber, already considered in the fourth and fifth records of the file, and labels of elements belonging to the scatterer or free space must not have the same value as those of PMA layers or corner elements. The last record of the input electromagnetic data file stores a character flag denoting the polarization of the incident plane wave: E for TM_z polarization, H for TE_z polarization.

The flow of the program CYL is then straightforward: first there is an initialization phase, in which a set of variables is defined depending of the polarization of the field and the type of ABC. If analytical ABC are chosen, an estimate of the radius ρ_1 of the outer boundary Γ is made in this phase, and the coefficients $a(\rho_1)$ and $\mathrm{b}(\rho_1)$ of (6.11) are computed. Then, the code CYL assembles and solves the linear system of equations. The routine calls for assembling the FEM matrix depending on the choice made for the ABC to be used, since analytical ABC and PMA require different treatment. Although the FEM matrix is built in different ways, the construction of the right-hand side of the linear system of equations is the same for both kinds of ABC, and also the solution process is common and is performed by the same subroutine of the software library LAPACK and by Matlab solver.

Whichever the type of boundary conditions used to solve the problem, the code prompts the user for the output file name, which should be given without extensions. This is automatically attached by the code depending on the content of the file: the extension ".TOT" is used for the file containing the nodal values of the total field, while ".BCS" is assigned to the file storing the nodal values of the scattered field only. Both files can be used as input of the postprocessor code described in Chapter 2 to produce field maps. As usual Matlab, providing a more flexible structure has slightly different file names for outputs created by adding suitable strings to the output file name. In particular `.scatterA.out` and `.scatterP.out` are postpended to the file name for files containing the scattered field amplitude and phase, respectively, whereas `.totalA.out` and `.totalP.out` are postpended to the file name for files containing the total field amplitude and phase.

6.8 Code CYL: Some Examples

To validate the code, the simple configuration of a perfectly conducting circular cylinder is considered first. The analytical solution of such problem is well-known for both TE_z and TM_z polarization [16], and will be used to asses the accuracy of the FEM solution. The cylinder has a radius of 0.5λ and the incident plane wave has TM_z polarization and direction of propagation characterized by $\phi_i = 0$. The problem is solved using both kinds of absorbing boundary conditions to terminate the computational domain. To employ analytic ABC, the fictitious contour Γ must be circular and is taken in this example with its center along the axes of the cylinder and with radius equal to 1λ. On the other hand, the usage of PMA does not require a circular boundary to truncate the domain, and in such case Γ is chosen as a square with side $l = 3\lambda$. Furthermore, in this latter case Γ is a perfectly conducting box internally coated with a perfectly matched anisotropic absorber 0.4λ thick. The PMA has been discretized using 3 layers of finite elements, as it can be seen from the geometric input file enclosed in the CD-ROM. The parameter a of the absorber has been chosen equal to $(0.85, -0.85)$, a value which has provided good performances, as the reader can experiment by running the code on this same example with different values.

Figure 6.5 shows the total field, exact [16] and numerically computed using the code CYL, along the x axis, that is along a line characterized by $\phi = 180^\circ$ and values of ρ ranging from 1λ to 0.5λ for the left half, and $\phi = 0^\circ$ and ρ ranging from 0.5λ to 1λ for the right half. Only 6 numerical values are reported for each side of the graph, because only 5 elements lies between the perfectly conducting surface of the cylinder at $\rho = 0.5\lambda$, and the surface Γ or the beginning of the PMA at $\rho = 1\lambda$. Intermediate values could be obtained by resorting to a finer mesh or by interpolating the nodal values computed. The numerical results agree very well with the analytic one on the right side of the cylinder ($\phi = 0^\circ$), which is the illuminated one, while some discrepancies can be observed in the shadow region behind the cylinder (left side of the graph, $\phi = 180^\circ$). It is worth noting that, as expected, the PMA approach provides overall better results than ABC, when the two kind of ABC are used at the same distance from the scatterer.

The simple case analyzed, exhibiting an axial symmetry, is particularly well-suited to the usage of analytic ABC: as a matter of fact, the axial symmetry of the scatterer produces a cylindrical outward traveling field that has a structure extremely well-matched to the analytic ABC. Although in this situation the analytic absorbing boundary conditions perform at their best, they are nonetheless outperformed by the generally more accurate PMA. This is particularly evident in the shadow region where the total field has low absolute value. The absorption provided by the PMA might be even improved in a number of ways that are left to the reader to explore. The parameters to play with are, for instance, the value of the constant a that characterizes the absorber and its profile variation [10]. It is also worth noting that the numerical accuracy could be improved by modifying the size of the mesh inside the PMA accordingly to the value given to the parameter a: the higher the real and imaginary part of a, the finer the mesh, so that the

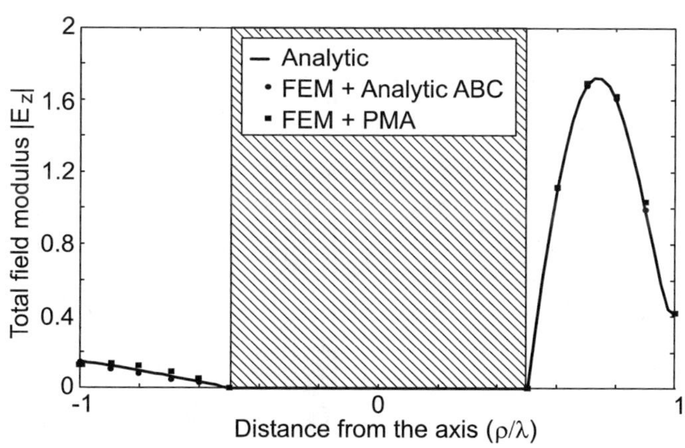

Figure 6.5 Comparison between the analytical and numerical solution for the problem of scattering from a perfectly conducting cylinder. Both analytic ABC and PMA have been used in the FEM solution.

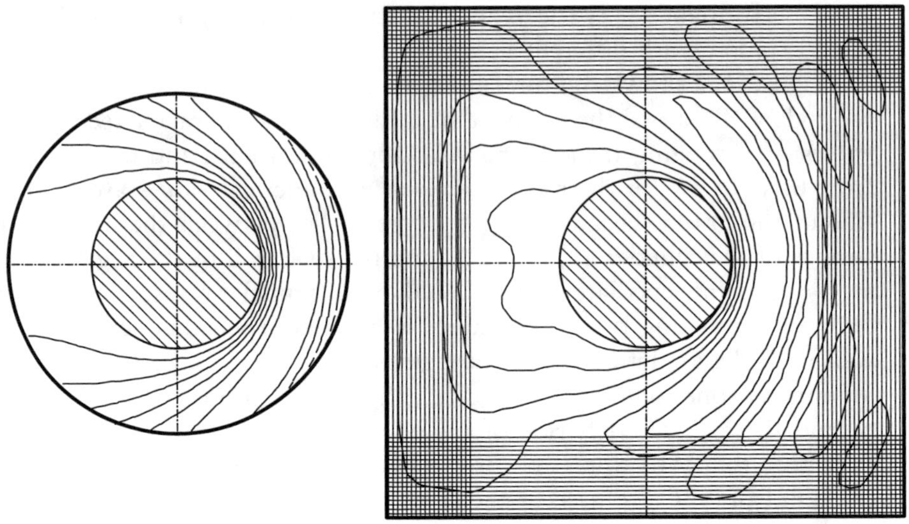

(a) (b)

Figure 6.6 Contour maps of the total field computed with FEM and the analytic ABC (a) and with FEM and PMA (b), for the problem of TM_z scattering from a perfectly conducting cylinder.

wavelength as well as the decay rate of the field inside the PMA are accurately modeled by the FEM approximation.

Figure 6.6 reports the field contour maps for the same problem of Figure 6.5 when the two different kinds of absorbing boundary conditions are used. Needless to say that, when the PMA is employed, the physically meaningful field is that extending between the perfectly conducting cylinder and the PMA layer itself.

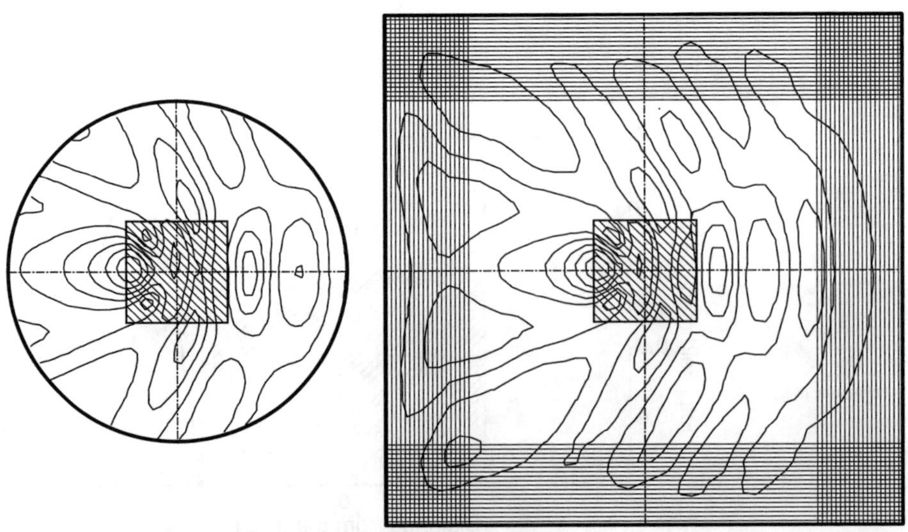

(a) (b)

Figure 6.7 Contour maps of the total field computed with FEM and analytic ABC (a) and with FEM and PMA (b), for the problem of TM_z scattering from a dielectric cylinder.

The code CYL has been used also to analyze the problem of scattering from a square dielectric cylinder with relative dielectric constant $\epsilon_r = 4$, for which no analytic solution is available. The polarization and the direction of propagation of the incident field are the same as in the previous example. The side of the square cross-section is chosen to be 0.3λ, and the enclosing boxes have the same dimensions as in the previous case. The contour maps of the total field for the two kinds of ABC are sketched in Figure 6.7.

6.9 The Code OWG

The code OWG is designed to deal with the problem of radiation from an open-ended waveguide and it is essentially created by merging the previously described code CYL with the code EHDEV presented in Chapter 4. In particular, the code EHDEV, which was designed for the analysis of N-port devices in rectangular waveguides, and implemented with a hard limit of 2 ports, is modified so as to be able to deal with device exhibiting at least one port having also an absorbing boundary to simulate radiation into free space. The overall structure of the code EHDEV is unchanged: the block-structured matrix and the block solver routine are retained, but the number of ports might be either one or two. The major change regards the assembling of the FEM matrix, that now must also account for radiation in open space. That is, it must allow the usage of analytic ABC or PMA at a portion of the boundary of the computational domain.

This approach of course makes OWG inherit EHDEV limitations too. In particular, that of being not suitable to treat inhomogeneous media when an E-plane device is considered. Furthermore, careful considerations must be done on what is really analyzed using this code. As a matter of fact, EHDEV can solve two-dimensional problem as E- or H-plane devices but cannot deal with a three-dimensional free-space region. The structure analyzed with the code OWG is then not the real three-dimensional problem, but rather its approximation in two-dimension. For example, when analyzing radiation from an H-plane horn opened on a ground plane the free-space region is modeled as a bidimensional space. In other words, the dimension a of the waveguide becomes infinite, while the dimension b remains finite, thus actually studying the propagation between two infinite p.e.c. planes. When the horn is fed with the fundamental TE_{10} mode, since the E_y component of the field is uniform in the direction perpendicular to the H-plane, the problem solved is equivalent to that of a parallel plate waveguide fed by the TE_1 mode, opened on a ground plane. This is not a major limitation, since the radiation pattern on the H-plane under study does not vary significantly from the real three-dimensional problem. Anyway, attention must be paid to the fact that the fields do decrease with a $1/\sqrt{r}$ law, rather than with the usual $1/r$ law of tridimensional propagation. Similar considerations apply when analyzing the E-plane.

Having said this, it is apparent that the structure of the code OWG is very similar to that of the code EHDEV. The subroutines used to build three of the four blocks of the solving matrix are indeed the same. The main changes have been done in the routine that assembles the FEM block of the solving matrix. This new routine is a variation of that used in the code CYL. The program OWG is actually an example of how easily the reader can modify all the provided codes to suit his practical needs.

The inputs needed by OWG are, as usual, a mesh data file and an electromagnetic data file. The mesh data file must complain to a mix of both the standards of the CYL and EHDEV codes, that is

- Dimensions are in millimeters;
- Although the EHDEV code can exploit all kind of elements, the ABC are implemented only for first-order triangular elements, which thus should be used;
- Nodes at p.e.c. and p.m.c. contours (Contours Γ_0 in Figure 6.4) must be identified by their own label: we will denote p.e.c. nodes with label 1 and p.m.c. nodes with label 2;
- Nodes at the input port (Contour Γ_1 in Figure 6.4) must be identified by their own label, here a label of 11 is used;
- The same standards of the CYL code applies for the ABC implementation (Contour Γ_A in Figure 6.4).

It is also possible to assign a further label to a set of nodes, to have the value of the field at those nodes stored in a separate file. This is extremely useful if, for example, the field at the aperture (line Γ_{ap} in Figure 6.4) is needed to compute the radiation pattern of the aperture, as shown later on.

Example 6.2

```
1              ! Number of ports
18.35, 9.175   ! Waveguide dimensions, in mm
29.175         ! Radius of analytical ABC circumference, in mm
11             ! Port label(s)
1,2            ! PEC and PMC labels
20             ! Radiating aperture label
1              ! ABC TYPE 1= Analytic 1st order, 2=PMA
10             ! Analytic ABC Label
5,50,55        ! PMA elements label X-parallel, Y-parallel, Corners
(1.,-1.)       ! PMA coefficient A
10             ! Number of modes for the expansion
H              ! Plane
1              ! Number of different dielectric materials
1.0            ! Epsilon(s)
1.0            ! Mu(s)
1              ! Label(s)
5              ! No. of Frequency point(s)
10.00 10.25 10.5 10.75 11       ! Frequency point(s), in GHz
```

The structure of the electromagnetic input file (Example 6.2) should by now be familiar to the reader. The first record specifies the number of ports connected to the waveguide circuits. The code OWG allows for one or two ports only, but it can be easily modified to take into account multiport structures using the same strategy adopted in the code EHDEV. On the other hand, the presence of more than one aperture in the ground plane is more tricky. It does not require any modification to the part of the code where the FEM matrix is assembled. Anyhow, since the radiation aperture is treated as the last port, it must be simply connected otherwise the enforcing of p.e.c. conditions at extreme nodes might fail at internal boundary nodes. The second record in the input file contains the waveguide dimension, in millimeters. It is very important that the width of the port in the mesh matches the right dimension a (for an H-plane device) or b (for an E-plane one) specified in this line.

The third line of the file provides the radius of the circular, or most likely semicircular, boundary Γ_A at which analytic ABC must be enforced. This input was not needed for the CYL code because in that case the radius can be easily computed within the code itself. The choice of the center and the radius of Γ_A is indeed a major point in the mesh generation step and can be not so easy when the aperture is fairly large or when multiple apertures are present. In these cases employing the PMA instead of the analytic ABC leads to substantial improvements in the accuracy of results.

The fourth line defines the label of the nodes at the input port of the horn. This label must be given also to the two nodes at the ends of the ports that belong also to the p.e.c. wall. The code automatically treats these two nodes as p.e.c nodes. The following record contains the labels of nodes at p.e.c. and p.m.c. walls, which, as usual, are set to 1 and 2, respectively. The sixth line gives the label of the nodes at the opening of the radiating structure. These nodes are treated as standard nodes in the solution process, but they are recognized and ordered as any other port. Only at the end of the analysis, they are used to plot the field distribution at the aperture. These data can also be used to compute the radiation pattern of the antenna, as shown later on.

The records from the seventh to the tenth contain flags and data related to the ABC implementation and are perfectly equal to the corresponding four lines needed by the CYL code. The next record contains the number of modes to be used in the modal expansion at the feeding port. This latter is a very important parameter and its influence on numerical results has been analyzed in Chapter 4, thus it will not be discussed here again. The flag stored in record twelve indicates if the case of E- or H-plane has to be considered.

The subsequent group of four lines contains: the number of different dielectric materials included in the computational domain, a list of the different permittivity, permeability, and the labels which characterize each element belonging to a given material. These four lines must be present also in the E-plane case, even if only homogeneous devices can be treated in such situation. The last record in the input electromagnetic data file stores the number of analysis to be performed and the frequency values, in GHz, to be used.

The code flow is straightforward and is quite similar to that of the EHDEV code of Chapter 4. After a general rescaling to MKS units, the nodes on the feeding port are recognized, the matrices needed to match the FEM solution to the modal expansion at that port are built. This task is accomplished using the same routines used by the code EHDEV. The FEM matrix is assembled in the usual way adding the terms related to the Bayliss and Turkel ABC if analytical ABC are used, or employing the subroutine ELEA.F to account for the presence of anisotropic material if PMA are chosen. The global system of equations has the same block structure as the global matrix related to the problem of waveguide discontinuities described in Chapter 4. It is then solved using the same routine.

The main output of the code is a file, whose name is keyed in by the user when prompted, containing all the reflection coefficients of all the modes of the modal expansion, grouped for each frequency, that is a GSM file akin to the one created by the EHDEV code and hence easily handled by `SeeGSM.m` function. Two sets of secondary files are also created. Each set is comprised of a number of files equal to the number of frequency points to be considered in the analysis. Files belonging to the first set are named FIELD_P#_f###.out, where ### is an incremental integer number indicating the frequency point while the first # is either 1 or 2 depending on which port is fed. These files contains the modulus of the nodal values of the field and can be post-processed to realize field maps. On the other hand, files belonging to the second set are named APERTURE_P#_f###.DAT and contain the field values, real part, imaginary part, modulus and phase, at the nodes at the aperture Γ_{ap}. This second set of file is created only if the user has given those node their unique label.

6.10 Code OWG: Some Examples

The first structure analyzed with the code OWG is a two-dimensional model of the classic problem of a rectangular waveguide opened on a ground plane, which is probably the simplest of aperture antennas. The ground plane is usually introduced to ease the analytic treatment of the problem. As a matter of fact, its presence allows the application of the equivalence theorem and the image principle, that permit to analytically express the radiated field [17]. Of course the FEM analysis can be carried out also without the ground plane, as will be shown in a later example.

A rectangular waveguide guide with internal dimensions $a = 18.35$ mm and $b = 9.175$ mm is used for this example, considering its radiation on the H-plane. The feeding port has been placed at a distance of 20 mm from the radiating aperture. Two different meshes have been built and used with different boundary conditions. When analytic ABC are employed, the computational domain in the free-space region is terminated with a circular contour Γ_A with radius 29.175 mm and with its center at the middle point of the aperture. In the second mesh, used with PMA, the free-space region is terminated with a half-square contour. The side of the square is 73.4 mm long. This p.e.c. box is internally coated with a 9.175 mm-thick PMA layer. Again, while the analytic ABC needs to be placed far enough from the radiating aperture, the PMA could be placed much closer. The computational domain in the two cases has been kept of similar size to have a fair comparison on field values.

The unimodal frequency band of the waveguide is in the range 10–15 GHz, thus a frequency sweep on this interval has been performed and ten modes have been used to expand the field at the input port of the radiating structure.

To investigate the PMA reliability, various different values for its a parameter (not to be confused with the waveguide width) have been used, namely $a = 1 - j, 1.5 - 1.5j, 2 - 2j, 2.5 - 2.5j, 3 - 3j$. The results are quite insensitive to the value assigned to this parameter. Some small variations can be noted in the input reflection coefficient (Figure 6.8) but differences on the field maps are unnoticeable. Results of Figures 6.8 and 6.9 show that both kinds of ABC gives good results for such a small aperture and such large distance of the ABC surface. This is the reason why the PMA behavior is very good for all the chosen values of parameter *a*. It anyway remains good even if the PMA is placed closer.

The graphs of Figures 6.8 and 6.9 show consistency between the results provided by the two kind of ABC, but do not validate the code. To this end, the field pattern at the frequency of 10 GHz is computed and compared with that obtained analytically using the approximation of the unperturbed TE_{10} field distribution at the radiating aperture. Figure 6.10 shows a remarkably good agreement among the approximated analytic solution and the numerical solutions obtained with both kind of ABC. These results not only validate the code but also mean that the approximation of considering an unperturbed TE_{10} field distribution at the aperture is fairly good in this case for the computation of the radiation pattern. What the approximate analytic treatment cannot provide is the standing wave ratio (SWR) of such a device. This quantity, of main interest for the designer, can be numerically evaluated by the OWG code that can provide the reflection coefficient of the fundamental mode at the input port. As seen in Figure 6.8, its magnitude is around 0.19 for the lowest frequency and for both types of ABC.

The next example is of more practical importance and it also shows some numerical issues that should be always carefully considered when applying the FEM. It consists of a rectangular H-plane sectorial horn fed with the same rectangular waveguide analyzed earlier. The horn is 40 mm long and its aperture is 36.7 mm wide, that is two-times wider than the broad dimension of the feeding rectangular waveguide. The input port of the device is taken at a distance of 10 mm from the beginning of the horn, but all other quantities involved in the electromag-

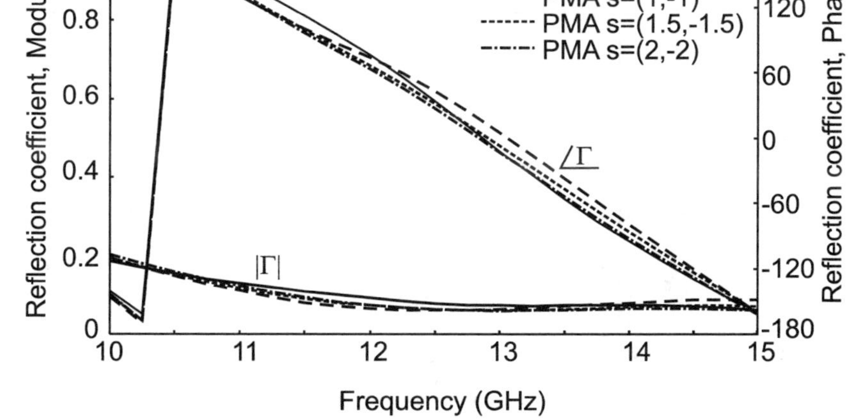

Figure 6.8 Reflection coefficient at the input port of a rectangular waveguide opened on a ground plane.

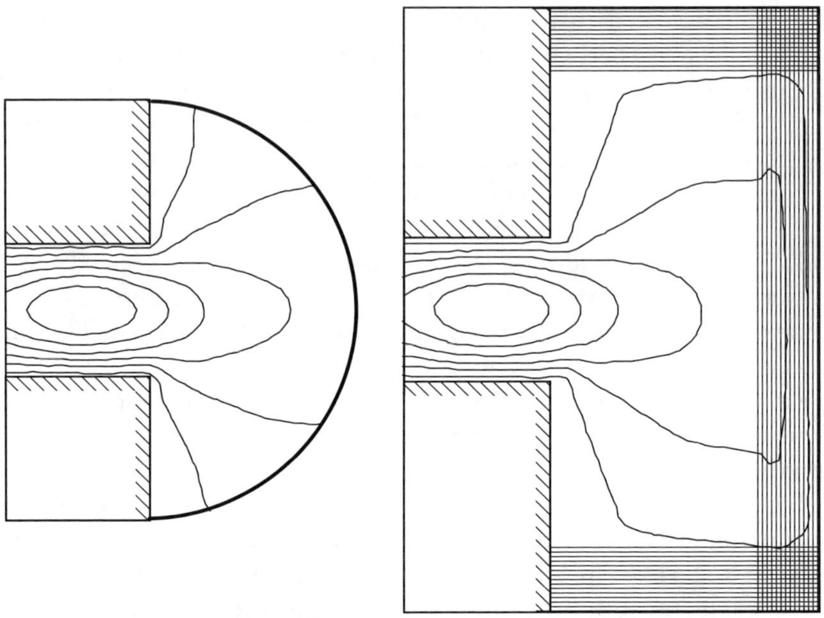

Figure 6.9 H-plane electric field map for a rectangular waveguide opened on a ground plane. Analytic ABC (a), and PMA (b), have been used to terminate the computational domain.

netic analysis (frequency, number of modes, etc.) are the same as in the previous example.

The first analysis is carried out using second-order Bayliss and Turkel ABC. The horn is again opened over a ground plane and the fictitious circular surface Γ_A, with radius 58.35 mm, is centered in the middle point of the horn aperture. If the mesh is built using the HORN.TXT mesh description file provided, the resulting

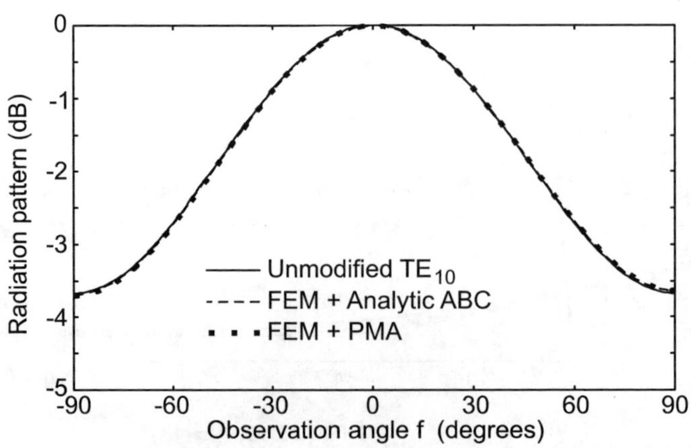

Figure 6.10 H-plane radiation pattern for a rectangular waveguide opened on a ground plane. Frequency $f = 10$ GHz.

subdivision in finite elements is sketched in Figure 6.11(a). The field map obtained by running the code OWG using this mesh is shown in Figure 6.12(a).

Figure 6.11(b) shows the mesh obtained after applying the Delaunay algorithm to the mesh of Figure 6.11(a). Figure 6.12(b) shows the electric field map computed by the OWG code using the mesh of Figure 6.11(b).

It can be seen that the field map of Figure 6.12(b) has an higher symmetry than the field map of Figure 6.12(a).

This behavior may be justified by noting that the Delaunay regularized mesh has an higher degree of symmetry with respect to the unregularized one. The sensitivity of results to the mesh structure can be ascribed to a sort of *numerical anisotropy* arising in computations that is exacerbated in this case by the coarseness of the mesh. As a matter of fact, to allow this example to run on low-memory computers, a rather coarse mesh has been used. The length of the element sides is thus somewhat critical, and the relative small difference between the oblique edges in the upper and lower part of the unregularized mesh makes the difference.

A slightly different, more realistic configuration has been analyzed using the PMA layer termination. It consists of a an H-plane sectorial horn in free space. No ground plane is present. The horn has the same dimension as in the previous case and is enclosed in a 100 mm × 73.4 mm p.e.c. box, internally coated with a PMA layer 9.175 mm thick. All the other parameters used for the analysis are the same as in the previous example.

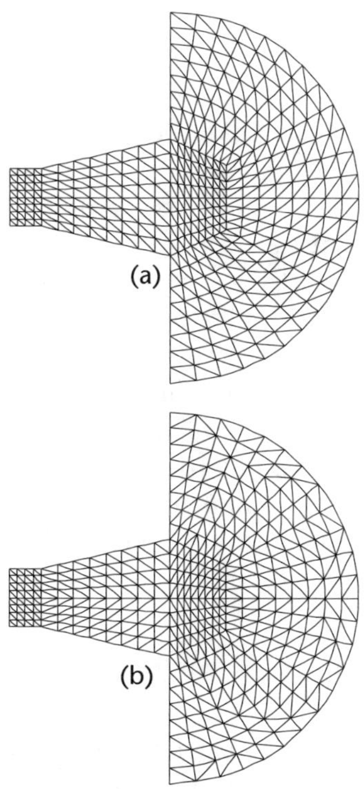

Figure 6.11 Meshes for a rectangular H-plane sectorial horn before (a), and after (b) Delaunay regularization.

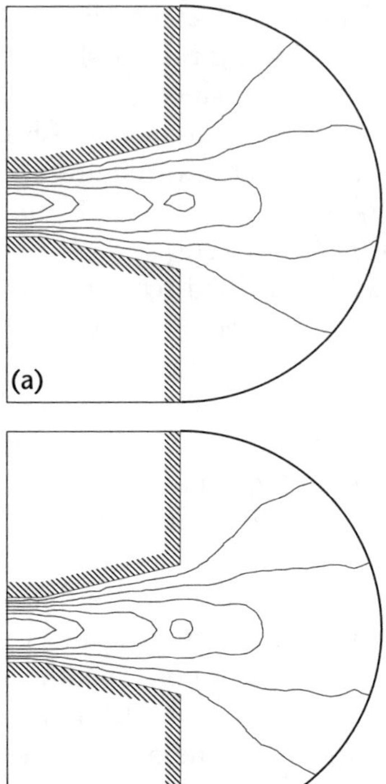

Figure 6.12 Computed electric field maps for the H-plane sectorial horn, using the mesh without (a), and with the Delaunay regularization (b).

Although this time, thanks to the general better behavior of the PMA, the field map appears regular and symmetric even for a nonregularized mesh (Figure 6.13) there is still the numerical anisotropy problem seen above. This becomes evident when plotting the field at the horn aperture (Figure 6.14). It is seen that the nonregularized mesh yields slightly asymmetric values for the field, while the regularized one (with Delaunay algorithm) yields better results.

The sensitivity to the mesh structure is still due to the coarseness of the mesh itself. The asymmetry in the field at the horn aperture is negligible when SWR is the quantity of interest, but the relatively large asymmetry on the phase that gives inaccurate results when the radiation pattern is sought.

The radiation pattern computed from the FEM analysis can be compared with the approximate analytic results. Following the standard ray approach, the E_z component of the field at the horn aperture can be approximated as

$$E_z = \cos\left(\frac{\pi}{A}y\right) e^{-jk\frac{y^2}{2R}} \tag{6.64}$$

In this equation, A is the aperture width, y is the coordinate along the aperture itself, the origin being placed in the middle point, and R is the distance between the aperture and the imaginary point where the two oblique sides of the horn meet. For our configuration, A is 36.7 mm and R is 80 mm.

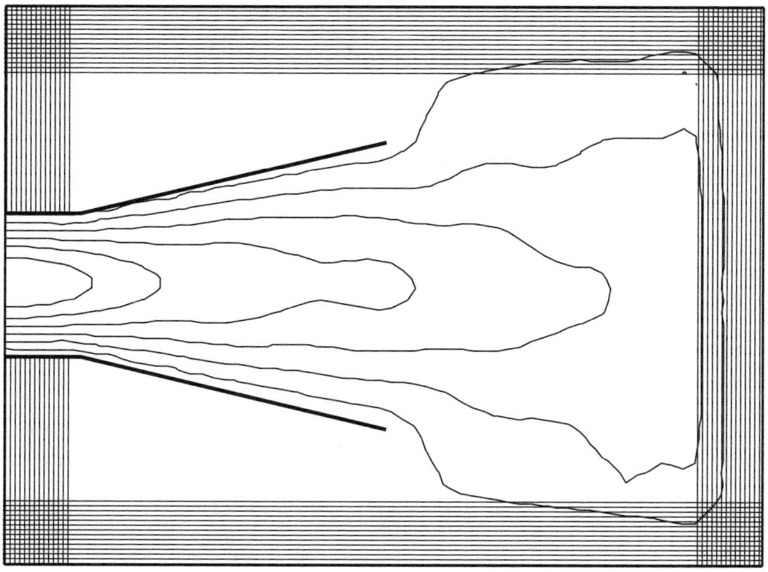

Figure 6.13 Electric field map for an H-plane sectorial horn at $f = 10$ GHz. The computational domain is terminated using a PMA layer.

Figure 6.15 shows the radiation pattern computed with approximated analytic solution, and with the two FEM models using Bayliss and Turkel ABC or PMA termination. As expected, the directivity of the horn is higher that of the open-ended waveguide. It is also seen that the approximated analytical approach gives accurate results in the direction of maximum radiation, but fails for different directions.

The reflection coefficient for this structure at the feeding port is much lower that that of the open waveguide, being estimated to $|\Gamma| = 0.0634$ with the Bayliss and Turkel ABC, and to $|\Gamma| = 0.0863$ when the PMA with $a = 1 - j$ is used.

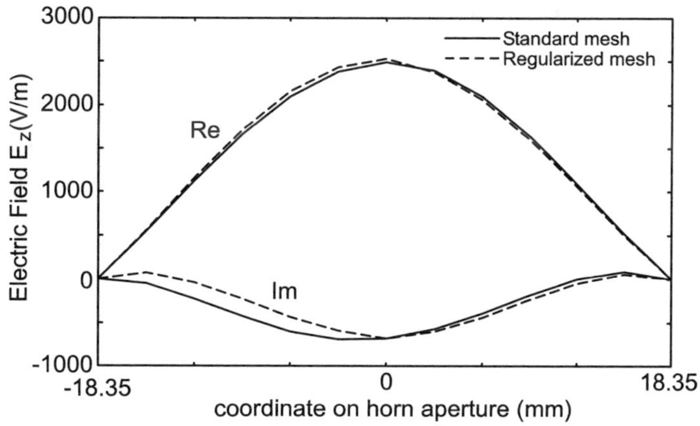

Figure 6.14 Real and imaginary part of the E_z field distribution at the aperture of the horn.

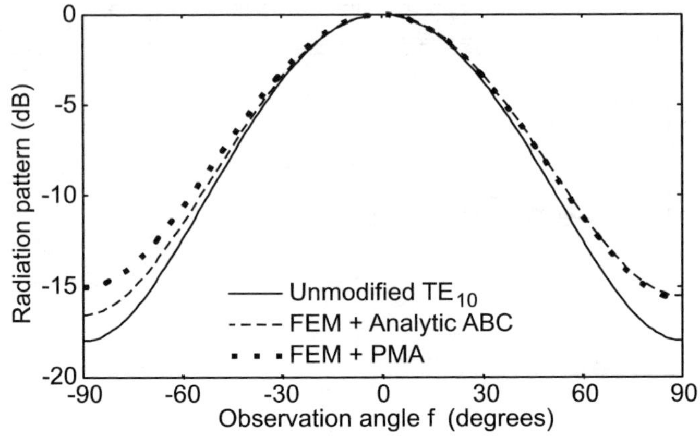

Figure 6.15 Radiation pattern for the H-plane sectorial horn under analysis at $f = 10$ GHz.

The difference between this two values can be explained considering the different configurations analyzed in the two cases.

As a very last example a couple of waveguides opened on a ground plane is analyzed. The configuration is similar to that analyzed at the beginning of this section (Figure 6.9, but with two waveguides open on the ground plane. Waveguides are identical ($a = 18.35$ mm, $b = 9.157$ mm). The problem is still studied on the H-plane and the two waveguides axes are $2a$ apart, that means that the portion of ground plane between the two apertures is a long.

Figure 6.16 shows the geometry and the field map, represented as contour lines, for port 1 fed with a TE_{10}. There is no noticeable field within the second waveguide in Figure 6.16. This is by itself an interesting and expected result since it is fairly well-known that coupling between two apertures on a ground plane is limited, especially on H-plane.

To have a confirmation of this, Figure 6.17 presents the corresponding generalized scattering matrix, as a function of frequency, limited to first mode. It is apparent how the symmetry of the device leads to $S_{11} \simeq S_{22}$, the slight differences being due to the asymmetry introduced by the mesh, and how the device reciprocity leads to $S_{12} = S_{21}$, this time the equality holds since it is bound to reciprocity, which is mesh-independent, and not to symmetry.

It is also apparent how coupling is very low, S_{12} being always lower than -30 dB, and this explains why there is no noticeable field within the second waveguide in Figure 6.16. The field can become more apparent if it is plot in dB rather than in linear scale. The reader is encouraged to change `SeeContours.m` function in this sense.

6.11 Disc Content

The software described in this chapter is stored in the subdirectory ABC of the directory CODES in the enclosed CD-ROM. The folder ABC has two further

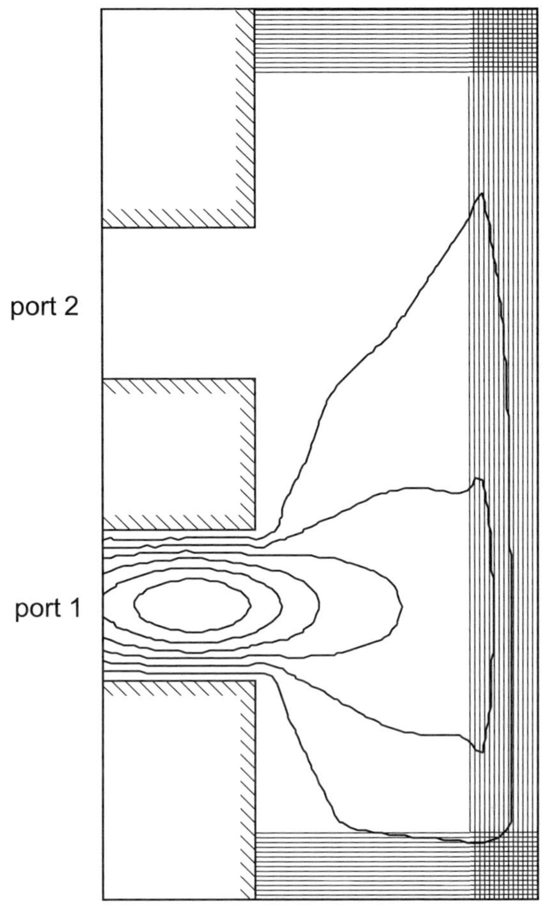

Figure 6.16 Field map for the two open-waveguides problem with lower waveguide (port 1) fed at $f = 10$ GHz.

subdirectories, named CYL and OWG, containing the files relative to the codes CYL and OWG, respectively.

6.11.1 FORTRAN

In the CYL subdirectory there are the FORTRAN 77 files CYL.F, CYL_SUBS.F, and ELEA.F, that must be compiled and linked all together with the library described in Chapter 2. The geometric and electromagnetic input data files relative to the examples discussed in Section 6.8 are stored in the two subdirectories PECYL and DIECYL of the folder CYL.

The FORTRAN 77 files OWG.F and OWG_SUBS.F are stored in the subdirectory OWG of the directory ABC. For the user convenience, there are the two duplicate files ELEA.F and EHD_SUBS.F, that are exactly the same as those in the directory CYL and EHDEV, respectively. They have been included here because must be compiled with all the other files stored in the directory OWG and linked all together with the library described in Chapter 2. The geometric and electro-

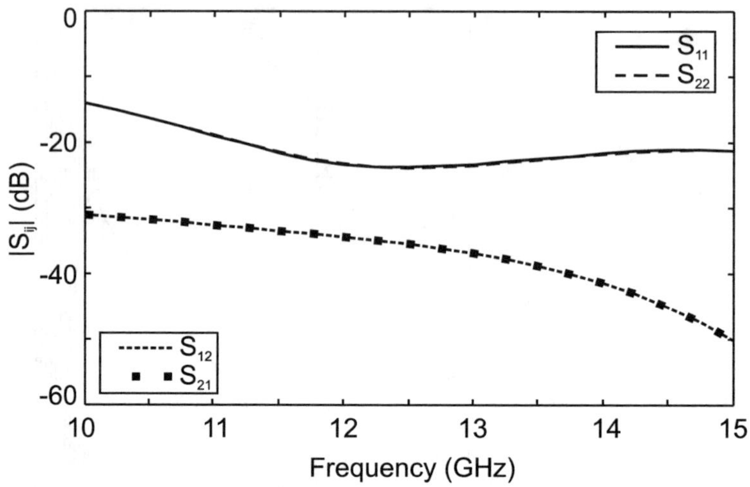

Figure 6.17 Scattering parameters, return loss and coupling, for the two open-waveguides problem with port 1 fed with TE_{10} at $f = 10$ GHz.

magnetic input data files relative to the examples discussed in Section 6.10 can be found in the three subdirectories OWG, HORN, and OWG2.

A more complete description of each file is given in the README file in the directory ABC itself.

6.11.2 Matlab

The Matlab subdirectory tree mimics the FORTRAN one. The CYL directory contains the `cyl.m` function, which must be called according to

```
cyl(fnamemesh,fnameaux,fnameout,fieldflag)
```

while OWG subdirectory contains the `owg.m` function, which must be called according to

```
owg(fnamemesh,fnameaux,fnameout,fieldflag)
```

In both cases the first three inputs are the mesh file, the auxiliary file and the word to prepend to all output file names, respectively, while the last flag, if set to 1, directs the code compute and save field maps.

The Matlab version contains the same subdirectories for the examples as in the FORTRAN version.

References

[1] Engquist, B. and Majda, A., "Absorbing Boundary Conditions for the Numerical Simulation of Waves," *Mathematics of Computation*, Vol. 31, pp. 629–651, 1977.

[2] D'Angelo, J. and Mayergoyz, I. D., "Finite Element Methods for the Solution of RF Radiation and Scattering Problems," *Electromagnetics*, Vol. 10, pp. 177–199, 1990.

[3] Mur, G., "Absorbing Boundary Conditions for the Finite-Difference Approximation of the Time-Domain Electromagnetic-Field Equation," *IEEE Transactions on Electromagnetic Compatibility*, Vol. EMC-23, pp. 377–382, 1981.

[4] Bayliss, A. and Turkel, E., "Radiation Boundary Conditions for Wave-Like Equations," *Communications on Pure and Applied Mathematics*, Vol. 33, pp. 707–725, 1980.

[5] Bayliss, A., Gunzburger, M., and Turkel, E., "Boundary Conditions for the Numerical Solution of Elliptic Equations in Exterior Regions," *SIAM Journal of Applied Mathematics*, Vol. 42, pp. 430–451, 1982.

[6] Peterson, A. F., "Absorbing Boundary Conditions for the Vector Wave Equation," *Microwave and Optical Technology Letters*, Vol. 1, pp. 62–64, 1988.

[7] Stupfel, B. and Mittra, R., "A Theoretical Study of Numerical Absorbing Boundary Conditions," *IEEE Transactions on Antennas and Propagation*, Vol. 43, pp. 478–487, 1995.

[8] Stupfel, B. and Mittra, R., "Numerical Absorbing Boundary Conditions for the Scalar and Vector Wave Equations," *IEEE Transactions on Antennas and Propagation*, Vol. 44, pp. 1015–1022, 1996.

[9] Botha, M. M. and Davidson, D. B., "Rigorous, Auxiliary Variable-Based Implementation of a Second-Order ABC for the Vector FEM," *IEEE Transactions on Antennas and Propagation*, Vol. 54(11), pp. 3499–3504, Nov. 2006.

[10] Berenger, J. P., "A Perfectly Matched Layer for the Absorption of Electromagnetic Waves," *Journal of Computational Physics*, Vol. 114, pp. 185–200, 1994.

[11] Karp, S. M., "A Convergent Farfield Expasion for Two-Dimensional Radiation Functions," *Comm. Pure Appl. Math.*, Vol. 14, pp. 427–434, 1961.

[12] Abramowitz, M. and Stegun, I. A., *Handbook of Mathematical Functions with Formulas, Graphs, and Mathematical Tables*, New York: Dover Publications, 1972.

[13] Webb, J. P., "Absorbing Boundary Conditions for the Finite Element Analysis of Planar Devices," *IEEE Transactions on Microwave Theory and Techniques*, Vol. 38, pp. 1328–1332, 1990.

[14] Pearson, L. W., et al., "Inward-Looking and Outward-Looking Formulations for Scattering from Penetrable Objects," *IEEE Transactions on Antennas and Propagation*, Vol. 40, pp. 714–720, 1992.

[15] Sacks, Z. S., et al., "A Perfectly Matched Anisotropic Absorber for Use as an Absorbing Boundary Condition," *IEEE Transactions on Antennas and Propagation*, Vol. 43, pp. 1460–1463, 1995.

[16] Harrington, R. F., *Time-Harmonic Electromagnetic Fields*, New York: McGraw-Hill, 1961.

[17] Balanis, C. A., *Advanced Engineering Electromagnetics*, New York: John Wiley & Sons, 1989.

PART II

Three Dimensions

CHAPTER 7

Finite Elements in Three Dimensions

As already pointed out in Chapter 2, the implementation of a finite element analysis relies on a series of standard steps which can be conveniently implemented by a series of standard tools and routines forming a library. This allows both for code reusability and easy maintenance.

In this chapter the essential tools for a three-dimensional FEM implementation will be presented, together with the basic theory needed to understand its functioning. Since the work flow is generally divided into three phases: *preprocessing*, *linear system solving*, and *postprocessing* the description will follow this same subdivision.

It is worth remarking that the three-dimensional codes are here presented only in FORTRAN version. This choice is due to the fact that FORTRAN-optimized codes are noticeably faster than Matlab in solving linear system of equations. Since three-dimensional codes usually lead to quite larger linear systems than those generated by two-dimensional codes, any increase in speed is essential.

On the other hand some Matlab routines for pre- and postprocessing are provided, since for those speed is a minor issue for those.

7.1 Preprocessing

If there are many choices for elements in two-dimensional, the three-dimensional case presents even more possibilities. The most used three-dimensional elements are *tetrahedra*, but also *hexahedra* and *prisms* or *pyramids* are possible. Using tetrahedra allows for easy coverage of very complex-shaped geometry having very general curved boundary. It is generally accepted that the use of tetrahedral elements is an optimal compromise between accuracy and computational efficiency for a general finite element implementation. The concept of three-dimensional tetrahedral mesh will be introduced both for nodal and vector elements, considering the mathematical form of interpolating function bases used in scientific literature and commercial softwares.

Being that the implementation of three-dimension meshing code is rather complex, we choose not to develop a new one for this book but rather to resort to a freely available software: TetGen[1]. As it will be described later, TetGen output

1. TetGen is developed by H. Si at Numerical Mathematics and Scientific Computing, Weierstrass Institute for Applied Analysis and Stochastics, Mohrenstr. 39, 10117 Berlin, Germany. TetGen is free for research and noncommercial uses. For any commercial utilization, a commercial license is available upon request.

needs to be converted for electromagnetic FEM usage. This is accomplished by a FORTRAN format converter. At the same time, for Matlab users, TetGen has also been encapsulated in a Matlab-coded wrapper designed both to run TetGen and to transform the nodal TetGen mesh into a suitable vector mesh for electromagnetic analysis.

It is worth remarking that much research has been carried out on three-dimensional meshing and that several commercial software packages that permit high-quality meshes for FEM three-dimensional electromagnetic problems do exist. Among the commercial software it is worth mentioning PATRAN, which is a MSC NASTRAN pre- and postprocessor. Besides commercial softwares, several freely available or even open-source mesh generators with very attractive performance exist, among which, TetGen, NETGEN, and GMSH. Furthermore commercial analysis softwares, like the leader in electromagnetic FEM analysis Ansoft HFSS, embed their own mesh generator.

TetGen has been chosen here for the simplicity of its interface and for the possibility to be compiled on a variety of different platforms. Furthermore, its input file is based on the piecewise linear complex (PLC) syntax, which is a standard used also by other mesh generators (such as TRIANGLE[2], which works on two-dimensional geometries and can be considered as a valid alternative to the two-dimensional mesh generator of this book and to Matlab PDE toolbox mesh generator).

TetGen, as the large part of the freely available meshing software, is specifically designed in order to obtain nodal meshes to be used inside finite-element framework solving mechanical or fluid dynamics problems. However physical problems concerning the propagation of electromagnetic fields are better discretized adopting a pure vectorial approach, in other words, being in general the problem of an unknown vector field, the use of a vector base inside each element is more natural and allows for easier inclusion of boundary conditions.

From this reasons it is evident that the need to expand the information related to the nodal mesh provided by TetGen including also information related to the edges of each element and their labels.

As already mentioned, TetGen takes as an input geometry description file a PLC file; whose syntax will be described in detail, in the next subsection. TetGen output is constituted by a set of text files that will be described in Section 7.1.2 together with the file format converter routine used to transform them into a single file in quick FEM format.

7.1.1 Input Geometry Description File

The input geometry file must describe completely the problem's domain with the possibility to define features and assigning labels for each subdomain and boundary faces.

The format of the input geometry file used by TetGen is known as piecewise linear complex file. For a complete description of TetGen input file format we

2. http://www.cs.cmu.edu/ quake/triangle.html.

suggest to look on the dedicated documentation at [1]. We will have a look at the input file syntax only for what concerns the need to run the examples included in the CD.

A PLC input file is a simple text file having an easy to grasp syntax (Table 7.1); it is divided into four consecutive parts describing geometry points, faces, holes, and regions, respectively (Figure 7.1). More in depth:

- The first block include information on the points used to define the geometrical three-dimensional domain, which are called *nodes*. In particular the first uncommented row reports four space-separated integers:
  - `NNODE` is the number of nodes in the geometry;
  - `DIM` is the dimensionality of the geometrical space which is set to 3 because we are working in a general *three-dimensional* space;
  - `0` indicates the number of attribute per node;
  - `1` makes the TetGen program take into account node labels.

Table 7.1 PLC Input Geometry Description File Syntax

Input Data	*Data Type*
`NNODE DIM 0 1`	`4 Integers`
⋮	*NNODE* lines of node definitions, each containing:
`IN X Y Z NL`	`1 Integer, 3 reals, 1 integer`
⋮	
`NFACE 1`	`2 Integers`
⋮	*NFACE* blocks of face definitions, each containing:
`NPOLY NHOLE LABEL`	`3 Integers`
⋮	*NPOLY* lines, each defyning a polygon:
`NN N(1), N(2), ..., N(NN)`	$n+1$ `Integers`
⋮	*NHOLE* lines, each defining a 2D hole:
`NH2 X Y Z`	`1 Integer, 3 reals`
⋮	
⋮	
`NHOLES`	`1 Integers`
⋮	*NHOLES* blocks of 3D hole definitions, each containing:
`NH3 X Y Z`	`1 Integer, 3 reals`
⋮	
`NREGIONS`	`1 Integers`
⋮	*NREGIONS* blocks of region definitions, each containing:
`NR X Y Z LABEL ATTR`	`1 Integer, 3 Reals, 2 Integers`
⋮	

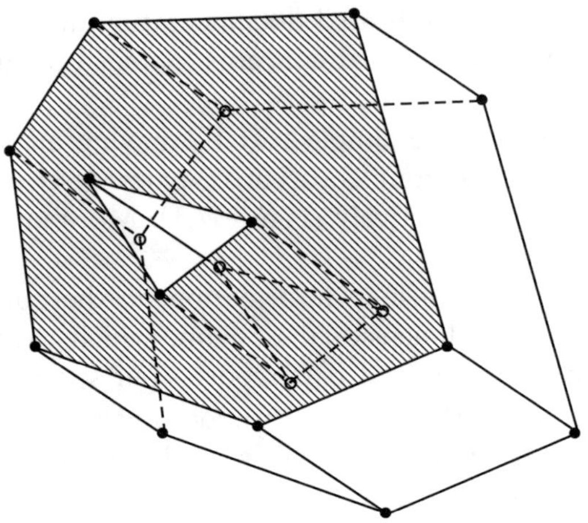

Figure 7.1 A generic volume, which can be described by a PLC file. Bullets are nodes and the hatched front face is a polygonal face defined by 6 nodes in which a triangular hole is cut.

- This first row is followed by NNODE additional rows each of which containing the data of a single node:
  - IN the unique, progressive, index identifying the node;
  - X Y Z three floating-point coordinates;
  - NL node label.
- The second block lists all the information defining face topology. The first line of this section contains two integers:
  - NFACE the number of faces defining the volumes;
  - 1 makes the TetGen program take into account face labels.
- Then a number of blocks follows, one for each face. Each block contains:
  - A first line with:
    NPOLY number of polygonal, closed, lines representing the face;
    NHOLE number of holes to be cut in a face (if a large polygon inscribes a smaller one then the smaller can be used to cut a hole in the larger);
    LABEL the label to be assigned to the face.
  - Then, for each polygon in the face there is a line containing:
    NN number of nodes defining the polygon;
    N(i) $i = 1,\ldots,NN$ integers pointing to the nodes that are used in the polygon definitions.
  - Then, for each hole in the face there is a line containing:
    NH2 the unique index identifying the two-dimensional hole;
    X Y Z three floating-point coordinates defining a point within the polygon which should be used to cut the hole in the face.
- A set of faces delimiting a closed-volume automatically define a three-dimensional domain. As polygons can be used to cut holes in faces, similarly volumes can be used to cut three-dimensional holes in larger volumes. The first line of the third section defines:
  NHOLES the number of closed volumes used to cut three-dimensional holes in other volumes. For each of these there is a line defining:

NH3 the unique index identifying the three-dimensional hole;
X Y Z three floating-point coordinates defining a point within the volume which should be used to cut the hole in a larger one.

- A volume, as defined automatically by a set of faces enclosing it, is not automatically assigned a label. If a label is to be assigned, then a volume must be defined as a region. The first line of the fourth block defines:
  - NREGIONS The number of closed volumes to which a label is to be assigned. For each of these there is a line defining:
    NR the unique index identifying the region;
    X Y Z three floating-point coordinates defining a point within the volume which should be used to define the region;
    LABEL the label assigned to the region.

To better understand the PLC file format, an example of a straight waveguide section is reported in Example 7.1; it is also included in the CD examples.

Example 7.1

```
#
# wr90.poly
#
# Sample file for standard waveguide description.
# this simple file describe a WR90 waveguide phase shifter at
# 10 GHz frequency.
#
# Part 1 - node list
8   3  0  1
1   -11.43  -5.08  0.00  11
2   -11.43   5.08  0.00  11
3    11.43   5.08  0.00  11
4    11.43  -5.08  0.00  11
5   -11.43  -5.08  9.93  12
6   -11.43   5.08  9.93  12
7    11.43   5.08  9.93  12
8    11.43  -5.08  9.93  12
# Part 2 - face list
6  1
1  0  11
4  1  2  3  4  1
1  0  1
4  1  2  6  5  1
1  0  1
4  2  3  7  6  1
1  0  1
4  3  4  8  7  1
1  0  1
4  1  4  8  5  1
1  0  12
4  5  6  7  8  1
# Part 3 - hole list
0
# Part 4 - region list
1
1 0.0 0.0 5.0 1 0.1
```

7.1.2 Output Mesh Description File

The geometry input file explained in the previous section is the input of the TetGen program which is executed with the command:

```
tetgen -pqaA plcfile.poly
```

where switches means:

- `-p` produce tetrahedra;
- `-q` produce high-quality mesh;
- `-a` applies a maximum volume constraint to tetrahedra;
- `-A` assigns labels to regions.

TetGen then generates a set of different text files having the same name, without extension, of the original input file but several different extensions: `.1.node`, `.1.ele`, `.1.face` that contains information about mesh nodes, elements, and boundary faces, respectively. TetGen also produces other files which are not necessary to us.

The content of the three files of interest is described below:

- `.1.node` file (Example 7.2): A first line defining the number of nodes (here 26), the number of coordinates per node (3, of course) and a couple of flags; then as many lines as there are nodes follow, starting with node index, followed by the three coordinates and, lastly, the label.
- `.1.ele` file (Example 7.3): A first line defining the number of elements (here 44), the number of nodes per element (4, in a tetrahedron) and an additional flag; then as many lines as there are elements follow, starting with element index, followed by the four indexes of the four nodes defining the tetrahedron, and at last the label of the tetrahedron.
- `.1.face` file (Example 7.4): A first line defining the number of faces (here 112), and an additional flag; then as many lines as there are faces follow, starting with face index, followed by the three indexes of the three nodes defining the triangular face, and at last the label of the face.

Example 7.2

```
26  3  0  1
    1  0.0  0.0  0.0  1
    2  0.0  1.0  0.0  1
    3  1.0  1.0  0.0  1
:
:
```

Example 7.3

```
44  4  1
    1     8  18  17  25  1
    2    21  16  15  25  1
```

```
     3    1  10  20  23  1
:
:
```

Example 7.4

```
112  1
     1   8  18  17  1
     2   8  25  18  0
     3  18  25  17  0
:
:
```

These files are read by the PLC2QF.F routine, accepting the base file name as an input and converting the mesh in the quick FEM format described below as output.

It must be noted that, in three-dimension, p.e.c. boundaries are reserved label -1, whereas waveguide ports, as they will be used in Chapter 9, are assigned positive labels corresponding to the port numbering scheme. If positive labels are present an additional mesh file, whose extension is `.mm` is also created. This latter file contains a reduced set of informations useful to define the two-dimensional meshes of the ports as they can be extracted from the tetrahedral mesh of the device. Since this file is only used in Chapter 9, its description can be found there.

PCL2QF.F can be wrapped within the DO_PCL2QF.F driver, which interactively asks the user for the base file name, without extensions, and writes the mesh in the quick FEM format by calling up the WRITE3DMESH.F subroutine.

Alternatively, the `plc2qf.m` program presents a single wrapper for both TetGen and mesh conversion. It carries out the following tasks:

- Take the `.poly` file as geometrical description input;
- Execute TetGen program using correct command line flags and generate `.1.node`, `.1.ele` and `.1.face` files;
- Complete the mesh information providing a list of composition for all the mesh edges with relative boundary labels, if it is necessary;
- At last it writes a `.fem`, and optionally a `.mm` file; containing all the information about the mesh that the computation program needs to represent the mesh from the numerical point of view.

The output mesh description file in quick FEM format is a file containing all the information required by processing FEM programs to create an internal mesh representation containing also label and attributes for subregions and boundaries of the spatial computation domain. All the electromagnetic codes described in the following chapters take as input a file with the `.fem` extension that contains the position and label of all the mesh nodes, the composition and region-dependent label of all the mesh tetrahedra, and the composition and boundary-dependent label for all the mesh edges. All these information are stored into a text file in a way very similar to what has been done previously for two-dimensional.

The mesh output file format has the general pattern listed in Table 7.2. Its explanation is so similar to that given for the two-dimensional meshes that it is

Table 7.2 Output Mesh Description File Syntax

Input Data	*Data Type*
`Header`	`Character*80`
`Elements = NELE`	`Integer`
`Nodes = NNODE`	`Integer`
`Edges = NSPIG`	`Integer`
`Elements`	a one-line tag of the element definition block
⋮	*NELE* lines of element definition, each containing:
`IE NN N(1) ... N(NN) LE`	`NN + 2 Integers`
⋮	
`Nodes`	a one-line tag of the node definition block
⋮	*NODE* lines of node definition, each containing:
`IN X Y Z LN`	`1 Integer, 2 Reals, 1 Integer`
⋮	
`Edges`	a one-line tag of the edge definition block
⋮	*NELE* lines of edge definition,
`IE NS N(1) ... N(NS)`	`NS + 1 Integers`
⋮	
`Edge Labels`	a one-line tag of the edge label definition block
⋮	*NSPIG* lines of element label definition, each containing:
`IS LS`	`2 Integers`
⋮	

useless to provide it again. The only evident difference is the presence of three coordinates (`X Y Z`) for each node. Furthermore, tetrahedrons are defined on the basis of 4 nodes each, so `NN`=4, and each tetrahedron has 6 edges, so `NS` is 6.

It is worth to remark that while the first and second blocks (elements and nodes) are strictly necessary to represent uniquely the given mesh, the edges and edge-labels blocks are auxiliary and are needed only for edge-based codes.

7.2 Element Matrices

After the preprocessing phase, and based on it, the solution procedure requires building the global coefficient matrix which, as shown in Part I of this book, is made up of contributions coming from each single element in the form of one or more local matrices.

The first step of the solution process is hence to calculate the local matrices needed for the elements. The form of local matrices is very dependent from the type of element used, the type of bases used, and obviously the type of equation which has to be solved. There are two main classes of elements used for electromagnetic simulation purposes: nodal elements and vector elements, both types were already presented in the chapters of Part I.

Nodal elements make use of scalar basis function and are hence suitable to approximate scalar fields. Nodal elements have been also used for a long time to

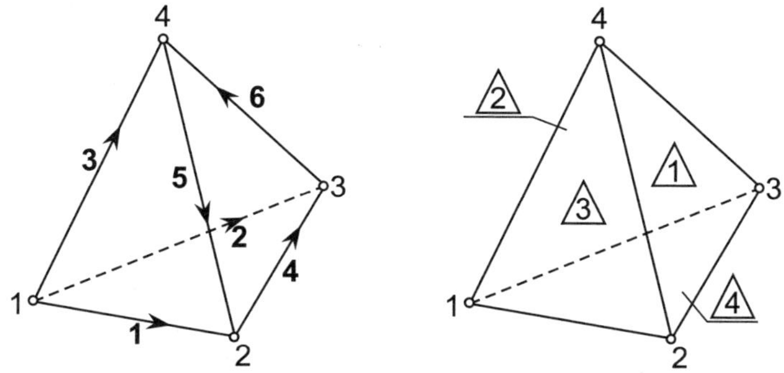

Figure 7.2 Tetrahedron with node (plain); edge (bold); and face (inset in triangles) enumeration.

simulate a vector field by separately approximating each component along orthogonal directions. This approach presents numerical problems which can be overcome in some ways but at the cost of increasing the global problem complexity, and of degenerating the solution accuracy.

Edge elements, on the other hand, make use of vector basis functions which are more appropriate to approximate a vector field and allow enforcing boundary conditions in a straightforward manner.

For what concerns shape, the most used geometrical forms are *tetrahedra* and *hexahedra*; the first one is the most popular and is widely used in commercial FEM-based software. Hexaedra, on the other hand, are less flexible but easier to generate. In this book only tetrahedra will be considered.

A tetrahedron is a geometrical three-dimensional structure defined by 4 non-coplanar vertexes. By joining each vertex with all the others a volume delimited by 4-triangular faces is obtained (Figure 7.2). Tetrahedra have many attracting properties, one of the most useful is its adaptability in discretizing geometrically complex structures. From the computational point of view, a tetrahedra is uniquely defined assigning its four vertexes in the three-dimensional space and defining a rule to enumerate its six edges and four triangular faces. This can be done as described in Tables 7.3 and 7.4.

An hexaedron, on the other hand, is a geometrical three-dimensional structure defined by eight points in the space and isomorphic to a cube, that is enclosed by six nonoverlapping quadrangular faces. Hexahedral elements are more suitable to discretize regular geometries, on these domains they permit smaller meshes, that is meshes with fewer degrees of freedom compared to tetrahedral meshes.

Table 7.3 Edge Numbering Scheme in a Tetrahedron

Start Point	*End Point*	*Edge Number*
1	2	1
1	3	2
1	4	3
2	3	4
4	2	5
3	4	6

Table 7.4 Face Numbering Scheme in a Tetrahedron

1st Point	*2nd Point*	*3rd Point*	*Face Number*
2	3	4	$\triangle_1$
1	3	4	$\triangle_2$
1	2	4	$\triangle_3$
1	2	3	$\triangle_4$

The cases of tetrahedral elements will be consider in the following. The hexahedral elements are not discussed since they are not exploited in the codes in the enclosed CD-ROM. Hexahedrons, and also other types of element geometries, are described widely in the literature, among others in [2–4].

7.2.1 Nodal Elements

Given a set of four noncoplanar vertexes $P = (x, y, z)$ a tetrahedron $\mathcal{S}$ in three-dimensional space is defined. Nodal basis for this kind of elements can be defined by means of the polynomial forms (2.8) defined in Chapter 2. In three-dimensional simplex coordinates ζ_j are defined as the ratio between the volume of the sub-elements the generic internal point P defines and the volume of the the element itself (Figure 7.3):

$$\zeta_j = \frac{V(P, \triangle_j)}{V} \qquad \text{with } j = 1, 2, 3, 4 \tag{7.1}$$

being V the volume of the whole tetrahedron $\mathcal{S}$ and $V(P, \triangle_j)$ the volume of the subtetrahedron defined by inner point P and the points belonging to the face $\triangle_j$ which is opposed of the jth vertex (Figure 7.3).

It is evident that the number of simplex coordinates associated to point P inside a tetrahedron is one unit grater than the geometrical dimension of the space, as it was in two dimensions, therefore the four three-dimensional simplex coordinates are linearly dependent as they were on triangles and their relation is

$$\zeta_1 + \zeta_2 + \zeta_3 + \zeta_4 = 1 \tag{7.2}$$

condition, which is due to the fact that the union of the subtetrahedra, defined by the internal point P, spans the whole volume of the tetrahedron $\mathcal{S}$.

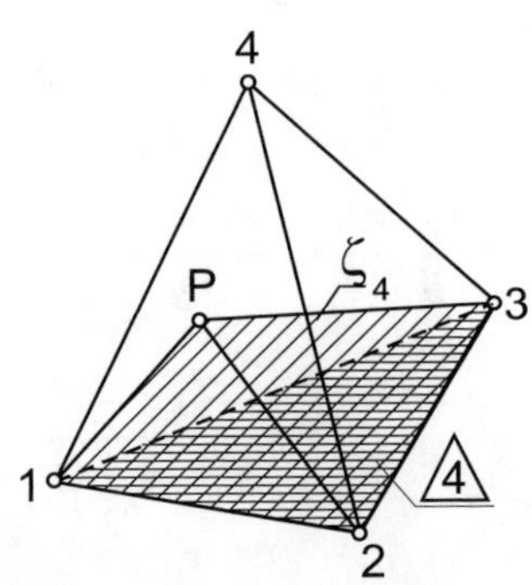

Figure 7.3 Geometrical interpretation of the simplex coordinates.

Using the simplex coordinates it is possible to retrieve the Cartesian coordinates of each point $P = (x, y, z)$, within the tetrahedron, using simple matrix-vector multiplication

$$\begin{bmatrix} 1 \\ x \\ y \\ z \end{bmatrix} = \begin{bmatrix} 1 & 1 & 1 & 1 \\ x_1 & x_2 & x_3 & x_4 \\ y_1 & y_2 & y_3 & y_4 \\ z_1 & z_2 & z_3 & z_4 \end{bmatrix} \begin{bmatrix} \zeta_1 \\ \zeta_2 \\ \zeta_3 \\ \zeta_4 \end{bmatrix} \tag{7.3}$$

Equation (7.3) describes the transformation from simplex to Cartesian coordinates. By naming $\bar{\mathbf{x}}$ the left-most vector in (7.3), $[Z^{(e)}]$ the matrix and $\bar{\boldsymbol{\zeta}}$ the right-most vector a more compact notation can be attained: $\bar{\mathbf{x}} = [Z^{(e)}]\bar{\boldsymbol{\zeta}}$. Furthermore, considering (7.2) it is possible to recast (7.3) as:

$$\begin{bmatrix} x \\ y \\ z \end{bmatrix} = \begin{bmatrix} (x_1 - x_4) & (x_2 - x_4) & (x_3 - x_4) \\ (y_1 - y_4) & (y_2 - y_4) & (y_3 - y_4) \\ (z_1 - z_4) & (z_2 - z_4) & (z_3 - z_4) \end{bmatrix} \begin{bmatrix} \zeta_1 \\ \zeta_2 \\ \zeta_3 \end{bmatrix} + \begin{bmatrix} x_4 \\ y_4 \\ z_4 \end{bmatrix} \tag{7.4}$$

which can be used to calculate partial derivatives or integrals in simplex space, and in particular the derivative of a generic function $f(\zeta_1, \zeta_2, \zeta_3)$, leading to the Jacobian matrix:

$$\begin{bmatrix} \frac{\partial f}{\partial \zeta_1} \\ \frac{\partial f}{\partial \zeta_2} \\ \frac{\partial f}{\partial \zeta_3} \end{bmatrix} = \begin{bmatrix} \frac{\partial x}{\partial \zeta_1} & \frac{\partial y}{\partial \zeta_1} & \frac{\partial z}{\partial \zeta_1} \\ \frac{\partial x}{\partial \zeta_2} & \frac{\partial y}{\partial \zeta_2} & \frac{\partial z}{\partial \zeta_2} \\ \frac{\partial x}{\partial \zeta_3} & \frac{\partial y}{\partial \zeta_3} & \frac{\partial z}{\partial \zeta_3} \end{bmatrix} \begin{bmatrix} \frac{\partial f}{\partial x} \\ \frac{\partial f}{\partial y} \\ \frac{\partial f}{\partial z} \end{bmatrix} = \mathbf{J} \begin{bmatrix} \frac{\partial}{\partial x} \\ \frac{\partial}{\partial y} \\ \frac{\partial}{\partial z} \end{bmatrix} f(x, y, z) \tag{7.5}$$

The Jacobian matrix allows, as viewed in Chapter 2, to take into account coordinate system changes inside integrals as in:

$$\iiint_\Omega f(x, y, z)\,dx\,dy\,dz = \iiint_\Omega f(\zeta_1, \zeta_2, \zeta_3)|\mathbf{J}|\,d\zeta_1 d\zeta_2 d\zeta_3 \tag{7.6}$$

The latter formula is very useful to calculate the element matrices in simplex coordinates (ζ_i) rather than in the real space (x, y, z).

The inverse relation between simplex and Cartesian coordinate is also useful to compute FEM integrals; it is evident that $\bar{\boldsymbol{\zeta}} = [Z^{(e)}]^{-1}\bar{\mathbf{x}}$, where the expanded inverse matrix is:

$$[Z^{(e)}]^{-1} = \begin{bmatrix} a_1 & b_1 & c_1 & d_1 \\ a_2 & b_2 & c_2 & d_2 \\ a_3 & b_3 & c_3 & d_3 \\ a_4 & b_4 & c_4 & d_4 \end{bmatrix} \tag{7.7}$$

it is well-known that

$$a_i = (-1)^{i-1}|[Z^{(e)}]_{1,i}|\ |[Z^{(e)}]|^{-1} \tag{7.8}$$

$$b_i = (-1)^{i}|[Z^{(e)}]_{2,i}|\ |[Z^{(e)}]|^{-1} \tag{7.9}$$

$$c_i = (-1)^{i+1}|[Z^{(e)}]_{3,i}|\ |[Z^{(e)}]|^{-1} \tag{7.10}$$

$$d_i = (-1)^{i+2}|[Z^{(e)}]_{4,i}|\ |[Z^{(e)}]|^{-1} \tag{7.11}$$

where the $|[Z^{(e)}]_{j,i}|$ is minor j,i of matrix $[Z^{(e)}]$ obtained by computing the determinant of the submatrix obtained excluding the jth row and the ith column from matrix $[Z^{(e)}]$ itself. It is worth remembering how $|[Z^{(e)}]| = 6V$, being V the volume of the tetrahedron.

7.2.1.1 First-Order Tetrahedron Elements

The simplest basis functions on a tetrahedron are first-order nodal functions defined by the simplex coordinates themselves:

$$\alpha_i = \zeta_i \qquad i = 1,2,3,4 \tag{7.12}$$

Using these bases, four degrees of freedom are assigned to each tetrahedron, one for each vertex. The local element matrices are defined, as for the two-dimensional case, by the following integrals on the tetrahedron volume Ω_e:

$$S_{ij}^{(e)} = \int_{\Omega_e} \nabla\alpha_i^{(e)} \cdot \nabla\alpha_j^{(e)} d\Omega \tag{7.13}$$

$$T_{ij}^{(e)} = \int_{\Omega_e} \alpha_i^{(e)} \alpha_j^{(e)} d\Omega \tag{7.14}$$

The elements of matrix (7.14) can be evaluated using the general formula [5]:

$$\int_{\Omega_e} \left(\zeta_1^{(e)}\right)^m \left(\zeta_2^{(e)}\right)^n \left(\zeta_3^{(e)}\right)^p \left(\zeta_4^{(e)}\right)^q d\Omega = \frac{m!n!p!q!}{(m+n+p+q+3)!} 6V^{(e)} \tag{7.15}$$

where m, n, p, and q are integers. For first-order bases, integrals (7.15) are either

$$\int_{\Omega_e} \alpha_i^2 d\Omega = \frac{2!}{5!} 6V^{(e)} \tag{7.16}$$

or

$$\int_{\Omega_e} \alpha_i \alpha_j d\Omega = \frac{1!1!}{5!} 6V^{(e)} \tag{7.17}$$

hence local matrix $\left[T^{(e)}\right]$ will be:

$$T_{ij}^{(e)} = \begin{cases} V^{(e)}/10 \text{ if } & i = j \\ V^{(e)}/20 \text{ if } & i \neq j \end{cases} \tag{7.18}$$

or in matrix form:

$$\left[T^{(e)}\right] = \frac{V^{(e)}}{20} \begin{bmatrix} 2 & 1 & 1 & 1 \\ 1 & 2 & 1 & 1 \\ 1 & 1 & 2 & 1 \\ 1 & 1 & 1 & 2 \end{bmatrix} \tag{7.19}$$

The entries of matrix $[S^{(e)}]$ can be obtained by resorting to (7.7). It is easy to see that, in global Cartesian coordinates, the gradient of the simplex coordinates appearing in (7.13) are constant vectors:

$$\nabla \zeta_j = \frac{1}{6V^{(e)}} \left[b_j \hat{\mathbf{x}} + c_j \hat{\mathbf{y}} + d_j \hat{\mathbf{z}} \right] \tag{7.20}$$

Hence the local matrix $\left[S^{(e)}\right]$ can be evaluated easily according to

$$S_{ij}^{(e)} = \frac{1}{(6V^{(e)})^2} \left\{ \int_{\Omega_e} b_i b_j \, d\Omega + \int_{\Omega_e} c_i c_j \, d\Omega + \int_{\Omega_e} d_i d_j \, d\Omega \right\} \tag{7.21}$$

If we define new matrices $[B]$, $[C]$, and $[D]$ so that it is

$$B_{ij}^{(e)} = \frac{1}{(6V^{(e)})^2} \int_{\Omega_e} b_i b_j d\Omega = \frac{b_i b_j}{36V^{(e)}} \tag{7.22}$$

$$C_{ij}^{(e)} = \frac{1}{(6V^{(e)})^2} \int_{\Omega_e} c_i c_j d\Omega = \frac{c_i c_j}{36V^{(e)}} \tag{7.23}$$

$$D_{ij}^{(e)} = \frac{1}{(6V^{(e)})^2} \int_{\Omega_e} d_i d_j d\Omega = \frac{d_i d_j}{36V^{(e)}} \tag{7.24}$$

then it is possible to attain the compact matrix form:

$$\left[S^{(e)}\right] = \left[B^{(e)}\right] + \left[C^{(e)}\right] + \left[D^{(e)}\right] \tag{7.25}$$

7.2.1.2 Higher Order Tetrahedron Elements

Nodal bases can be defined up to arbitrarily high order in terms of Silvester polynomials (2.8) [5]. In general a n-order scalar interpolating function is defined as a product of Silvester polynomial of various orders, each function of one of the four simplex coordinates:

$$\alpha_i(\zeta_1, \zeta_2, \zeta_3, \zeta_4) = R_{m_1}(n, \zeta_1) R_{m_2}(n, \zeta_2) R_{m_3}(n, \zeta_3) R_{m_4}(n, \zeta_4) \tag{7.26}$$

where $m_1 + m_2 + m_3 + m_4 = n$.

The number of nodes, and hence degrees of freedom, for the element is given by the formula:

$$N^{(e)} = \frac{(n+1)(n+2)(n+3)}{6} \tag{7.27}$$

which indeed coincides with the dimension of the vector space of n-order polynomials in three-dimensional space. The node locations are depicted in Figure 7.4.

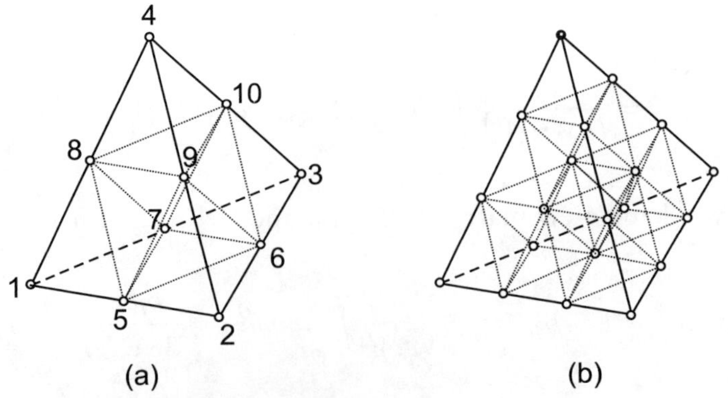

Figure 7.4 Positions of the nodes in the physical space for a second-order tetrahedron element (a), and for a third-order tetrahedron element (b).

Second-order scalar bases are explicitly given in the following. The total number of degrees of freedom per tetrahedron, from (7.27), is 10; therefore also 10-interpolating basis functions are to be defined on the basis of (7.26) and (2.8):

$$\begin{aligned}
\alpha_1 &= R_2(2,\zeta_1) = \zeta_1(2\zeta_1 - 1) \\
\alpha_2 &= R_2(2,\zeta_2) = \zeta_2(2\zeta_2 - 1) \\
\alpha_3 &= R_2(2,\zeta_3) = \zeta_3(2\zeta_3 - 1) \\
\alpha_4 &= R_2(2,\zeta_4) = \zeta_4(2\zeta_4 - 1) \\
\alpha_5 &= R_1(2,\zeta_1)R_1(2,\zeta_2) = 4\zeta_1\zeta_2 \\
\alpha_6 &= R_1(2,\zeta_2)R_1(2,\zeta_3) = 4\zeta_2\zeta_3 \\
\alpha_7 &= R_1(2,\zeta_1)R_1(2,\zeta_3) = 4\zeta_1\zeta_3 \\
\alpha_8 &= R_1(2,\zeta_1)R_1(2,\zeta_4) = 4\zeta_1\zeta_4 \\
\alpha_9 &= R_1(2,\zeta_2)R_1(2,\zeta_4) = 4\zeta_2\zeta_4 \\
\alpha_{10} &= R_1(2,\zeta_3)R_1(2,\zeta_4) = 4\zeta_3\zeta_4
\end{aligned} \tag{7.28}$$

In the following, local matrices entries will be given. For a complete development of these calculations the reader can refer to [5].

Local FEM matrices are built, as in the first-order case, from (7.13) and (7.14). Considering (7.14), the integrals involving products of basis functions can be recast as products of simplex coordinates and hence the calculation can be performed using (7.15), yielding, for example:

$$\begin{aligned}
T_{12}^{(e)} &= \int_{\Omega_e} \alpha_1^{(e)}\alpha_2^{(e)} d\Omega \\
&= \int_{\Omega_e} \zeta_1(2\zeta_1 - 1)\zeta_2(2\zeta_2 - 1) d\Omega \\
&= 4\int_{\Omega_e} \zeta_1^2\zeta_2^2 d\Omega - 2\int_{\Omega_e} \zeta_1^2\zeta_2 d\Omega - 2\int_{\Omega_e} \zeta_1\zeta_2^2 d\Omega + \int_{\Omega_e} \zeta_1\zeta_2 d\Omega
\end{aligned} \tag{7.29}$$

each of the four integrals in (7.29) is in the form given in (7.15) and their numerical values are

$$T_{ij}^{(e)} = 6V^{(e)}\left(4\frac{2!2!}{7!} - 2\frac{2!}{6!} - 2\frac{2!}{6!} + \frac{1}{5!}\right) = \frac{12V^{(e)}}{5040} \tag{7.30}$$

Following this procedure it is possible to obtain all terms $T_{ij}^{(e)}$ of matrix $[T^{(e)}]$ for second-order scalar interpolating basis

$$\left[T^{(e)}\right] = \frac{V^{(e)}}{420}\begin{bmatrix} 6 & 1 & 1 & 1 & -4 & -6 & -4 & -4 & -6 & -6 \\ 1 & 6 & 1 & 1 & -4 & -4 & -6 & -6 & -4 & -6 \\ 1 & 1 & 6 & 1 & -6 & -4 & -4 & -6 & -6 & -4 \\ 1 & 1 & 1 & 6 & -6 & -6 & -6 & -4 & -4 & -4 \\ -4 & -4 & -6 & -6 & 32 & 16 & 16 & 16 & 16 & 8 \\ -6 & -4 & -4 & -6 & 16 & 32 & 16 & 8 & 16 & 16 \\ -4 & -6 & -4 & -6 & 16 & 16 & 32 & 16 & 8 & 16 \\ -4 & -6 & -6 & -4 & 16 & 8 & 16 & 32 & 16 & 16 \\ -6 & -4 & -6 & -4 & 16 & 16 & 8 & 16 & 32 & 16 \\ -6 & -6 & -4 & -4 & 8 & 16 & 16 & 16 & 16 & 32 \end{bmatrix} \quad (7.31)$$

For what concerns the entries of $[S^{(e)}]$ matrix (7.13), these can be obtained considering that the partial derivatives of the basis functions with respect to the simplex coordinates can be recast as:

$$\begin{aligned} \frac{\partial \alpha_i}{\partial x} &= \sum_{k=1}^{4} \frac{\partial \alpha_i}{\partial \zeta_k}\frac{\partial \zeta_k}{\partial x} = \sum_{k=1}^{4} \frac{b_k}{6V^{(e)}}\frac{\partial \alpha_i}{\partial \zeta_k} \\ \frac{\partial \alpha_i}{\partial y} &= \sum_{k=1}^{4} \frac{\partial \alpha_i}{\partial \zeta_k}\frac{\partial \zeta_k}{\partial y} = \sum_{k=1}^{4} \frac{c_k}{6V^{(e)}}\frac{\partial \alpha_i}{\partial \zeta_k} \\ \frac{\partial \alpha_i}{\partial z} &= \sum_{k=1}^{4} \frac{\partial \alpha_i}{\partial \zeta_k}\frac{\partial \zeta_k}{\partial z} = \sum_{k=1}^{4} \frac{d_k}{6V^{(e)}}\frac{\partial \alpha_i}{\partial \zeta_k} \end{aligned} \quad (7.32)$$

Hence substituting (7.32) into (7.13) one obtains

$$\begin{aligned} S_{ij}^{(e)} &= \int_{\Omega_e} \nabla \alpha_i^{(e)} \cdot \nabla \alpha_j^{(e)} d\Omega \\ &= \frac{1}{(6V^{(e)})^2}\sum_{k=1}^{4}\sum_{h=1}^{4}(b_k b_h + c_k c_h + d_k d_h)\int_{\Omega_e}\frac{\partial \alpha_i}{\partial \zeta_k}\frac{\partial \alpha_j}{\partial \zeta_h}\, d\Omega \end{aligned} \quad (7.33)$$

The partial derivatives of basis functions can be conveniently represented by means of the basis function themselves. Being indeed α_i second-order polynomials in ζ_k it is apparent how these derivatives can be represented by an expansion over the α_j function themselves, since they span a space larger than that spanned by their derivatives.

It is possible to write

$$\frac{\partial \alpha_i}{\partial \zeta_k} = \sum_{s=1}^{10} D_{is}^{(k)} \alpha_s \qquad \text{being:} \qquad D_{is}^{(k)} = \left.\frac{\partial \alpha_i}{\partial \zeta_k}\right|_{P_s} \quad (7.34)$$

were P_s is the interpolation point of the tetrahedron where function α_s is unitary. Hence the integral in (7.33) becomes:

$$\int_{\Omega_e} \frac{\partial \alpha_i}{\partial \zeta_k} \frac{\partial \alpha_j}{\partial \zeta_h} d\Omega = \sum_{s=1}^{10} \sum_{r=1}^{10} D_{is}^{(k)} \left\{ \int_{\Omega_e} \alpha_i \alpha_j d\Omega \right\} D_{jr}^{(h)} \tag{7.35}$$

Substituting (7.35) in (7.33) the local matrix $[S^{(e)}]$ can be written in terms of the already known entries of matrix $[T^{(e)}]$, of the coefficients b_i, c_i, and d_i, and of four new matrices $[D^{(k)}]$, $k = 1, \ldots, 4$ constructed by assembling coefficients $D_{is}^{(k)}$

$$\left[S^{(e)}\right] = \frac{1}{(6V^{(e)})^2} \sum_{k=1}^{4} \sum_{h=1}^{4} (b_k b_h + c_k c_h + d_k d_h) \left[D^{(k)}\right] \left[T^{(e)}\right] \left[D^{(h)}\right]^t \tag{7.36}$$

The terms $D_{is}^{(k)}$ are named *differentiation coefficients* and $[D^{(k)}]$ are the differentiation matrices. For the sake of completeness, the form of one of the four differentiation matrices is reported here below, the other three being trivial to reconstruct

$$\left[D^{(1)}\right] = \begin{bmatrix} 3 & -1 & -1 & -1 & 1 & -1 & 1 & 1 & -1 & -1 \\ 0 & 0 & 0 & 0 & 0 & 0 & 0 & 0 & 0 & 0 \\ 0 & 0 & 0 & 0 & 0 & 0 & 0 & 0 & 0 & 0 \\ 0 & 0 & 0 & 0 & 0 & 0 & 0 & 0 & 0 & 0 \\ 0 & 4 & 0 & 0 & 2 & 2 & 0 & 0 & 2 & 0 \\ 0 & 0 & 0 & 0 & 0 & 0 & 0 & 0 & 0 & 0 \\ 0 & 0 & 4 & 0 & 0 & 2 & 2 & 0 & 0 & 2 \\ 0 & 0 & 0 & 4 & 0 & 0 & 0 & 2 & 2 & 2 \\ 0 & 0 & 0 & 0 & 0 & 0 & 0 & 0 & 0 & 0 \\ 0 & 0 & 0 & 0 & 0 & 0 & 0 & 0 & 0 & 0 \end{bmatrix} \tag{7.37}$$

Obviously, the higher the order of the element, the larger the local element matrices are. Hence, the increase in accuracy gained with higher order is paid in terms of higher memory requirements and CPU times, both bounded to the lower sparsity of the global matrix. three-dimensional scalar elements have been presented here mainly to introduce some fundamental aspects of three-dimensional finite elements, being the solution of electromagnetic three-dimensional problems better performed by resorting to the vector elements which will be the subject of the next section.

7.2.2 Vector Elements

The use of nodal elements for the discretization of vector fields in electromagnetic problems is usually not the best choice, as already pointed out for the two-dimensional case. The use of scalar bases requires the separate use of a complete set of bases for each of the 3 three-dimensional field components, hence, using a first-order element, the internal field will be discretized by 12 degrees of freedom, that is 4 for each component. At the same time, the continuity condition for the fields across the element faces must be explicitly enforced. If all three components are enforced continuously, this is too strong a condition, if only the tangent component is enforced continuously, this is the correct condition but its enforcement is awkward in a generic tetrahedral mesh. The size of the problem tends to increase due to the large number of continuity conditions between field components needed at each node of each elements in assembling the global matrix. This, as a side effect, also leads to nonconforming solutions to the electromagnetic model, also known as spurious modes.

As demonstrated in [6–8], spurious mode in three-dimensional electromagnetic vector problems are originated by the incorrect continuity condition at the element boundaries. This is equivalent to find the solution of the discretized problem in a function space which is *greater* than the function space spanned by the solution of Maxwells equations.

Hence the use of vector bases is preferable than the use of scalar components discretization. A lot of papers are present in technical literature that report particular types of bases for tetrahedron, hexahedron, or covariant elements [9–12]. Here are reported only the type of elements that are used in the application described in the following chapters and included on the CD-ROM.

An electromagnetic field **E** or **H** must satisfy, the complete set of Maxwells equations that, in a sourceless region, can be recast as single-field Helmholtz homogeneous equations, with the additional condition that either $\nabla \cdot \mathbf{E} = 0$ or $\nabla \cdot \mathbf{H} = 0$. These two partial differential equation problems are equivalent, hence only the first one will be considered here. The condition stated by the vanishing of the E-field divergence is equivalent to require that the solution of the vector Helmholtz equation must be curl-conforming or, in a more formal way, that the solution of the partial differential equation will belong to the Sobolev functions set $\mathcal{H}(\nabla\times,\Omega)$ [13], that is:

$$\mathcal{H}(\nabla\times,\Omega) = \left\{ \mathbf{E} \in \left[\mathcal{L}^2(\Omega)\right]^3 : \nabla \times \mathbf{E} \in \left[\mathcal{L}^2(\Omega)\right]^3 \right\} \tag{7.38}$$

Searching the approximated solution of the vector Helmholtz equation via finite elements, this last condition must be stated explicitly in either one of two different ways: including a separate set of conditions obtained from a weak form drawn by $\nabla \cdot \mathbf{E} = 0$ into the geometrical domain Ω, or using a set of basis functions belonging to $\mathcal{H}(\nabla\times,\Omega)$. In the latter case the curl-conforming condition is naturally imposed, and the finite element approximation of the electric field, in the Ω domain, is a linear finite summation of basis functions defined on each $\Omega^{(e)} \in \Omega$ which satisfy the curl-conforming condition. This latter option is usually preferred.

In the following subsection the vector bases most widely used, as far as the authors know, are considered and their local element matrices derived for the tetrahedral elements.

7.2.2.1 First-Order Tetrahedral Elements

Considering a tetrahedral element, the simplest, yet widely exploited, type of basis functions are the so called *Nedelec* functions [2, 7, 8, 14]. A basis function is associated to each of the tetrahedron edges, that leads to six bases and to the following field expansion within each element:

$$\mathbf{E}^{(e)} = \sum_{i=1}^{6} E_i^{(e)} \boldsymbol{\alpha}_i^{(e)} \tag{7.39}$$

where the basis function associated to the ith edge is given by:

$$\boldsymbol{\alpha}_i^{(e)} = l_i \left(\zeta_{i_1} \nabla \zeta_{i_2} - \zeta_{i_2} \nabla \zeta_{i_1} \right) \tag{7.40}$$

being l_i the length of the ith edge and i_1 and i_2 are the local number of the two nodes defining the ith edge, as indicated in Table 7.3.

It is worth to remark that the function $\nabla \zeta_j$ is a constant vector normal to the jth face and going inward. Such a property leads to important geometrical features for this kind of bases, besides the main one of curl-conformity within $\Omega^{(e)}$. By using Green's formulas:

$$\nabla \cdot \boldsymbol{\alpha}_i^{(e)} = l_i \nabla \cdot \left(\zeta_{i_1} \nabla \zeta_{i_2} - \zeta_{i_2} \nabla \zeta_{i_1} \right) = l_i \left(\zeta_{i_1} \nabla \cdot \nabla \zeta_{i_2} - \zeta_{i_2} \nabla \cdot \nabla \zeta_{i_1} \right) = 0 \tag{7.41}$$

where the vanishing value of the divergence within the entire element is due to the vanishing of the divergence for the two constant vectors $\nabla \zeta_{i_1}$ and $\nabla \zeta_{i_2}$. This feature means that the field expressed by (7.40), within the tetrahedral element, satisfies the Gauss theorem for all the possible values of the degrees of freedom, hence the generic field represented as an expansion on these bases naturally belongs to the space spanned by the solutions of Maxwells equations.

From the vanishing of ζ_i at the ith face follows another interesting feature: $\boldsymbol{\alpha}_i^{(e)}$ is orthogonal at faces $\triangle_{i_1}$ and $\triangle_{i_2}$. This implies that it is also orthogonal to all tetrahedron edges except at the ith one where it exhibits a constant tangent unitary value. On this latter edge the basis also exhibits a normal component which is linear.

Indeed one can visualize the generic $\boldsymbol{\alpha}_i^{(e)}$ easily if a cylindrical reference with the z-axis coincident with the edge opposite to the ith is considered; the edge opposite to the ith being that edge of the tetrahedron which does not share any vertex with edge i^{th}. In this reference $\boldsymbol{\alpha}_i^{(e)}$ in a generic point $P(r, \varphi, z)$ is directed along $\hat{\varphi}$ and its amplitude is proportional to r.

Figure 7.5 shows $\boldsymbol{\alpha}_1^{(e)}$. It is evident how the vector basis is normal at faces $\triangle_1$ and $\triangle_2$ and how it "curls" around edge 6, opposite to 1, presenting a constant tangential component on edge 1. This latter edge is indeed the only one exhibiting a tangential component for the basis function.

It is apparent that, by associating a degree of freedom to the value of the tangential component at the ith edge, the component of the field along that edge direction is fixed. Neighboring tetrahedra sharing that edge share also the degree of freedom and, hence, the continuity of the tangential component of the field is guaranteed, while the normal component of the field at the same edge may be discontinuous. Of course since on a given face only the bases relative to the

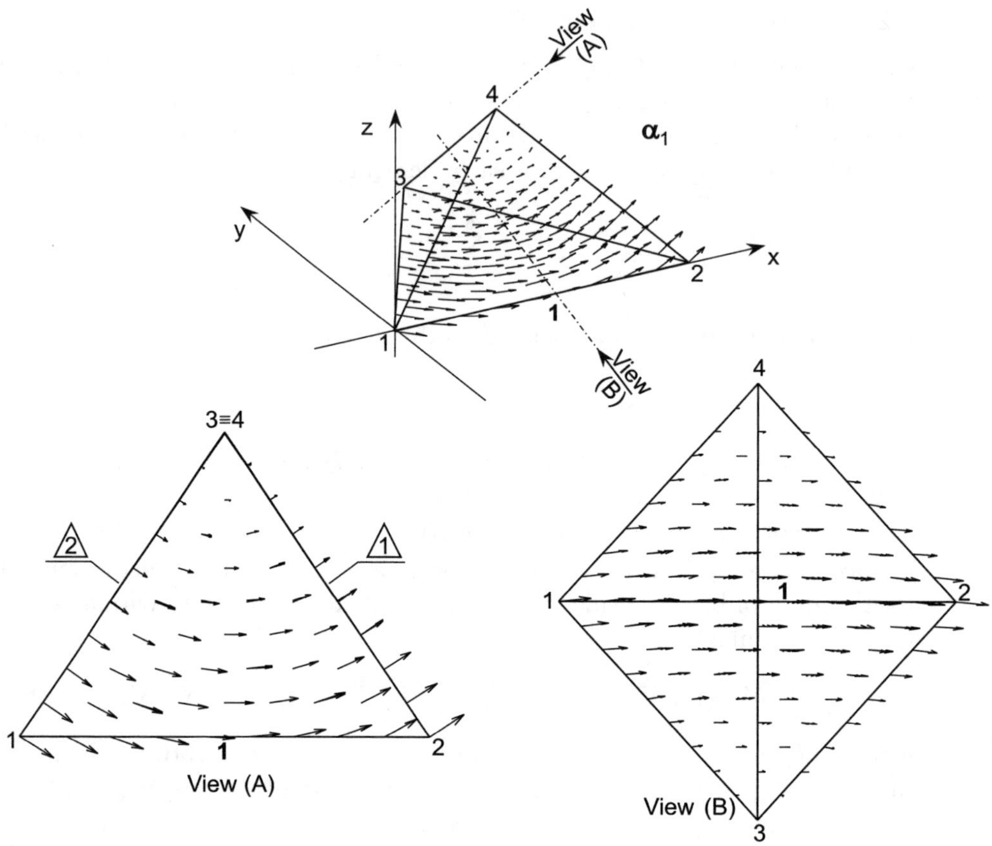

Figure 7.5 Edge first-order basis functions number 1 represented in a generic three-dimensional view (top), and on 2 views, one aligned with edge 6 (a), and one normal to both edges 1 and 6 (b).

edges of that face exhibit a nonzero tangential component on the face itself, from the continuity of the tangential components at the edges it descends naturally the continuity of the tangential component at the whole face, whereas the normal component can be discontinuous.

Applying $\boldsymbol{\alpha}_i^{(e)}$, the curl-differential operator allows us to highlight some other fundamental properties of this basis:

$$\nabla \times \boldsymbol{\alpha}_i^{(e)} = l_i \nabla \times \left(\zeta_{i_1} \nabla \zeta_{i_2} - \zeta_{i_2} \nabla \zeta_{i_1}\right) = 2 l_i \nabla \zeta_{i_1} \times \nabla \zeta_{i_2} \tag{7.42}$$

being the gradients constant, the field curl too is a constant vector within the tetrahedral element for first-order basis.

The above-mentioned properties are useful in order to compute the local element matrices defined in Chapter 2 for the two-dimensional case:

$$E_{ij}^{(e)} = \int_{\Omega_e} \nabla \times \boldsymbol{\alpha}_i^{(e)} \cdot \nabla \times \boldsymbol{\alpha}_j^{(e)} d\Omega \tag{7.43}$$

$$F_{ij}^{(e)} = \int_{\Omega_e} \boldsymbol{\alpha}_i^{(e)} \cdot \boldsymbol{\alpha}_j^{(e)} d\Omega \tag{7.44}$$

To compute (7.43), one can resort to (7.42). The integral of these constants over element volume $V^{(e)} = ||[Z^{(e)}]||/6$ is

$$E_{ij}^{(e)} = 4V^{(e)} l_i l_j \left(\nabla\zeta_{i_1} \times \nabla\zeta_{i_2}\right) \cdot \left(\nabla\zeta_{j1} \times \nabla\zeta_{j2}\right) \tag{7.45}$$

where, by applying (7.20)

$$\begin{aligned} E_{ij}^{(e)} = \frac{4 l_i l_j}{36^2 \left(V^{(e)}\right)^3} \{ & \left[b_{i_1}\hat{\mathbf{x}} + c_{i_1}\hat{\mathbf{y}} + d_{i_1}\hat{\mathbf{z}}\right] \times \left[b_{i_2}\hat{\mathbf{x}} + c_{i_2}\hat{\mathbf{y}} + d_{i_2}\hat{\mathbf{z}}\right] \\ & \left[b_{j_1}\hat{\mathbf{x}} + c_{j_1}\hat{\mathbf{y}} + d_{j_1}\hat{\mathbf{z}}\right] \times \left[b_{j_2}\hat{\mathbf{x}} + c_{j_2}\hat{\mathbf{y}} + d_{j_2}\hat{\mathbf{z}}\right] \} \end{aligned} \tag{7.46}$$

or

$$\begin{aligned} E_{ij}^{(e)} = & \frac{4 l_i l_j}{36^2 \left(V^{(e)}\right)^3} \{ \left[c_{i_1} d_{i_2} - d_{i_1} c_{i_2}\right] \left[c_{j_1} d_{j_2} - d_{j_1} c_{j_2}\right] \\ & + \left[d_{i_1} b_{i_2} - b_{i_1} d_{i_2}\right] \left[d_{j_1} b_{j_2} - b_{j_1} d_{j_2}\right] \\ & + \left[b_{i_1} c_{i_2} - c_{i_1} b_{i_2}\right] \left[b_{j_1} c_{j_2} - c_{j_1} b_{j_2}\right] \} \end{aligned} \tag{7.47}$$

The computation of $[F^{(e)}]$ matrix is performed by substituting (7.40) into (7.44)

$$\begin{aligned} \int_{\Omega_e} \boldsymbol{\alpha}_i^{(e)} \cdot \boldsymbol{\alpha}_j^{(e)} d\Omega = l_i l_j \Big[& \left(\nabla\zeta_{i_2} \cdot \nabla\zeta_{j2}\right) \int_{\Omega_e} \zeta_{i_1} \zeta_{j1} d\Omega - \left(\nabla\zeta_{i_2} \cdot \nabla\zeta_{j1}\right) \int_{\Omega_e} \zeta_{i_1} \zeta_{j2} d\Omega \\ & - \left(\nabla\zeta_{i_1} \cdot \nabla\zeta_{j2}\right) \int_{\Omega_e} \zeta_{i_2} \zeta_{j1} d\Omega + \left(\nabla\zeta_{i_1} \cdot \nabla\zeta_{j1}\right) \int_{\Omega_e} \zeta_{i_2} \zeta_{j2} d\Omega \Big] \end{aligned} \tag{7.48}$$

that is, by recognizing that the integrals are the same as those appearing in the definition of nodal scalar elements

$$\begin{aligned} F_{ij}^{(e)} = l_i l_j \Big[& \left(\nabla\zeta_{i_2} \cdot \nabla\zeta_{j2}\right) T_{i_1,j1}^{(e)} - \left(\nabla\zeta_{i_2} \cdot \nabla\zeta_{j1}\right) T_{i_1,j2}^{(e)} \\ & - \left(\nabla\zeta_{i_1} \cdot \nabla\zeta_{j2}\right) T_{i_2,j1}^{(e)} + \left(\nabla\zeta_{i_1} \cdot \nabla\zeta_{j1}\right) T_{i_2,j2}^{(e)} \Big] \end{aligned} \tag{7.49}$$

Using again (7.20) and (7.19) one obtains:

$$\begin{aligned} F_{ij}^{(e)} = \frac{l_i l_j}{720 V^{(e)}} \Big[& \left(\delta_{i_1 j_1} + 1\right) f_{i_2 j_2} \\ & - \left(\delta_{i_1 j_2} + 1\right) f_{i_2 j_1} - \left(\delta_{i_2 j_1} + 1\right) f_{i_1 j_2} + \left(\delta_{i_2 j_2} + 1\right) f_{i_1 j_1} \Big] \end{aligned} \tag{7.50}$$

where δ_{ij} is Kronecker's delta, whose value is 1 if $i = j$ and zero otherwise, and f_{ij} is a matrix accounting for the scalar products of the gradients

$$f_{ij} = b_i b_j + c_i c_j + d_i d_j \tag{7.51}$$

7.3 Global Matrices

Assembling the global matrices from the local ones is a straightforward procedure once the reader has mastered the two-dimensional case. The main difference is that, in the three-dimensional case, only local vector matrices are used, where for two-dimensional or two-and-a-half-dimensional analysis also nodal elements were exploited.

Similarly to the two-dimensional vector elements scheme, the assembling of global matrices starts from the computed local edge matrices in which each entry on column and row is related to one of the degrees of freedom of the element. As it was in two-dimensional edge elements, each edge has a conventional direction which, in three-dimensional is summarize in Table 7.3.

For each element, its local matrix is inserted in the global matrix by adding or subtracting it to the pertinent entry, which is identified by the global numbering of the element edges. The choice of adding or subtracting is linked to the relative orientation of the edge in the global and local numbering scheme. If orientations are the same the contribution is added. Otherwise it is subtracted.

From the numerical point of view, the storage of the global matrices is performed as in the two-dimensional cases, but the matrix dimension are generally larger. It is worth noticing that usually compressed band storage is less efficient for three-dimensional problems than for two-dimensional problems because three-dimensional tetrahedrons have more complex and numerous adjacency links than two-dimensional triangles. For this reasons the sparse matrix representation would be preferred over the compressed band one. The numerical implementation of this kind of data structure for three-dimensional finite elements framework is exactly the same as explained for the two-dimensional case in the Chapter 2.

7.4 Solving the Linear System of Equations

The dimension of linear systems of equations for finite elements analysis, of three-dimensional problems, increases very quickly with the size of the domain to be considered for the solution of the electromagnetic problems. A very important property of the matrices involved in such numerical problems is its high degree of sparsity, generally indicated as the ratio N_{nz}/N^2 between the number of nonzero entries in the matrix, N_{nz}, and the number of entries in the matrix which, if N is the number of degrees of freedom, is N^2.

Many techniques have been developed for the solution of linear systems of equations with a large sparse matrix.

It is beyond the scope of this book to treat these algorithms and discuss their features. Furthermore it is relatively of little use to select a single solver and embed it in a FEM code. The most flexible approach would be that of dumping the system matrix to the disc after assembling, then selecting one of the many solvers compiled as stand-alone softwares.

In this section a brief overview of existing solvers is given; there are essentially two kinds of solving methods: direct solvers, and iterative solvers, to deal with a linear system in the form

$$[A]\mathbf{x} = \mathbf{b} \tag{7.52}$$

The first kind requires fully knowing either the factorization of matrix $[A]$ or its inverse $[A]^{-1}$. The idea of performing the inverse of the matrix being of course impractical. Usage of these methods is really scarcely efficient for large sparse matrices because matrix factorization typically leads to substantial filling-up of the matrix. This deteriorates the sparsity of the factorized matrix, hence increasing dramatically the memory required for its allocation. However, very efficient direct solvers based on the multifrontal unbalanced method have been developed to partially overcome this issue and are able to efficiently factorize large sparse matrices. Among the packages implementing this method UMFPACK [15] is quite widely used in research and commercial softwares implementing finite elements. One should anyway note that obtaining a factorization of the system matrix can be very important, especially when the linear system has to be solved several times with different right-hand sides, since factorization is performed only once and the solution is then reduced to fast matrix-vector products.

The second class of solving methods are the iterative ones, that allow to solve a linear sparse systems by knowing matrices $[A]$ and $[A]^T$ only. Indeed this knowledge needs to be at a functional level, meaning that the code must be able to perform matrix-vector products, with the matrix and its transposition not necessarily effectively stored in memory, if needed matrix entries can be computed on the fly. The theoretical foundation of these methods is rooted into the theory of Krylov subspaces for which many excellent references exist, in particular [16, 17]. Also in this case there are many different algorithms to obtain the solution iteratively, and several freely available numerical packages that can be used and included in finite element software. The most basic algorithm is the conjugate gradient method (CG) that is very powerful, fast and appropriate for large, sparse, symmetric and positive definite systems. For unsymmetric sparse systems the biconjugate gradient method (BiCG) has been developed; however, also other methods like the conjugate gradient squared method (CGS), the quasiminimal residual method (QMR), and the generally minimal residual method (GMRES) are very efficient and widely used. The advantages and drawbacks of each method are explained with implementation details in many dedicated literature works, a very complete detailed treatment can be found in [17].

It is however important to remark that, when iterative methods are used, special care must be taken to the conditioning of the matrix A, measured by its condition number $\mathcal{K}_A = \|[A]\| \times \left\|[A]^{-1}\right\|$. Usually $\mathcal{K}_A$ tends to increase by increasing the matrix dimension and the accurate solution of an ill-conditioned system requires multiplying both the left- and right-hand side of the system by a suitable preconditioning matrix $[P]$

$$[P][A]\mathbf{x} = [P]\mathbf{b} \tag{7.53}$$

In order to lead to a faster converging iterative scheme the matrix $[P]$ must satisfy two semiempirical conditions:

- It must be a "good" approximation of the inverse of system matrix $[A]$, actually if it were $[P] = [A]^{-1}$, then the solution will be trivially determined by $\mathbf{x} = [P]\mathbf{b}$.
- It must be easy and fast to compute, this condition is clearly conflicting with the first one.

The preconditioning matrix for the given problem is not unique and it can be sought using different strategies. The most simple and well-known is that of using a diagonal preconditioner, obtained by inverting the diagonal elements of the $[A]$ matrix

$$P_{ij} = \begin{cases} A_{ij}^{-1} & \text{if } i = j \\ 0 & \text{if } i \neq j \end{cases} \tag{7.54}$$

Greater advantages are obtained via more sophisticated precontitioners, like the block diagonal preconditioners, where $[P]$ is a block diagonal matrix in which each of its blocks is obtained by inverting the correspondent diagonal block of the original matrix $[A]$.

The incomplete LU (ILU) preconditioner can be regarded as an improvement of the block diagonal one and generally provides excellent speed-up in convergence [17].

7.5 Disc Content

The accompanying CD-ROM is organized, as mentioned in Chapter 2, in a FORTRAN framework and in a Matlab framework. Within the FORTRAN framework there is a directory 3D-PART II holding the three-dimensional FEM codes:

7.5.1 3D-PART II

- TOOLS3D: The codes of this chapter. Basically there is just the driver routine DO_PLC2QF.F
  - LIB: The subroutines described in this chapter:
    ELEE3D.F: Edge elements, tetrahedral, first order;
    PLC2QF.F: The format converter;
    READ3DMESH.F: Subroutine to read a three-dimensional mesh file;
    WRITE3DMESH.F: Subroutine to write a three-dimensional mesh file;
    WRITE3DMM.F: Subroutine to write the port MM file;
    RENUMBER3D.F: Band optimization renumbering subroutine;
    RENUMBERPORTS3D.F: Band optimization renumbering subroutine for ports MM file.

- MATLAB: The Matlab codes for preprocessing:
  plc2qf.m: Mesh format converter with TetGen driver;
  SeeLabeledEdges.m: Draws a three-dimensional representation of the mesh, drawing only the edges of a tetrahedron which have a nonzero label.
- CAVITY: This directory contains all the programs of the following chapter.
- WDEV: This directory contains all the programs of Chapter 9.

The single subdirectories of CAVITY and WDEV will be discussed in the relative chapters, but all share the same organization:

- A top-level subdirectory containing the source code(s);
- A series of subdirectories, one for each example discussed in the text.

References

[1] Hang, S., *TetGen, A Quality Tetrahedral Mesh Generator and Three-Dimensional Delaunay Triangulator, User's Manual*, http://tetgen.berlios.de, 2006.

[2] Graglia, R.D., Wilton, D.R., and Peterson, A.F., "Higher Order Interpolatory Vector Bases for Computational Electromagnetics," *IEEE Transactions on Antennas and Propagation*, Vol. 45, pp. 329-332, 1997.

[3] Graglia, R. D., et al., "Higher Order Interpolatory Vector Bases on Prism Elements," *IEEE Transactions on Antennas and Propagation*, Vol. 46(3), pp. 442–450, March 1998.

[4] Graglia, R. D. and Gheroma, I.-L., "Higher Order Interpolatory Vector Bases on Pyramidal Elements," *IEEE Transactions on Antennas and Propagation*, Vol. 47(5), pp. 775-782, May 1999.

[5] Silvester, P. P. and Ferrari, R. L., *Finite Elements for Electrical Engineers*, Cambridge, UK: Cambridge University Press, 3rd edition, 1996.

[6] Wong, S.H. and Cendes, Z. J., "Combined Finite Element-Modal Solution of Three-Dimensional Eddy Current Problems," *IEEE Trans. Magnetics*, Vol. 24(6), pp. 2685–2687, Nov. 1988.

[7] Bossavit, A., "Whitney Forms: A Class of Finite Elements for Three-Dimensional Computations in Electromagnetics," *IEEE Proceedings*, Vol. 135(A)(8), pp. 493–500, Nov. 1988.

[8] Bossavit, A., "Solving Maxwell's Equations in a Closed Cavity and the Question of Spurious Modes," *IEEE Transaction on Magnetics*, Vol. 26(2), pp. 702–705, 1990.

[9] van Welij, J. S., "Calculation of Eddy Currents in Terms of H on Hexahedra," *IEEE Transaction on Magnetics*, Vol. 21, pp. 2239-2241, 1985.

[10] Özdemir, T. and Volakis, J. L., "Triangular Prisms for Edge-Based Vector Finite Element Analysis of Conformal Antennas," *IEEE Transaction on Antennas and Propagation*, Vol. 45(5), pp. 788-797, May 1997.

[11] Jin, J . M. and Volakis, J. L., "Electromagnetic Scattering by and Transmission Through a Three Dimensional Slot in a Thick Conducting Plane," *IEEE Transaction on Antennas and Propagation*, Vol. 39(4), pp. 543–550, 1991.

[12] Jin, J. M. and Volakis, J. L., "Scattering and Radiation from Microstrip Patch Antennas and Arrays Residing in a Cavity," *IEEE Transaction on Antennas and Propagation*, Vol. 39, pp. 1598–1604, 1991.

[13] Volakis, J. L, Kempel, L. C., Chatterjee, A., "*Finite Element Method for Electromagnetics*," IEEE Press, 1998.

[14] Savage, J. S. and Peterson, A. F., "Higher-Order Vector Finite Elements for Tetrahedral Cells," *IEEE Transactions on Microwave Theory and Techniques*, Vol. 44(6), pp. 874–879, June 1996.

[15] Davis, T. A. and Duff, I. S., "A Combined Unifrontal/Multifrontal Method for Unsymmetric Sparse Matrices," *ACM Transactions on Mathematical Software*, Vol. 25(1), pp. 1–19, 1999.

[16] van der Vorst, H., "*Iterative Krylov Methods for Large Linear Systems*," Cambridge, U.K.: Cambridge University Press, 2003.

[17] Saad, Y., "*Iterative Methods for Sparse Linear Systems*," SIAM, 2nd edition, 2000.

CHAPTER 8

Resonant Cavities

The first and most simple three-dimensional application of finite elements for wave electromagnetics is the quest for a metallic cavity, possibly inhomogeneously filled, resonant modes.

This is naturally a closed and sourceless electromagnetic problem, considering that the cavity is a region in three-dimensional space completely enclosed in metallic walls. The electromagnetic problem, as will be shown below, has to be solved in this closed domain without any electric nor magnetic field sources. The resulting electric and magnetic field solution constitutes an infinite set of modes existing at discrete frequency points. From the mathematical point of view, solving an electromagnetic cavity problem is equivalent to finding the set of bases for the kernel of the differential operator represented by Maxwells equations with suitable boundary conditions. This kind of problem, akin but somewhat simpler than the one treated in Chapter 3, thanks to the fully three-dimensional framework, will be introduced and discussed in this chapter.

Indeed the FEM solution of cavity problems is quite old, starting from early solutions which were three-dimensional extensions of two-dimensional problem, thanks to rotational symmetry [1] up to true three-dimensional solutions [2], where hexaedral nodal elements are exploited in a penalty framework, and to [3, 4] where the same approach is extended to dielectrically loaded cavities. Subsequent papers dealt with divergence-free elements [5, 6] or exploited a potential-based approach [7] and, finally, vector elements [8, 9] to avoid spurious solutions. Indeed the cavity problem has been widely used as a benchmark to test newly developed element performances, as in [10–16]. Finite element books investigating this problem are, among others [17, 18].

8.1 Formulation of the Three-Dimensional Eigenvalue Problem

From the theoretical point of view, a cavity is a closed region of space bounded by impenetrable walls. This implies that the electromagnetic field inside the cavity is completely unrelated to whatever happens outside the cavity. Furthermore, there are no sources either electric or magnetic inside the cavity. Such a theoretical device can be obtained by enclosing the cavity volume with a perfect conducting material either electric (p.e.c.) or magnetic (p.m.c.). The internal region can include inhomogeneous dielectrics as well as p.e.c. or p.m.c. regions, as sketched in Figure 8.1. As it will be shown later the solution for the field within the cavity constitutes an infinite but discrete set of frequencies at which a nonnull field distribution within the cavity is possible. These frequencies are called resonant frequencies and the corresponding field distributions are called resonant modes.

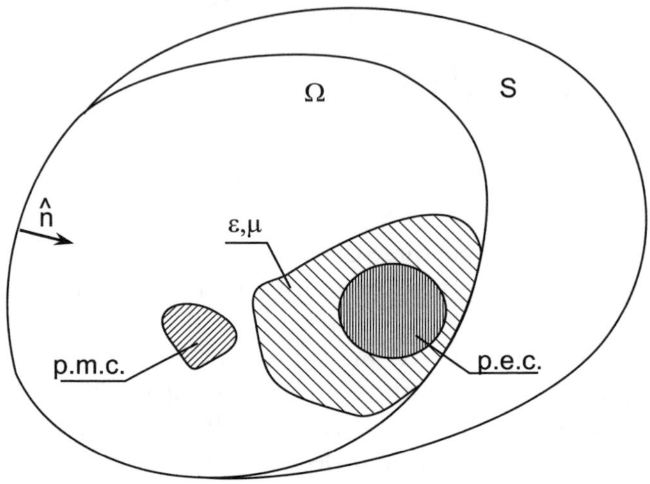

Figure 8.1 Example of inhomogeneous cavity spanning domain Ω of boundary *S* and with inhomogeneities.

It is evident that a similar device would have little if no practical interest, being the field within it completely isolated from the external environment, and hence no possible mutual influence between the internal and external field is possible. Practical devices are coupled to the exterior via some hole, expressly made small so as to perturb as little as possible the resonant fields and frequencies, hence the study of a closed cavity is of practical interest. Indeed, resonators in general, and cavity resonators in particular, have many important applications in microwave networks where the cavity characteristic of exhibiting only a discrete set of resonant frequencies is exploited. Cavities can provide frequency references, while sets of interconnected cavities resonating at slightly different frequencies can implement a waveguide passband filter [19–23] or can be exploited to measure electromagnetic properties of materials [24–26].

The volume encompassed by the cavity will be denoted Ω. Its boundary will be denoted S and its inward looking normal $\hat{\mathbf{n}}$. Within the cavity the field distribution is given by the solution of a partial differential homogeneous operator, while the field on the boundary S must satisfy a set of Dirichlet and/or Neumann homogeneous boundary conditions. In particular, the two-curl Maxwells equations are sufficient in order to find electric and magnetic fields within Ω:

$$\begin{cases} \nabla \times \mathbf{E} = -j\omega\mu\mathbf{H} \\ \nabla \times \mathbf{H} = j\omega\varepsilon\mathbf{E} \end{cases} \tag{8.1}$$

In (8.1) field sources of both electric, **J**, and magnetic, **M**, type are expressly disregarded. Furthermore the internal dielectric materials are considered to be linear and isotropic, which leads to permittivities and permeabilities which are, in presence of losses, frequency dependent scalar complex values. In this case it is convenient to define relative permittivity and permeability as ε_r and μ_r so that it is: $\varepsilon = \varepsilon_0\varepsilon_r = \varepsilon_0(\varepsilon_1 - j\varepsilon_2)$ and $\mu = \mu_0\mu_r = \mu_0(\mu_1 - j\mu_2)$, being ε_0 and μ_0 the permittivity and permeability of the free space. Then (8.1) can be recast as:

$$\begin{cases} \dfrac{1}{\mu_r}\nabla \times \mathbf{E} = -j\omega\mu_0\mathbf{H} \\ \dfrac{1}{\varepsilon_r}\nabla \times \mathbf{H} = j\omega\varepsilon_0\mathbf{E} \end{cases} \tag{8.2}$$

After some straightforward calculations it is possible to recast the system of two first-order partial differential equations (8.2) as a single second-order partial differential equation:

$$\nabla \times \frac{1}{\mu_r}\nabla \times \mathbf{E} - k_0^2\varepsilon_r\mathbf{E} = 0 \tag{8.3}$$

being $k_0^2 = \omega^2\varepsilon_0\mu_0$ the free-space wavenumber.

The Helmholtz equation (8.3) can be obtained also in terms of the magnetic field **H**. Strictly speaking the two forms of the Helmholtz equation are equivalent, but since (8.3) presents the curl of a term containing $1/\mu_r$ it is preferable for problems where μ_r is homogeneous, whereas the magnetic-field Helmholtz equation is dually preferable where ε_r is homogeneous.

In any case, having the two equations in the same mathematical form, the analysis will be carried out on (8.3). On the boundary S of domain Ω, and on internal p.e.c. or p.m.c., if any, suitable boundary conditions are to be imposed. As already seen in two-dimensional the conditions for a perfect electric conductor boundary S^{pec} are:

$$\hat{\mathbf{n}} \times \mathbf{E} = 0; \qquad \text{or} \qquad \hat{\mathbf{n}} \times \frac{1}{\varepsilon_r}\nabla \times \mathbf{H} = 0 \qquad \text{on } S^{pec} \tag{8.4}$$

For a perfect magnetic conductor sheet S^{pmc} the condition to satisfy is given on the surface tangent magnetic field component:

$$\hat{\mathbf{n}} \times \frac{1}{\mu_r}\nabla \times \mathbf{E} = 0; \qquad \text{or} \qquad \hat{\mathbf{n}} \times \mathbf{H} = 0 \qquad \text{on } S^{pmc} \tag{8.5}$$

More complex and general nonpenetrable sheets S^c may be defined in terms of the linear homogeneous Robin relation:

$$\alpha\hat{\mathbf{n}} \times \mathbf{E} + \beta\hat{\mathbf{n}} \times \frac{1}{\mu_r}\nabla \times \mathbf{E} = 0 \qquad \text{on } S^c \tag{8.6}$$

where the two constants α and β are determined to accomplish physical properties of the boundary surface itself. Another very useful nonpenetrable boundary condition is that holding on a resistive sheets, also known as Leontovich boundary condition. This is a relation between the tangential components of the electric and magnetic fields:

$$\hat{\mathbf{n}} \times \hat{\mathbf{n}} \times \mathbf{E} = Z_s\hat{\mathbf{n}} \times \mathbf{H} \qquad \text{on } S^r \tag{8.7}$$

being Z_s the surface impedance of the sheet S^r. In this case the cavity is not isolated in a strict sense, inasmuch as there is a net flux of power flowing out through the surface, but since the fields are confined anyway within the cavity itself, this latter can still be considered isolated.

The set of (8.3) and the boundary conditions (8.4) to (8.7) on the surface S represent the eigenvalue problem. Finite elements are applied to this problem via the Galerkin procedure already described in previous chapters. If $\mathcal{F}$ is the space of C^1 vector functions defined on Ω satisfying the prescribed boundary conditions on S, then the possible solutions of (8.3) defines a subset $\mathcal{K}$ of $\mathcal{F}$ which is known as the kernel of the homogeneous operator (8.3). Once an inner product is defined over $\mathcal{F}$, test functions $\mathbf{W}$ are chosen in this same function space.

The weak formulation of the eigenvalue problem is obtained projecting the operator (8.3) on the $\mathcal{F}$ space:

$$\int_{\Omega} \mathbf{W} \cdot \nabla \times \frac{1}{\mu_r} \nabla \times \mathbf{E} \, dV - k_0^2 \int_{\Omega} \mathbf{W} \cdot \varepsilon_r \mathbf{E} \, dV = 0 \tag{8.8}$$

By applying Green's theorem the previous equation can be recasted as:

$$\begin{aligned} &\int_{\Omega} (\nabla \times \mathbf{W}) \cdot \left(\frac{1}{\mu_r} \nabla \times \mathbf{E} \right) dV - k_0^2 \int_{\Omega} \mathbf{W} \cdot \varepsilon_r \mathbf{E} \, dV \\ &\quad + \int_{S} \mathbf{W} \cdot \left(\hat{\mathbf{n}} \times \frac{1}{\mu_r} \nabla \times \mathbf{E} \right) dS = 0 \end{aligned} \tag{8.9}$$

The weak form equation of the problem is hence constituted by three terms: the first two are related to the internal domain field distribution; while the third takes into account the boundary conditions, therefore if the boundary conditions of the original electromagnetic problem are defined on subsets of the total closed boundary surface, on each part the appropriate condition is to be enforced separately. For perfect magnetic boundaries on which the electric field have to satisfy condition (8.5), the third term in (8.9) vanishes, therefore they are also called natural conditions for such kinds of problems.

More complex is the case of p.e.c. boundaries. In this case the tangential component of the electric field must vanish, but since the weight functions $\mathbf{W}$ belong to the $\mathcal{F}$ space, weights are vanishing at p.e.c. boundaries, so they are unable to correctly test the unknown field. For this reason a set of auxiliary conditions are to be explicitly imposed on S^{pec} to guarantee that the electric field expansion there will have a vanishing tangential component:

$$\int_{S^{pec}} \bar{\mathbf{W}} \cdot (\hat{\mathbf{n}} \times \mathbf{E}) \, dS = 0 \tag{8.10}$$

being the $\bar{\mathbf{W}}$ appropriate weights defined on S^{pec}.

For the sake of completeness the boundary conditions (8.6) and (8.7) become, in weak terms

$$-\int_{S^c} \frac{\alpha}{\beta} \mathbf{W} \cdot (\hat{\mathbf{n}} \times \mathbf{E}) \, dS \tag{8.11}$$

and

$$jk_0 \zeta_0 \int_{S^r} \frac{1}{Z_s} (\hat{\mathbf{n}} \times \mathbf{W}) \cdot (\hat{\mathbf{n}} \times \mathbf{E}) \, dS \tag{8.12}$$

leading to the complete weak operator

$$\int_{\Omega} (\nabla \times \mathbf{W}) \cdot \left(\frac{1}{\mu_r} \nabla \times \mathbf{E} \right) dV - k_0^2 \int_{\Omega} \mathbf{W} \cdot \varepsilon_r \mathbf{E} \, dV - \int_{S^c} \frac{\alpha}{\beta} \mathbf{W} \cdot (\hat{\mathbf{n}} \times \mathbf{E}) \, dS \\ + jk_0 \zeta_0 \int_{S^r} \frac{1}{Z_s} (\hat{\mathbf{n}} \times \mathbf{W}) \cdot (\hat{\mathbf{n}} \times \mathbf{E}) \, dS = 0 \tag{8.13}$$

The code included in the CD-ROM deals only with cavity with p.e.c. or p.m.c. boundaries, hence (8.13) simplifies to

$$\begin{cases} \int_{\Omega} (\nabla \times \mathbf{W}) \cdot \left(\frac{1}{\mu_r} \nabla \times \mathbf{E} \right) dV - k_0^2 \int_{\Omega} \mathbf{W} \cdot \varepsilon_r \mathbf{E} \, dV = 0 \\ \int_{S^{pec}} \mathbf{W}' \cdot (\hat{\mathbf{n}} \times \mathbf{E}) \, dS = 0 \end{cases} \tag{8.14}$$

To compute with the solution of the Galerkin weak form of the original electromagnetic problem in a finite element framework, a mesh of the domain must be considered. A tetrahedral mesh is usually preferred. If Ω_e is a generic tetrahedron of the mesh and the Nedelec basis functions are chosen to expand the field, then

$$\mathbf{E}^{(e)} = \sum_{j=1}^{N_e} e_j^{(e)} \, \boldsymbol{\alpha}_j^{(e)} \tag{8.15}$$

being N_e the number of bases $\boldsymbol{\alpha}_j^{(e)}$ pertaining to the single element and $e_j^{(e)}$ the unknown expansion coefficients. Here e is used in place of E in order to avoid possible confusion with the elements of matrix $[E]$. The weak formulation within the element is as usual obtained by using, as weights $\mathbf{W}$ the same bases $\boldsymbol{\alpha}_j^{(e)}$, which of course belong to $\mathcal{F}$:

$$\sum_{j=1}^{N_e} \frac{1}{\mu_r^{(e)}} \left(\int_{\Omega_e} \nabla \times \boldsymbol{\alpha}_i^{(e)} \cdot \nabla \times \boldsymbol{\alpha}_j^{(e)} \, dV \right) e_j^{(e)} \\ - k_0^2 \sum_{j=1}^{N_e} \varepsilon_r^{(e)} \left(\int_{\Omega} \boldsymbol{\alpha}_i^{(e)} \cdot \boldsymbol{\alpha}_j^{(e)} \, dV \right) e_j^{(e)} = 0 \tag{8.16}$$

where $j = 1, \ldots, N_e$. Within each element permittivity and permeability are considered constant. This allows carrying them out of the integrals and reverting to the element matrices already defined in Chapter 7:

$$E_{ij}^{(e)} = \int_{\Omega_e} \nabla \times \boldsymbol{\alpha}_i^{(e)} \cdot \nabla \times \boldsymbol{\alpha}_j^{(e)} \, dV \tag{8.17}$$

and

$$F_{ij}^{(e)} = \int_{\Omega_e} \boldsymbol{\alpha}_i^{(e)} \cdot \boldsymbol{\alpha}_j^{(e)} \, dV \tag{8.18}$$

The weak form within the element can hence be written in matrix form as

$$\frac{1}{\mu_r^{(e)}} \left[E^{(e)} \right] \left[e^{(e)} \right] - k_0^2 \varepsilon_r^{(e)} \left[F^{(e)} \right] \left[e^{(e)} \right] = 0 \tag{8.19}$$

being $\left[e^{(e)}\right]$ the vector of the unknown expansion coefficients in the element (e). In three-dimensional problems the case of anisotropic dielectric or magnetic media is of relevant interest. In this case permittivities and permeabilities are described by tensors $[\varepsilon]$ and $[\mu]$, which are 3×3 matrices. In this case it is not possible to extract these quantities from within the integrals and hence material tensors must be taken into account when computing $\left[E^{(e)}\right]$ and $\left[F^{(e)}\right]$ matrices.

By assembling all the local matrices, a global generalized eigenvalue problem is obtained, whose unknowns are all the electric field expansion coefficients which are internal to Ω domain or on the p.m.c, type boundaries. These will be arranged in vector $[e]_I$. On the other hand, coefficients relative to edges lying on p.e.c surfaces can be arranged in a $[e]_B$ vector and these must be explicitly set to zero. By assuming that the $\mathbf{W}'$ weights are Dirac's functions centered on the edge midpoint, integral (8.10) reduces to the expansion coefficient itself and hence the problems complete form is:

$$\begin{bmatrix} 0 & 0 \\ 0 & [E] \end{bmatrix} \begin{bmatrix} [e]_B \\ [e]_I \end{bmatrix} = k_0^2 \begin{bmatrix} [I] & 0 \\ 0 & [F] \end{bmatrix} \begin{bmatrix} [e]_B \\ [e]_I \end{bmatrix} \tag{8.20}$$

where $[E]$ and $[F]$ are the global matrices obtained assembling all the $[E^{(e)}]$ and $[F^{(e)}]$, respectively, for all the internal coefficients of the expansion, while $[I]$ is the identity matrix having a dimension equal to the number of unknowns, or degrees of freedom (DoF), on the p.e.c surfaces.

The matrix (8.20) represents a linear algebra generalized eigenvalue problem, whose solution always exists, since it can be proved that matrix $[F]$, and hence the whole right-hand side matrix, is hermitian and nonsingular. From the theory of eigenvalue linear problems follows that (8.20) has at least as many vanishing eigenvalues as there are unknowns associated to edges lying on the p.e.c. (equal to the dimension of vector $[e]_B$). It is also evident that only the solution of the sub-problem concerning internal degrees of freedom will lead to as many nonvanishing eigenvalues as there are internal unknowns. The generalized problem to be solved is then:

$$[E][e]_I = k_0^2[F][e]_I \tag{8.21}$$

The numerical eigenvalues k_0^2 are only an approximation of the infinite discrete set of eigenvalues of the original electromagnetic problem, but it is clear that, by increasing the number of degrees of freedom in the mesh, the calculated eigenvalues become more accurate and also their numbers increase including a greater number of higher and higher frequency resonant modes. Furthermore, since the lower resonant frequencies are relevant to larger wavelengths, the cavity mesh is much finer, in terms of wavelengths, for the lower modes than for the higher. Numerically computed lower modes are hence comparatively more accurate than higher modes, in a given mesh.

Since the matrices $[E]$ and $[F]$ are hermitian then the eigenvalues k_0^2 of the generalized eigenvalue problem are real numbers, for the linear operators spectral theorem [27]. Then solution of the numerical problem (8.21) allows to compute the real-valued eigenvalues for the given cavity problem. As explained in the chapter dedicated to waveguide modes, the eigenvalue magnitude is strictly related to the characteristics of the resonant mode itself: the minimum, physical, eigenvalue $\left(k_0^{(1)}\right)^2$ represents the first cavity mode relative to the frequency $f^{(1)}$, and, in general, to the ith eigenvalue $\left(k_0^{(i)}\right)^2$ corresponds the frequency $f^{(i)}$ according to the relation:

$$f^{(i)} = \frac{c_0 k_0^{(i)}}{2\pi} \tag{8.22}$$

being c_0 the light speed in vacuum.

The eigenvector $[e]_I^{(i)}$ relative to $\left(k_0^{(i)}\right)^2$ represents the numerical approximation of the field within the cavity or equivalently the field of the ith resonant mode.

It is important to stress that even if spurious modes are avoided by the vector formulation, there might exist several null or nearly null eigenvalues. These latter being eigenvalues which should be null, theoretically, but are nonnull, numerically, due to the finite machine precision. All these null and nearly-null eigenvalues are nonphysical and must be discarded. The strong point of edge elements being that no nonphysical eigenvalue is significantly nonnull and hence mixed with real physical eigenvalues, as it is the case with nodal elements where, to distinguish physical and nonphysical eigenvalue the corresponding eigenvectors, and relative field distribution, must be analyzed.

8.2 Numerical Implementation

The algorithm described in the previous section has been implemented using first-order Nedelec tetrahedral elements, as explained in Chapter 7.

The assembling of the local matrices $\left[E^{(e)}\right]$ and $\left[F^{(e)}\right]$ into a pair of global matrices $[E]$ and $[F]$ is performed according to the same rules outlined in Chapters 2 and 3 for triangular edge elements. Entry $E_{i,j}^{(e)}$ in local matrix $\left[E^{(e)}\right]$ is added to entry E_{i_g,j_g}, being i and j the edge numbers in the local numbering scheme of the tetrahedron and i_g and j_g the corresponding edge numbers in the mesh global numbering scheme. Of course edge direction is to be taken into account when passing from local to global numbering, as it was outlined in Chapters 2 and 3, by subtracting, rather than adding, contributions if the local and global edge directions do not match. Similar assembly rules hold for $[F]$.

For the sake of simplicity, the code presented here can deal only with isotropic lossless dielectrics. This leads to real and symmetric matrices and hence specialized solvers can be used. All the routines are written in FORTRAN code in order to guarantee high speed both in building up the matrices and in solving the generalized eigenvalue problem.

The assembled matrices are stored in full-matrix form, in order to simplify the code and make it easier to understand. For this reason the numerical solution is performed via eigensolvers that do not take advantage of the sparsity of the global matrices. The reader can improve the proposed code in order to exploit the sparsity and to increase the performance both in term of stored memory and computing speed, or to include anisotropic or lossy media with ease.

In particular Netlib[1] contains, besides LAPACK (already largely used in this book and even included in the accompanying CD-ROM), several routines for sparse matrices, among which `lanz`, which solves large sparse generalized eigenproblems and which can work on matrices stored in files. This allows for a very easy integration in FEM codes since the user can resort to any memorization scheme within the FEM code. Then, by integrating a suitable output routine writing the matrices in the appropriate format, the integration with `lanz` is completed.

In the provided code the numerical solution of (8.21), since matrices are real and symmetric, and both are nonsingular, is performed by an optimized `LAPACK` routines. The routine, `SGGEV`, accepts as inputs two matrices $[A]$ and $[B]$, and solves

$$[A][\bar{x}] = \lambda [B][\bar{x}] \tag{8.23}$$

by reducing it to the standard eigenvalue problem:

$$[C][\bar{y}] = \lambda [\bar{y}] \tag{8.24}$$

being $[\bar{y}] = [L^*][\bar{x}]$ and $[C] = [L]^{-1}[A][L^*]^{-1}$ and, finally, being $[B] = [L][L^*]$ the Cholesky hermitian factorization of matrix $[B]$. Such a factorization is possible only for hermitian positive definite matrix.

After the main processing the computed eigenvalues need to be sorted by magnitude, the same sorting is performed on the corresponding eigenvectors, in case postprocessing facilities are available. The null eigenvalues which are present are relative to eigenvectors which span the kernel subspace of the $\nabla \times \mathbf{E}$ operator, these can simply be discarded, since they correspond to nonphysical resonance modes, also known as spurious modes, as explained in the previous chapter.

8.3 The Code Cavity

The cavity code structure is easy to describe and it implements the already analyzed basic steps of an FEM code. Its required inputs are of course a mesh file, in the standard format described in the previous chapter and several auxiliary data included in the auxiliary input file.

The mesh can be defined in any unit, since the auxiliary file allows for anisotropic scaling of the geometry. It must be noted that the corresponding eigenvalues will be given in the inverse of the same unit. That is, if the cavity dimensions

1. Netlib: A repository of mathematical software, data, documents, address lists, and other useful items—http://www.netlib.no/

are in meters, $k_0^{(i)}$ will be m^{-1}, if the cavity dimensions are in millimeters, $k_0^{(i)}$ will be mm^{-1}, and so on. The first three numbers in the first line of the auxiliary input file (Example 8.1) allow for a separate scaling along the three coordinate directions x, y, and z of the mesh. This feature is particularly useful when seeking the location of a resonant mode as a function of the shape factor of a cavity, that is, the determination of resonance as cavity dimensions vary with a given law. The benefit is that of avoiding remeshing—which is a long and tedious procedure—and being able to perform several simulations on similar but different geometries with just one mesh. Of course rescaling should not produce too thin and slender tetrahedrons to maintain the accuracy of the FEM technique.

Example 8.1

```
1.0,1.0,1.0              !# Scaling factors sx,sy,sz
1                        !# Neglect border labels (label=-1) [0=no,1=yes]
2                        !# Number of different labels
2.0, 4.1                 !# Er1
100, 101                 !# Element label
```

The second row of the auxiliary input file states whether the code should not associate an unknown to certain edges. Metallic edges should exhibit a null tangent component of the electric field. This was enforced in some of the two-dimensional codes by setting the corresponding row and column of the global matrix to 0 and by setting 1 on the diagonal. By also setting to 0 the corresponding term in the right-hand side vector of the solving system we are guaranteed that the given unknown will vanish. This is of course suboptimal since the system exhibits many useless rows and columns, and is particularly true in three-dimensional. Hence, this procedure is not applied and, to optimize the code, edges are internally renumbered so as to eliminate those boundary edges on which no unknown is present. These edges are, by default, those with label -1. This means that, if the second line of the auxiliary input file holds a 1 then the edges with -1 label are discarded, and the cavity is p.e.c. bounded. If it is set to 0 then the edges are considered and we can think of the cavity as bounded by p.m.c. walls.

The following lines contain data regarding the physical materials within the analysis domain. First the number of dielectric materials besides free space is given. Then two lines presenting, respectively, the list of ε_r values and the corresponding list of element labels characterizing elements in the given material. In Example 8.1 there are two different dielectric materials plus empty space (which, by default, has label 0) and their dielectric constants and associated labels are `(2.0,100)` and `(4.1,101)`, respectively.

Once the files are read the code performs scaling and the renumbering of the edges if needed. Then the global matrices are assembled from the local ones. Finally the solution of the generalized eigenproblem is revealed in the routine `SGGEV` that is actually a driver routine able to solve more general problems in which the matrices `A` and `B` are unsymmetric and complex, or in other words in presence of anisotropic and lossy materials. The computed eigenvalues are stored in `ALPHAR`, `ALPHAI`, and `BETA` array which are the real part and the imaginary part of the numerator and the denominator of the generalized eigenvalues of the problem. The eigenvalues can then be obtained as (ALPHAR[i] + $\sqrt{-1}$ALPHAI[i])/BETA[i] and hence it is:

$$k_0^{(i)} = \sqrt{\frac{\text{ALPHAR}[i] + \sqrt{-1}\text{ALPHAI}[i]}{\text{BETA}[i]}} \tag{8.25}$$

of course, since all eigenvalues are real, vector `ALPHAI` is zero. The array of eigenvalues is finally ordered and zero or near-to-zero eigenvalues are discarded. Results are written to two files: (1) file `eig_raw.out` contains the three vectors `ALPHAR`, `ALPHAI`, and `BETA`, mainly for debugging purposes, while `eig_sorted.out` contains the eigenvalues after applying (8.25), sorting and eliminating nonphysical ones.

It is important to remark that, by exploiting a vector formulation, spurious modes are avoided. This means that there are no unphysical field configurations with a resonant frequency value which can be mistaken with the frequency value of a physical mode, but there does exist zero-frequency, hence obviously unphysical, solutions and, due to numerical errors, these frequencies might be small but nonzero.

By examining the values of `ALPHAR`, `ALPHAI`, and `BETA` it appears that `ALPHAI` is null, due to the problem symmetry which leads only to real eigenvalues, and that all the entries in `BETA` are of comparable order of magnitude. On the other hand, some entries in `ALPHAR` are very close to zero. The problem is that the square root in (8.25) smooths out this difference and makes the distinction between numerically nonzero and physically nonzero eigenvalue less evident, even if still clear.

For the reader's convenience, an automatic procedure discriminating nonphysical eigenvalues is implemented. The code computes the average value $\overline{\text{ALPHAR}}$ of all ALPHAR entries and then assumes exactly equal to zero all those entries for which ALPHAR[i] $< \overline{\text{ALPHAR}} \times 10^{-5}$. The reader can check the correctness of this assumption by analyzing the `eig_raw.out` file.

8.4 Code Cavity: Some Examples

The first test case is a simple rectangular cavity, enclosed by p.e.c. walls, with dimensions $a \times b \times c$ along the x, y, and z axis, respectively (Figure 8.2).

The resonant modes eigenvalues of this kind of closed cavity are well-known and given by the analytic expression:

$$k_{mnp} = \sqrt{\left(\frac{m\pi}{a}\right)^2 + \left(\frac{n\pi}{b}\right)^2 + \left(\frac{p\pi}{c}\right)^2} \tag{8.26}$$

The actual dimensions of the cavity are $a = 10$ mm, $b = 7$ mm, and $c = 12$ mm. The results attained via the `CAVITY` code are compared with analytical solutions from (8.26). A series of separate simulations is performed using different meshes for the same cavity, with an increasing number of elements, and hence of unknowns or, which is the same, degrees of freedom. The results obtained are reported in Table 8.1. As is shown, the calculated eigenvalues are quite accurate also in case

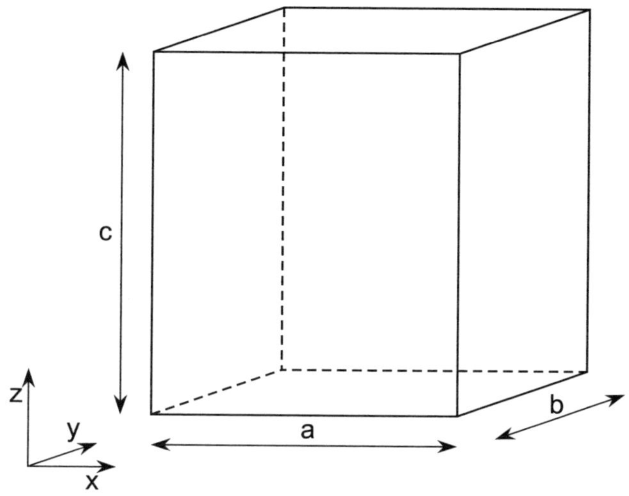

Figure 8.2 Homegeneus rectangular cavity with p.e.c boundary with size $a \times b \times c$.

of relatively poor discretizations, and the accuracy tends to increase as the number of degrees of freedom used for the analysis increases. The relative error is here, and in the following, computed as

$$\Delta f = 100 \frac{f_{\text{FEM}} - f_{\text{Analytic}}}{f_{\text{Analytic}}} \tag{8.27}$$

To better grasp the relation between the accuracy of the geometrical discretization and the accuracy obtained on the FEM computed eingenvalues the first eigenvalue is analyzed more in detail. Additional sets of simulations are performed on the cavity increasing the number of unknowns from 120 to 3,275. The subdirectory `RECT` of the `CAVITY` tree holds all these meshes, each of which is named

Table 8.1 Eigenmodes Comparison Between Analytical and Finite Elements Solutions

Mode Index		*Analytical*		*FEM (717 DoF)*			*FEM (2269 DoF)*		
i	mnp	f_i	$k_i\ (m^{-1})$	f_i	$k_i\ (m^{-1})$	Δf	f_i	$k_i\ (m^{-1})$	Δf
1	101	19.51 GHz	408.9	19.39 GHz	406.3	0.64%	19.43 GHz	407.3	0.39%
2	011	24.79 GHz	519.6	24.49 GHz	513.2	1.23%	24.63 GHz	516.1	0.67%
3	110	26.14 GHz	547.8	25.91 GHz	543.1	0.86%	25.96 GHz	544.0	0.69%
4	111	28.97 GHz	607.2	28.60 GHz	599.5	1.27%	28.73 GHz	602.1	0.84%
5	102	29.13 GHz	610.6	28.79 GHz	603.4	1.18%	28.78 GHz	603.1	1.23%
6	201	29.98 GHz	628.3	28.90 GHz	605.7	3.60%	28.82 GHz	604.1	3.85%
7	012	32.48 GHz	680.7	31.80 GHz	666.4	2.10%	32.06 GHz	671.9	1.29%
8	112	32.90 GHz	689.6	32.22 GHz	675.2	2.09%	32.46 GHz	680.3	1.35%
9	210	36.16 GHz	757.8	35.26 GHz	738.9	2.49%	35.59 GHz	745.8	1.58%
10	211	36.84 GHz	772.1	35.75 GHz	749.3	2.95%	35.84 GHz	751.1	2.72%

rect_####.fem, #### being the number of the pertinent degrees of freedom. The obtained first resonant frequency is compared with the analytical solution and the relative error computed. This error is plotted as a function of the number of degrees of freedom in Figure 8.3. It is evident the convergence of the first eigenvalue to the analytical solution.

As a second example the same rectangular cavity is partially filled with a dielectric slab exhibiting a permittivity ε_r placed parallel to the xz plane, lying on the cavity wall and of thickness t. The geometry is shown in Figure 8.4.

It is worth noticing that an approximate analytic solution still holds in the present case, since the cavity is a segment of $a \times b$ loaded waveguide of length c shortened at both ends. In the following it will be assumed, as in the previous example, $c > a > b$. In such a cavity two separate relations hold in each of the homogeneous regions. It is impossible in this problem to define TE and TM modes with respect to z but it is possible to do so with respect to y. In particular for TM_y mode it is $k_x = \frac{\pi}{a}$ but k_y is not zero, since the waveguide propagation constant must of course be unique but the two mediums propagation constants are different. The two relations are [28]:

$$k_0^2 = \omega^2 \varepsilon_0 \mu_0 = k_z^2 + \left(\frac{\pi}{a}\right)^2 + k_{y_0}^2 \tag{8.28}$$

$$k_1^2 = \omega^2 \varepsilon_1 \mu_1 = k_z^2 + \left(\frac{\pi}{a}\right)^2 + k_{y_1}^2 \tag{8.29}$$

By imposing the field continuity at the interface the following additional relation holds:

$$\frac{k_{y_1}}{\varepsilon_1} \tan(k_{y_1} t) = \frac{k_{y_0}}{\varepsilon_0} \tan[k_{y_0}(b - t)] \tag{8.30}$$

and for TE_y modes similar relations hold.

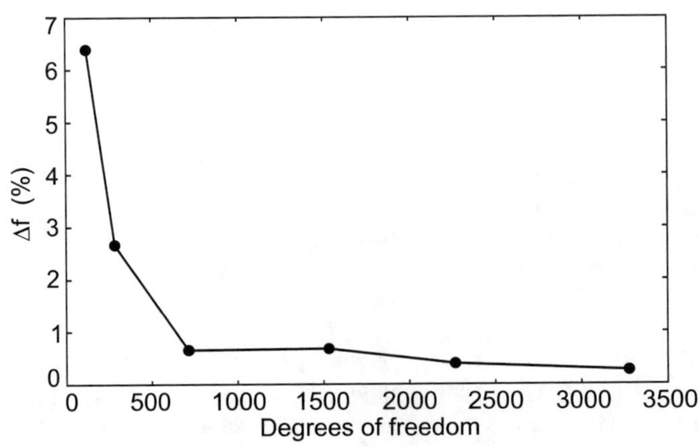

Figure 8.3 First eigenmode convergence as a function of the number of degrees of freedom for a 10 mm ×7 mm ×12 mm empy cavity.

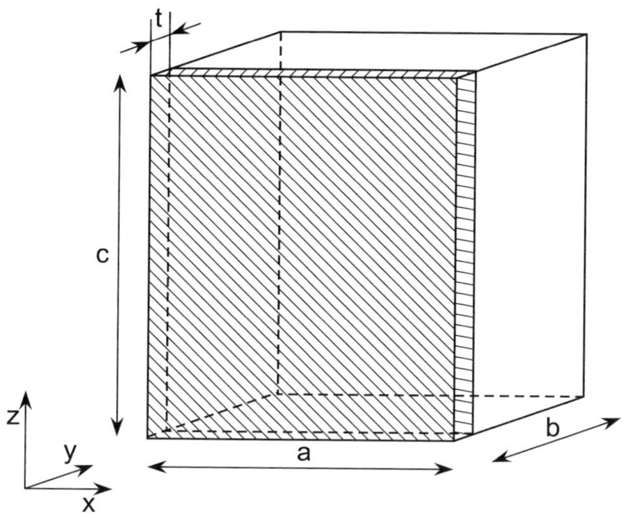

Figure 8.4 Rectangular cavity with p.e.c boundary with size $a \times b \times c$ loaded by a t-thick dielectric slab.

If k_1 is similar to k_0, it is expected that k_{y_0} and k_{y_1} are small, since the guide does not differ much from an empty one, where $k_{y_0} = k_{y_1} = 0$. In this case a Taylor expansion of (8.30) leads to

$$\frac{k_{y_1}^2 t}{\varepsilon_1} \simeq -\frac{k_{y_0}^2 (b-t)}{\varepsilon_0} \tag{8.31}$$

where it is interesting to note that, since both t and $b-t$ are positive, then if k_{y_1} is real k_{y_0} it must be imaginary, and vice versa. From this inevitably descends that k_z is somewhere between the propagation constants of the corresponding modes of the two homogeneous waveguides, one empty and the other filled with dielectric.

If a resonant cavity is considered, k_z cannot be arbitrary but c must be an integer number of half-guided wave wavelengths λ_g. Since $k_z = \frac{2\pi}{\lambda_g}$ it must be $k_z = \frac{p\pi}{c}$ being p a positive integer. If, as in our case, $c > a > b$ then the first mode is that for $p = 1$ and (8.28) becomes:

$$\omega_0^2 \varepsilon_0 \mu_0 = \left(\frac{\pi}{c}\right)^2 + \left(\frac{\pi}{a}\right)^2 + k_{y_0}^2 \tag{8.32}$$

$$\omega_0^2 \varepsilon_1 \mu_1 = \left(\frac{\pi}{c}\right)^2 + \left(\frac{\pi}{a}\right)^2 + k_{y_1}^2 \tag{8.33}$$

being $\omega_0 = 2\pi f_0$ and f_0 the resonant frequency of the loaded cavity. By recognizing that $k_0^{(e)} = \omega^{(e)}\sqrt{\varepsilon_0 \mu_0} = \sqrt{\left(\frac{\pi}{a}\right)^2 + \left(\frac{\pi}{c}\right)^2}$ is the resonance at $\omega_0^{(e)} = 2\pi f_0^{(e)}$ of the empty cavity of equal dimensions and by casting (8.31) into (8.32) and (8.33) one gets

$$\omega_0^2 \varepsilon_0 \mu_0 = \left(\omega^{(e)}\right)^2 \varepsilon_0 \mu_0 + k_{y_1}^2 \frac{\varepsilon_0 t}{\varepsilon_1 (b-t)} \tag{8.34}$$

$$\omega_0^2 \varepsilon_1 \mu_1 = \left(\omega^{(e)}\right)^2 \varepsilon_0 \mu_0 + k_{y_1}^2 \tag{8.35}$$

and solving for ω_0

$$\omega_0 = \omega^{(e)} \sqrt{\frac{\varepsilon_0 \mu_0 \varepsilon_1 (b - t) + \varepsilon_0^2 \mu_0 t}{\varepsilon_0 \mu_0 \varepsilon_1 (b - t) + \varepsilon_0 \varepsilon_1 \mu_1 t}} \tag{8.36}$$

For $\mu_0 = \mu_1$ and $\varepsilon_1 = \varepsilon_r \varepsilon_0$ (8.36) can be recast as

$$\omega_0 = \omega^{(e)} \sqrt{1 - \frac{\varepsilon_r - 1}{\varepsilon_r} \frac{t}{b}} \tag{8.37}$$

if $t \ll b$ a Taylor expansion $\sqrt{1 - x} = 1 - x/2 + o(x)$, up to the first term, can be used, leading to

$$2\pi f_0 \simeq 2\pi f_0^{(e)} \left[1 - \frac{1}{2} \frac{\varepsilon_r - 1}{\varepsilon_r} \frac{t}{b}\right] \tag{8.38}$$

or

$$\frac{f_0 - f_0^{(e)}}{f_0^{(e)}} \simeq -\frac{1}{2} \frac{\varepsilon_r - 1}{\varepsilon_r} \frac{t}{b} \tag{8.39}$$

which gives the relative shift in the resonant frequency of the loaded cavity with respect to the empty one. To the same result one arrives also by exploiting a perturbative approach based on stored energy [28]. The accuracy of this analytical perturbative approximation is of course strongly dependent on ε_r and t, which must be small.

Table 8.2 lists the resonant frequencies and the computed eigenvalues obtained with the `CAVITY` finite element code for rectangular cavity with a $a = 10$ mm, $b = 7$ mm, and $c = 12$ mm and loaded with a $t = 1$ mm-thick dielectric slab parallel to and touching the largest face of the cavity. The relative permittivity of the slab is $\varepsilon_r = 2$. The two data sets are relative to two different discretizations.

For the corresponding unloaded cavity $f_0^{(e)} = 19.51$ GHz and hence, from (8.39), $f_0 = 18.81$ GHz. This value too is reported in the table.

Since the analytical formula is itself approximate, and holds only for the first mode, the errors on the computed frequencies and eigenvalues are computed by considering the FEM solution on the finer mesh as a reference. It is worth noticing that some frequencies are shifted more than others with respect to the empty cavity. This is due to the fact that the dielectric effect is stronger for some modes than for others, as will be explained soon.

Finally, Figure 8.5 presents the resonant frequency of the fundamental mode for the above-mentioned problem as a function of $\varepsilon_r \in [1, 6]$. The dash-dot horizontal line at 19.51 GHz represents the resonance of the unloaded cavity. The dotted line is the frequency given by the approximated analytical formula (8.39) while solid lines are the first 5 resonant frequencies found via FEM. It is worth noticing that, for higher values of permittivity, the FEM numerical solution for the first mode

Table 8.2 Eigenvalues Convergence Increasing the Total Number of Degrees of Freedom for the Cavity with a Dielectric Slab

Mode Index	*Analytical*	*FEM (622 DoF)*		*FEM (2401 DoF)*		
i	f_i	f_i	k_i (m^{-1})	f_i	k_i (m^{-1})	Δf
1	18.81 GHz	18.08 GHz	379.1	18.57 GHz	389.3	2.65%
2		24.18 GHz	506.8	24.32 GHz	509.8	0.58%
3		25.33 GHz	531.0	25.64 GHz	537.5	1.21%
4		26.02 GHz	545.4	27.25 GHz	571.2	4.62%
5		27.20 GHz	570.2	27.62 GHz	578.8	1.51%
6		27.34 GHz	572.9	28.36 GHz	594.4	3.68%
7		28.16 GHz	590.1	30.13 GHz	631.5	6.78%
8		30.99 GHz	649.4	32.06 GHz	671.9	3.41%
9		31.48 GHz	659.8	33.96 GHz	711.7	7.56%
10		32.63 GHz	683.9	35.16 GHz	736.8	7.45%

and the analytical solution are quite different, and that the FEM solution is to be considered closer to the exact one.

It is interesting to note that in the present case the accuracy of the resonant frequency does not seem to deteriorate steadily as higher modes are considered. Indeed, second mode looks much more accurate than the first one. This phenomenon has a simple explanation. The first mode is derived from an $a \times b$ waveguide segment of length c shortened at both ends. This means that the mode is a standing TE_{10} wave (a TE_{101} in cavity notation) with an electric field directed along y which is constant in y and behaves like a sinusoid in x and z, hence there is a strong electric field in the dielectric, at the slab center. The second mode, on the other hand, can be thought of as deriving from a standing TE_{10} wave in a waveguide with cross-section $c \times a$ of length b and shortened at both ends. This time the electric field is x directed and, while uniform along x present a sinusoidal variation in y and z. This implies that the dielectric slab, which is close to the $y = 0$ face and thin, exhibits internally a relatively low field. In the first case the presence of the

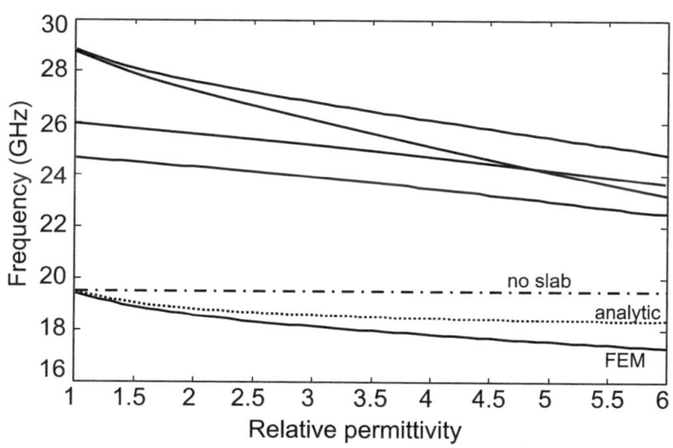

Figure 8.5 Resonant frequencies of a rectangular cavity loaded with a dielectric slab as a function of the ε_r of the slab.

dielectric has a greater effect on the electric field distribution inside the cavity and hence the resonant frequency is more affected. The accuracy of the eigenvalues corresponding to modes with a strong electric field in the dielectric slab can be increased by improving the mesh. Mesh should be finer within the slab since accuracy is linked to the element edge length in term of wavelength *within the medium* filling the element. It is hence clear that, if mesh is uniform, results exhibiting higher electric fields within the slab will be less accurate.

The third example is relative to a double-ridged cavity. This can be thought of as a $a \times b \times c$ rectangular cavity with two grooves, both w wide and t deep, on the $y = 0$ and $y = b$ faces, aligned along the z axis (Figure 8.6).

This is a cavity obtained from a double-ridged waveguide and, since the characteristic of such waveguides is to have a lower TE_{10} cutoff and a higher TE_{20} with respect to a waveguide of equal a and b dimension, we can expect the first resonance to be lower than that of the rectangular cavity. The numerical evaluation is carried out again on a $a = 10$ mm, $b = 7$ mm, and $c = 12$ mm cavity, the grooves are $w = 2$ mm wide and $t = 1$ mm deep.

Table 8.3 lists the eigenvalues computed in two different cases. The first one is relative to a very coarse mesh, leading to quite inaccurate results, while the second one is performed on a sufficiently dense mesh. The relative errors are calculated by considering the exact results attained on the finer mesh, as in the previous example. It is evident the large error affecting the results obtained on the coarser mesh. Note that the first and the third eigenvalues, that are the first two modes which are actually derived from a shortened ridged waveguide, are clearly affected by the presence of the grooves and both exhibit a lower resonant frequency than their equivalent in a rectangular empty box listed in Table 8.1.

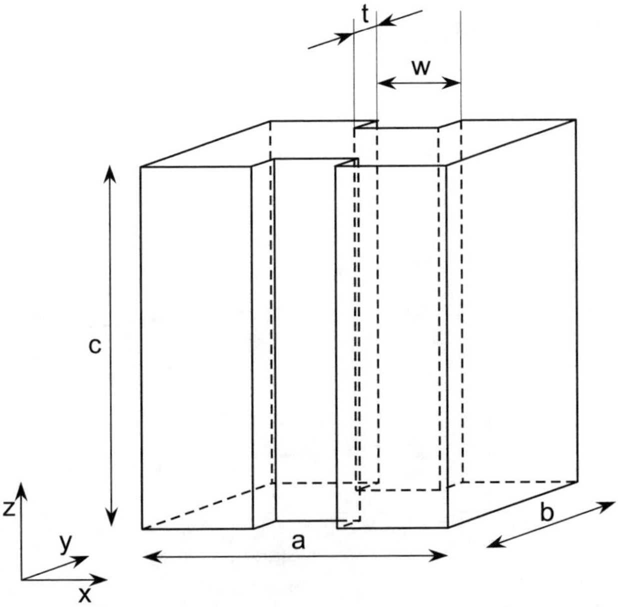

Figure 8.6 Example of homogeneous double-ridged p.e.c.-bounded cavity.

Table 8.3 Eigenvalues Convergence Increasing the Number of Degrees of Freedom for the Double Ridged Cavity

Mode Index	*FEM (254 DoF)*		*FEM (2038 DoF)*		
i	f_i	k_i (m^{-1})	f_i	k_i (m^{-1})	Δf
1	16.74 GHz	350.9	17.84 GHz	373.8	6.12%
2	24.12 GHz	505.6	25.29 GHz	530.1	4.61%
3	24.76 GHz	518.9	26.75 GHz	560.6	7.43%
4	26.23 GHz	549.8	27.69 GHz	580.4	5.26%
5	27.07 GHz	567.3	30.84 GHz	646.3	12.23%
6	29.30 GHz	614.0	32.41 GHz	679.2	9.59%
7	29.91 GHz	626.8	33.08 GHz	693.4	9.61%
8	30.63 GHz	642.0	33.16 GHz	694.9	7.62%
9	31.77 GHz	665.8	34.00 GHz	712.6	6.56%
10	32.03 GHz	671.2	37.82 GHz	792.6	15.32%

As a last example, the rectangular cavity already investigated ($a = 10$ mm, $b = 7$ mm, and $c = 12$ mm) is loaded with a dielectric cube of edge $d = 3$ mm placed at the center of the cavity floor (Figure 8.7). For this problem a series of 300 simulations has been carried out, varying the value of the relative permittivity of the dielectric cube from $\varepsilon_r = 1$ to $\varepsilon_r = 30$. The results are shown in Figure 8.8, which presents the values of the cavity resonant frequencies as a function of ε_r. In the present case the different effect on the various modes due to the dielectric sample is even more evident than for the slab-loaded cavity. There are modes nearly independent on ε appearing in Figure 8.8 as almost horizontal lines, and modes very sensitive to ε variations, with many mode crossovers.

It is apparent how, if the sample permittivity is unknown, it is fairly easy to recover its value by measuring the cavity first resonant frequency. It is also possible, by exploiting the knowledge of more than one resonant frequency, to determine the permittivities of an inhomogeneous sample made of several homogeneous

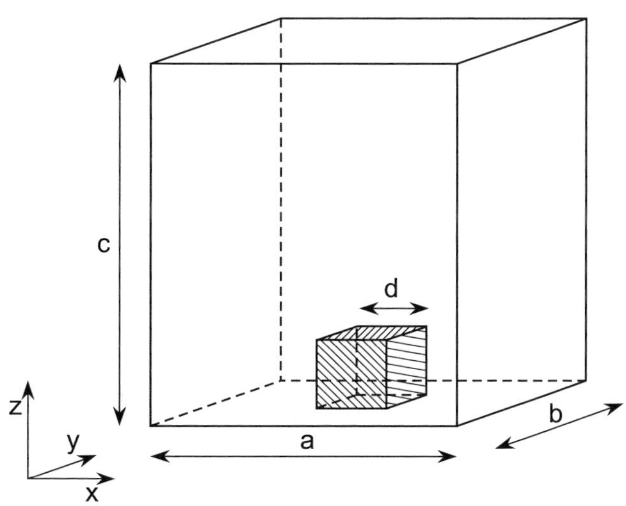

Figure 8.7 Example of loaded rectangular cavity.

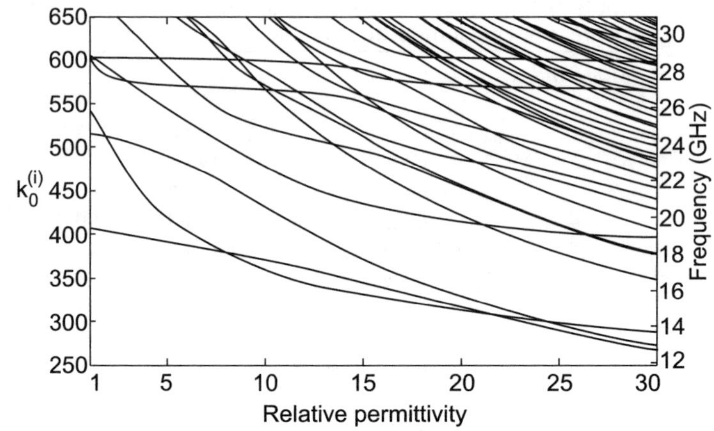

Figure 8.8 Resonant frequencies for the first modes as a function of sample ε_r.

materials, if its geometry is fully known and some sort of automatic optimization procedure is exploited [26].

8.5 Disc Content

The software described here is contained in the subdirectory CAVITY of the directory CODES3D in the enclosed CD-ROM. The program files are the main program CAVITY.F and its auxiliary subroutines file: CAVITYSUBS.F. The code needs to be linked with the three-dimensional library discussed in Chapter 7 containing elemental matrices and mesh reading routines. Of course it must be linked also with the `BLAS` and `LAPACK` libraries.

The three subdirectories RECT, SLAB, RIDGED, and DIEL, contain all the input and mesh files of the examples discussed in the previous section. A more complete description of each file is contained in the README file in the directory CAVITY itself.

References

[1] Davies, J. B., Fernandez, F. A., and Philippou, G. Y., "Finite Element Analysis of All Modes in Cavities with Circular Symmetry," *IEEE Transactions on Microwave Theory and Techniques*, Vol. 11(11), pp. 1975–1982, Nov. 1982.

[2] Hara, M., et al., "A Three-Dimensional Analysis of RF Electromagnetic Fields by the Finite-Element Method," *IEEE Transactions on Magnetics*, Vol. MAG-19(6), pp. 2417–2420, Nov. 1983.

[3] Webb, J. P., "The Finite-Element Method for Finding Modes of Dielectric-Loaded Cavities," *IEEE Transactions on Microwave Theory and Techniques*, Vol. 33(7), pp. 635–639, July 1985.

[4] Webb, J. P., "Efficient Generation of Divergence-Free Fields for the Finite-Element Analysis of Three-Dimensional Cavity Resonances," *IEEE Transactions on Magnetics*, Vol. 24(1), pp. 162–165, Jan. 1988.

[5] Konrad, A., "A Direct Three-Dimensional Finite-Element Method for the Solution of Electromagnetic Fields in Cavities," *IEEE Transactions on Magnetics*, Vol. MAG-21(6), pp. 2276–2279, Nov. 1985.

[6] Kobelansky, A. J. and Webb, J. P., "Eliminating Spurious Mode in Finite-Element Waveguide Problems by Using Divergence-Free Fields," *Electronics Letters*, Vol. 22(11), pp. 569–570, May 1986.

[7] Bardi, I., Biro, O., and Preis, K., "Finite-Element Scheme for Three-Dimensional Cavities Without Spurious Modes," *IEEE Transactions on Magnetics*, Vol. 27(5), pp. 4036–4039, Sep. 1991.

[8] Bossavit, A., "A Rationale for 'Edge Elements' in Three-Dimensional Fields Computations," *IEEE Transactions on Magnetics*, Vol. 24(1), pp. 74–79, Jan. 1988.

[9] Chatterjee, A., Jin, J. M., and Volakis, J. L., "Computation of Cavity Resonances Using Edge-Based Finite Elements," *IEEE Transactions on Microwave Theory and Techniques*, Vol. 40, pp. 2106–2108, 1992.

[10] Bardi, I., et al., "Nodal and Edge Element Analysys of Inhomogeneously Loaded Three-Dimensional Cavities," *IEEE Transactions on Magnetics*, Vol. 28(2), pp. 1142–1145, Mar. 1992.

[11] Savage, J. S. and Peterson, A. F., "Higher-Order Vector Finite Elements for Tetrahedral Cells," *IEEE Transactions on Microwave Theory and Techniques*, Vol. 44(6), pp. 874–879, June 1996.

[12] Wang, Y., Monk, P., and Szabo, B., "Computing Cavity Modes Using the *p*-Version of the Finite-Element Method," *IEEE Transactions on Magnetics*, Vol. 32(3), pp. 1934–1940, May 1996.

[13] Perepelitsa, S. G., Dyczij-Edlinger, R., and Lee, J.-F., "Finite-Element Analysis of Arbitrarily Shaped Cavity Resonators Using $H^1(curl)$ Elements," *IEEE Transactions on Magnetics*, Vol. 33(2), pp. 1776–1779, Mar. 1997.

[14] Martini, E., Selleri, S., and Pelosi, G., "Three-Dimensional FEM Robustness to Mesh Distortion Versus Basis Functions Order," *Microwave and Optical Technology Letters*, Vol. 27(1), pp. 17–20, Oct. 2000.

[15] Martini, E. and Selleri, S., "An Innovative Class of Curvilinear Tetrahedral Elements," *Electronics Letters*, Vol. 37, pp. 557–558, 2001.

[16] Ilić, M. M. and Notaroš, B. M., "Higher Order Hierarchical Curved Hexahedral Vector Finite Elements for Electromagnetic Modeling," *IEEE Transactions on Microwave Theory and Techniques*, Vol. 51(3), pp. 1026–1033, Mar. 2003.

[17] Silvester, P. P. and Ferrari, R. L., *Finite Elements for Electrical Engineers*, Cambridge, UK: Cambridge University Press, 3rd edition, 1996.

[18] Jin, J. M., *The Finite Element Method in Electromagnetics*, New York: John Wiley & Sons, 1993.

[19] Matthaei, G. L., Young, L., and Jones, E. M. T., *Microwave Filters, Impedance-Matching Networks, and Coupling Structures*, Dedham, MA: Artech House, 1980.

[20] Wang, J. S. and Mittra, R., "A Finite-Element Cavity Resonance Method for Waveguide and Microstrip Line Discontinuity Problems," *IEEE Transactions on Microwave Theory and Techniques*, Vol. 42, pp. 433–440, 1994.

[21] Groiss, S., et al., "Parameters of Lossy Cavity Resonators Calculated by the Finite-Element Method," *IEEE Transactions on Magnetics*, Vol. 32(3), pp. 894–897, May 1996.

[22] Coccioli, R., et al., "Design of Tolerance-Corrected Filters Employing Half-Cylinder Posts," *IEEE Transactions on Microwave Theory and Techniques*, Vol. 46(1), pp. 116–118, Jan. 1998.

[23] Cid, J. M. and Zapata, J., "CAD of Rectangular-Waveguide H-Plane Circuits by Segmentation, Finite Elements and Artificial Neural Networks," *Electronics Letters*, Vol. 37(2), pp. 98–99, Jan. 18, 2001.

[24] Vaid, J. K., Parkash, A., and Mansingh, A., "Measurement of Dielectric Parameters at Microwave Frequencies by Cavity Perturbation Technique," *IEEE Transactions on Microwave Theory and Techniques*, Vol. MTT-27(9), pp. 791–795, Sep. 1979.

[25] Meng, B., Booske, J., and Cooper, R., "Extended Cavity Perturbation Technique to Determine the Complex Permittivity of Dielectric Materials," *IEEE Transactions on Microwave Theory and Techniques*, Vol. 43(12), pp. 2633–2636, Dec. 1995.

[26] Coccioli, R., Pelosi, G., and Selleri, S., "Characterization of Dielectric Materials with the Finite-Element Method," *IEEE Trans. Microwave Theory Tech.*, Vol. 47(7), pp. 1106–1112, July 1999.

[27] Lang, S., *Linear Algebra*, Reading, MA: Addison-Wesley, 3rd edition, 1987.

[28] Harrington, R. F., *Time-Harmonic Electromagnetic Fields*, New York: McGraw-Hill, 1961.

CHAPTER 9

Waveguide Devices

In this chapter the problem of the finite element analysis of waveguide devices is revised and extended. The solution introduced in Chapter 4 was two-and-a-half-dimensional and limited to E-plane and H-plane devices, with limitations in the presence of dielectrics. Here this same problem will be extended to a full three-dimensional solution of waveguide devices containing general, lossy, anisotropic materials. The only true limitation being that of requiring the device to be fed via rectangular waveguides.

The algorithm described in the following sections belongs to the widely investigated class of hybrid finite element/modal expansion techniques. The basic theory of these techniques was already discussed in Chapter 4 and some references to open literature can be found there. Other interesting papers dealing with this topic and more specific to three-dimensional implementations are the very early papers [1–3] as well as more recent and advanced ones [4–8].

In principle, any kind of guiding structure can be used at ports, provided modal field distributions are known. Rectangular waveguides presents analytically known modes, and the same is true for circular waveguides, coaxial cables, and few other waveguides. If more general structures need to be considered then modes can be also obtained numerically. By extracting a two-dimensional triangular mesh pertinent to the waveguide section from the tetrahedral mesh of the whole device it is straightforward to obtain modal fields with an application of the Chapter 3 code WG.

9.1 Opening the Cavity: Formulation

Let's consider a microwave passive device which can be described as a multiport junction, that is, some sort of complex-shape domain enclosed in a perfectly conducting surface, which presents a finite number N of openings, each representing the cross-section of a semiinfinite uniform wave guiding structure extending normally from the surface to infinity. Each of these openings is a port of the device.

The electromagnetic wave problem defined over this limited three-dimensional domain Ω can include generic materials, possibly with losses and/or anisotropic behavior both in term of electric and magnetic properties.

The boundary S of the domain Ω is a surface which can be considered composed by the perfect electric conductor surface delimiting the device S^{pec}, by a possible perfect magnetic conductor surface S^{pmc}, and by the surfaces $S^{(k)}$ defining the $k = 1, \ldots, N$ ports:

$$S = S^{pec} \cup S^{pmc} \cup \left[\bigcup_{k=1}^{N} S^{(k)} \right] \tag{9.1}$$

Other types of boundary conditions can be considered, like those described in Chapter 8, but for the sake of simplicity only perfect electric conducting boundaries will be considered here.

A global Cartesian coordinate system x, y, z is defined on Ω, while on each port k a local Cartesian coordinate system is defined, $x^{(k)}, y^{(k)}, z^{(k)}$ where axis $x^{(k)}$ is aligned with the rectangular port larger dimension $a^{(k)}$ and axis $y^{(k)}$ is aligned with the port shorter dimension $b^{(k)}$. The axis $z^{(k)}$ is chosen so as to exit the device (Figure 9.1).

Assuming that in Ω no sources are present, Maxwells equations reduced to the two curl equations which can be recast as the single Helmholtz equation for the electric field:

$$\nabla \times \left([\mu_r]^{-1} \nabla \times \mathbf{E} \right) - k_0^2 [\varepsilon_r] \mathbf{E} = 0 \tag{9.2}$$

The magnetic field being related to the electric field through the equation:

$$- j k_0 \zeta_0 \mathbf{H} = [\mu_r]^{-1} \nabla \times \mathbf{E} \tag{9.3}$$

On S^{pec} the tangential component of the electric field must vanish, that is $\hat{\mathbf{n}} \times \hat{\mathbf{E}} = 0$, being $\hat{\mathbf{n}}$ the normal unit vector normal to S^{pec}.

Suitable boundary conditions enforcing the continuity of the transversal electric and magnetic field need to be enforced at each port (k):

$$\hat{\mathbf{z}}^{(k)} \times \mathbf{E} = \hat{\mathbf{z}}^{(k)} \times \mathbf{E}_{wg}^{(k)} \quad \text{on } S_k,\ k = 1, \ldots, N \tag{9.4}$$

$$\hat{\mathbf{z}}^{(k)} \times \mathbf{H} = \hat{\mathbf{z}}^{(k)} \times \mathbf{H}_{wg}^{(k)} \quad \text{on } S_k,\ k = 1, \ldots, N \tag{9.5}$$

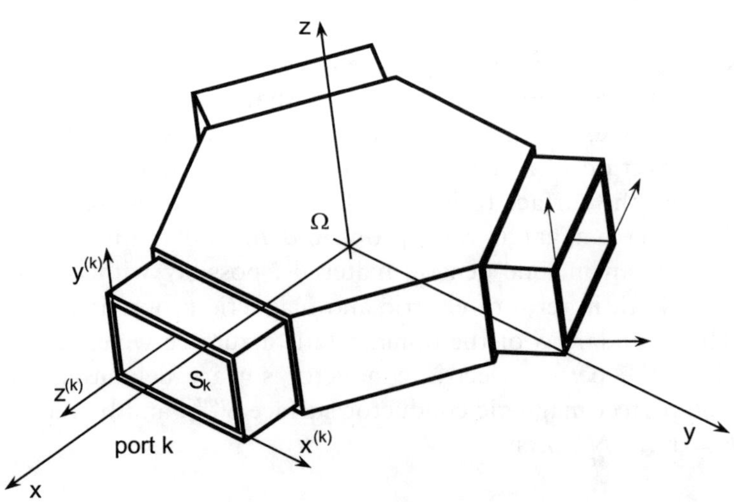

Figure 9.1 Generic three-dimensional device with rectangular ports and relative coordinate systems.

In (9.4) and (9.5) on the left-hand side there are the internal fields, while on the right-hand side there is the field at the port outside the device, that is in the semiinfinite feeding waveguide. In the semiinfinite waveguide, the electric field can be expressed as a modal expansion, similar to what was done in Chapter 4.

$$\mathbf{E}_{wg}^{(k)} = \sum_{m=1}^{\infty} \left[A_m^{(k)} \mathbf{e}_m^{(k)}(x^{(k)}, y^{(k)}) e^{j\beta_m^{(k)} z^{(k)}} + B_m^{(k)} \mathbf{e}_m^{(k)}(x^{(k)}, y^{(k)}) e^{-j\beta_m^{(k)} z^{(k)}} \right] \tag{9.6}$$

In (9.6) the notation used in (4.1) is expanded: $\mathbf{e}_m^{(k)}(x^{(k)}, y^{(k)})$ is the electric field tangential component of mode m at port k, being m a single index spanning both TE and TM modes in a suitable order and $\beta_m^{(k)}$ its pertinent wavenumber

$$\beta_m^{(k)} = \begin{cases} \sqrt{k_0^2 - \left(k_{tm}^{(k)}\right)^2} & \text{if } k_0 \geq k_{tm}^{(k)} \\ j\sqrt{\left(k_{tm}^{(k)}\right)^2 - k_0^2} & \text{if } k_0 < k_{tm}^{(k)} \end{cases} \tag{9.7}$$

with

$$k_{tm}^{(k)} = \sqrt{\left(k_{xm}^{(k)}\right)^2 + \left(k_{ym}^{(k)}\right)^2} = \sqrt{\left(\frac{i_m \pi}{a^{(k)}}\right)^2 + \left(\frac{j_m \pi}{b^{(k)}}\right)^2} \tag{9.8}$$

being $a^{(k)}$ and $b^{(k)}$ the larger and shorter dimensions of the rectangular port k, respectively, and i_m and j_m the two integer indexes determining mode m, either a $\text{TE}_{i_m j_m}$ or a $\text{TM}_{i_m j_m}$, with eigenvalue $k_{tm}^{(k)}$. Finally, $A_m^{(k)}$ and $B_m^{(k)}$ are the impressed complex modal amplitudes for the impinging mth mode on the kth waveguide port and the unknown complex modal amplitude for the scattered mth mode at the kth waveguide port, respectively.

The presence in the summation (9.6) of the impinging waves with amplitude $A_m^{(k)}$ formally represents the possibility to feed all the ports of the device with all modes at the same time rather than just the fundamental mode at a single ports as it was in Chapter 4. Of course to retrieve a GSM only one mode must be fed at a single port at a time, hence all $A_m^{(k)}$ will be zero except one, but in principle it can be very useful to be able to feed several ports at once in a multiport device, in particular when a device is designed to perform a summation or difference of signals coming from its input ports, like what may happen in monopulse radar comparators [9–11].

If the eigenfunctions

$$\mathbf{\Phi} = k_{ym}^{(k)} \cos\left(k_{xm}^{(k)} x^{(k)}\right) \sin\left(k_{ym}^{(k)} y^{(k)}\right) \hat{\mathbf{x}} - k_{xm}^{(k)} \sin\left(k_{xm}^{(k)} x^{(k)}\right) \cos\left(k_{ym}^{(k)} y^{(k)}\right) \hat{\mathbf{y}} \tag{9.9}$$

and

$$\mathbf{\Psi} = k_{xm}^{(k)} \cos\left(k_{xm}^{(k)} x^{(k)}\right) \sin\left(k_{ym}^{(k)} y^{(k)}\right) \hat{\mathbf{x}} + k_{ym}^{(k)} \sin\left(k_{xm}^{(k)} x^{(k)}\right) \cos\left(k_{ym}^{(k)} y^{(k)}\right) \hat{\mathbf{y}} \tag{9.10}$$

are defined, then the analytic expression for the transverse fields are, for the TE modes:

$$\mathbf{e}_m^{(k)} = \mathbf{e}_{m_H}^{(k)} = j\omega\mu\mathbf{\Phi}_m^{(k)} \tag{9.11}$$

$$\mathbf{h}_m^{(k)} = \mathbf{h}_{m_H}^{(k)} = \frac{1}{Z_{m_H}^{(k)}}(\mp\hat{\mathbf{z}}) \times \mathbf{e}_{m_H}^{(k)} = \mp j\beta_m^{(k)}\hat{\mathbf{z}} \times \mathbf{\Phi}_m^{(k)} \tag{9.12}$$

being $Z_{m_H}^{(k)} = \frac{k_0}{\beta_m^{(k)}}\zeta_0$ the modal impedance and $\zeta_0 = 120\pi$ the free-space wave impedance. In (9.12) the sign must be chosen according to the direction of propagation: minus if the wave propagates in the $z^{(k)}$ positive direction (outgoing wave); plus if the wave propagates in the $z^{(k)}$ negative direction (incoming wave).

For TM modes the transverse field expressions are:

$$\mathbf{e}_m^{(k)} = \mathbf{e}_{m_E}^{(k)} = -j\beta_m^{(k)}\mathbf{\Psi}_m^{(k)} \tag{9.13}$$

$$\mathbf{h}_m^{(k)} = \mathbf{h}_{m_E}^{(k)} = \frac{1}{Z_{m_E}^{(k)}}(\mp\hat{\mathbf{z}}) \times \mathbf{e}_{m_E}^{(k)} = \mp j\omega\varepsilon\hat{\mathbf{z}} \times \mathbf{\Psi}_m^{(k)} \tag{9.14}$$

with $Z_{m_E}^{(k)} = \frac{\beta_m^{(k)}}{k_0}\zeta_0$ the TM mode impedance and with the same convention on the sign of (9.14).

Using (9.3) the needed continuity condition for the tangential component of the magnetic field can be recast as a condition on the electric field, so the complete problem, comprehensive of boundary conditions, can be summarized as:

$$\begin{cases} \nabla \times \left([\mu_r]^{-1}\nabla \times \mathbf{E}\right) - k_0^2[\varepsilon_r]\mathbf{E} = 0 & \text{in } \Omega_T \\ \hat{\mathbf{n}} \times \mathbf{E} = 0 & \text{on } S^{pec} \\ \hat{\mathbf{n}} \times \nabla \times \mathbf{E} = 0 & \text{on } S^{pmc} \\ \hat{\mathbf{z}}^{(k)} \times \mathbf{E} = \hat{\mathbf{z}}^{(k)} \times \mathbf{E}_{wg}^{(k)} & \text{on } S_k \\ \dfrac{-1}{jk_0\zeta_0}\hat{\mathbf{z}}^{(k)} \times [\mu]^{-1}\nabla \times \mathbf{E} = \hat{\mathbf{z}}^{(k)} \times \mathbf{H} & \\ \qquad = \hat{\mathbf{z}}^{(k)} \times \mathbf{H}_{wg}^{(k)} = \dfrac{-1}{jk_0\zeta_0}\hat{\mathbf{z}}^{(k)} & \\ \qquad \times [\mu]^{-1}\nabla \times \mathbf{E}_{wg}^{(k)} & \text{on } S_k \end{cases} \tag{9.15}$$

To compute the solution of the proposed problem via finite elements, we need to recast (9.15) as a weak form by applying the usual Galerkin procedure.

Weight functions $\mathbf{W}$ are chosen within a space $\mathcal{V}$ which is a subspace of the Hilbert space of the curl-conforming vector functions $\mathcal{H}^{curl}(\Omega)$ for the electromagnetic operator defined on Ω. With the usual definition of a scalar product, it is possible to achieve the weak form of (9.15) for each equation separately. The first equation of the system (9.15) is the vector Helmholtz equation holding inside the domain and becomes:

$$\int_\Omega \nabla \times \mathbf{W} \cdot [\mu_r]^{-1} \nabla \times \mathbf{E}\, d\Omega - k_0^2 \int_\Omega \mathbf{W} \cdot [\varepsilon_r] \mathbf{E}\, d\Omega - \oint_S \mathbf{W} \cdot \hat{\mathbf{n}} \times [\mu_r]^{-1} \nabla \times \mathbf{E}\, dS = 0 \tag{9.16}$$

In (9.16) the boundary surface integral can be split, thanks to linearity, into three separate contributions, each pertaining to the portions of S which lie: (1) on a perfect electric conductor S^{pec}; (2) on a perfect magnetic conductor S^{pmc}; and (3) on the ports $S^{(k)}$.

For S^{pec} where Dirichlet boundary conditions hold, it is

$$\int_{S^{pec}} \mathbf{W} \cdot \hat{\mathbf{n}} \times [\mu_r]^{-1} \nabla \times \mathbf{E}\, dS = 0 \tag{9.17}$$

because the $\mathbf{W}$ test functions reduction to S vanishes on S^{pec} as it is the usual choice already presented in all the chapters.

Also for S^{pmc}, on which Neumann boundary conditions hold, the integral vanishes

$$\int_{S^{pmc}} \mathbf{W} \cdot \hat{\mathbf{n}} \times [\mu_r]^{-1} \nabla \times \mathbf{E}\, dS = 0 \tag{9.18}$$

because the factor $\hat{\mathbf{n}} \times [\mu_r]^{-1} \nabla \times \mathbf{E}$ is proportional to the tangential magnetic field and hence zero on p.m.c boundaries.

For these reasons the surface integral in (9.16), reduces only to the integral on the ports:

$$\int_\Omega \nabla \times \mathbf{W} \cdot [\mu_r]^{-1} \nabla \times \mathbf{E}\, d\Omega - k_0^2 \int_\Omega \mathbf{W} \cdot [\varepsilon_r] \mathbf{E}\, d\Omega - \sum_{k=1}^{N} \int_{S_k} \mathbf{W} \cdot \hat{\mathbf{z}}^{(k)} [\mu_r]^{-1} \times \nabla \times \mathbf{E}_{wg}^{(k)}\, dS = 0 \tag{9.19}$$

In (9.19) the surface integral implicitly accounts for the magnetic field continuity at the ports, hence the first, second, third, and fifth equations of the system (9.15) are accounted for. Since the code will be developed for empty rectangular waveguides as feeding structures $[\mu_r]^{-1}$ will be explicitly considered unitary in the surface integrals on ports in the following. On the other hand the electric field continuity conditions have not been imposed yet. Hence there is the need to explicitly consider the weak form of the remaining fourth equation of the system (9.16).

For such an equation the weight functions, $\bar{\mathbf{W}}$, are commonly chosen as the modal eigenfunctions $\boldsymbol{\Phi}$ and $\boldsymbol{\Psi}$ for the waveguide port itself, since they are solutions for the homogeneous Maxwells equations, they are guaranteed to belong to the $\mathcal{H}^{curl}(S^{(k)})$ function space. In this way for each port one can write:

$$\oint_{S_k} \bar{\mathbf{W}} \cdot \hat{\mathbf{n}} \times \mathbf{E} \times \hat{\mathbf{n}}\, dS = \oint_{S_k} \bar{\mathbf{W}} \cdot \hat{\mathbf{n}} \times \left\{ \sum_{m=1}^{\infty} \left[A_m^{(k)} \mathbf{e}_m^{(k)}(x^{(k)}, y^{(k)}) e^{j\beta_m^{(k)} z^{(k)}} + B_m^{(k)} \mathbf{e}_m^{(k)}(x^{(k)}, y^{(k)}) e^{-j\beta_m^{(k)} z^{(k)}} \right] \right\} \times \hat{\mathbf{n}}\, dS \tag{9.20}$$

where $\hat{\mathbf{n}} \times \mathbf{E} \times \hat{\mathbf{n}}$ is the component of the electric field tangent to the port surface.

Given the weak form of the problem defined by the pair (9.19) and (9.20) it is possible to apply the standard finite element approach. The next step is hence to define a mesh on the domain Ω. Within each mesh element the unknown field is expanded in terms of bases which are the element interpolating functions; in the present case, being the electric field a vector field, a vector mesh is considered. Elements are tetrahedra, and basis functions Whitney functions, as in Chapter 7 and indicated with $\boldsymbol{\alpha}_i^{(e)}$, being i the index of the tetrahedron (e) edge relative to the basis functions in the local numeration scheme of the element. The unknown electric field $\mathbf{E}$ within the single element (e) is hence:

$$\mathbf{E}^{(e)} = \sum_{j=1}^{N_e} e_j^{(e)} \boldsymbol{\alpha}_j^{(e)} \tag{9.21}$$

in which the unknowns are the N_e coefficients $e_j^{(e)}$ of the finite expansion of the electric field $\mathbf{E}$ on the tetrahedron. Lowercase e being used in place of uppercase E as it was in Chapter 8 to avoid possible confusion with element matrices.

Substituting (9.21) and (9.6) in (9.19), in a Galerkin framework in which weight functions $\mathbf{W}$ are the same basis function, $\boldsymbol{\alpha}_i$, used for field expansion, the following element-wise weak equation can be obtained

$$\begin{aligned}
&\sum_{j=1}^{N_e} \left\{ \int_{\Omega_e} \nabla \times \boldsymbol{\alpha}_i^{(e)} \cdot [\mu_r]^{-1} \nabla \times \boldsymbol{\alpha}_j^{(e)} \, d\Omega - k_0^2 \int_{\Omega} \boldsymbol{\alpha}_i^{(e)} \cdot [\varepsilon_r] \boldsymbol{\alpha}_j^{(e)} \, d\Omega \right\} e_j^{(e)} + \\
&\quad - \sum_{k=1}^{N} \sum_{m=1}^{\infty} B_m^{(k)} \int_{S_k} \boldsymbol{\alpha}_i^{(e)} \cdot \hat{\mathbf{z}}^{(k)} \times \left[\nabla \times \left(\mathbf{e}_m^{(k)} e^{-j\beta_m^{(k)} z^{(k)}} \right) \right]_{z^{(k)}=0} dS \\
&\quad = \sum_{k=1}^{N} \sum_{m=1}^{\infty} A_m^{(k)} \int_{S_k} \boldsymbol{\alpha}_i^{(e)} \cdot \hat{\mathbf{z}}^{(k)} \times \left[\nabla \times \left(\mathbf{e}_m^{(k)} e^{j\beta_m^{(k)} z^{(k)}} \right) \right]_{z^{(k)}=0} dS
\end{aligned} \tag{9.22}$$

for any $i = 1, 2, ..., N_e$ and for any element, (e), of the mesh. The surface integrals appear of course only on those elements having a face over a port. The following element-dependent contributions can be extracted from the integrals in (9.22):

$$E_{ij}^{(e)} = \int_{\Omega^{(e)}} \nabla \times \boldsymbol{\alpha}_i^{(e)} \cdot [\mu_r]^{-1} \nabla \times \boldsymbol{\alpha}_j^{(e)} \, d\Omega \tag{9.23}$$

$$F_{ij}^{(e)} = \int_{\Omega^{(e)}} \boldsymbol{\alpha}_i^{(e)} \cdot [\varepsilon_r] \boldsymbol{\alpha}_j^{(e)} \, d\Omega \tag{9.24}$$

The term $\hat{\mathbf{z}}^{(k)} \times \left[\nabla \times \left(\mathbf{e}_m^{(k)} e^{-j\beta_m^{(k)} z^{(k)}} \right) \right]_{z^{(k)}=0}$ in the surface integrals is, except for a multiplicative term $\frac{-1}{jk_0\zeta_0}$, the tangential component of the magnetic field [see the last of (9.15)]. By exploiting the expressions of the magnetic field tangential components provided in (9.12) and (9.14) one obtains

$$\begin{aligned}
P_{im}^{(e),(k)} &= \int_{S_k} \boldsymbol{\alpha}_i^{(e)} \cdot \hat{\mathbf{z}}^{(k)} \times \left[\nabla \times \mathbf{e}_m^{(p)} e^{-j\beta_m^{(k)} z^{(k)}} \right]_{z^{(k)}=0} dS \\
&= -jk_0\zeta_0 \int_{S_k} \boldsymbol{\alpha}_i^{(e)} \cdot \hat{\mathbf{z}}^{(k)} \times \left[\mathbf{h}_m^{(k)} \right]_{z^{(k)}=0} dS
\end{aligned} \tag{9.25}$$

where indices i,j are relative to the bases and span both the range $1,2,\ldots,N_e$, while m is the index of the mode on port (k). In a numerical implementation the number of modes considered must be finite and hence it will be $m = 1,\ldots,M^{(k)}$ at each port k.

It is also worth noting that (9.25) sign changes if incoming or outgoing waves are considered according to the signs in (9.12) and (9.14).

By introducing (9.12) and (9.14) into (9.25), and by considering that $\hat{\mathbf{z}} \times \hat{\mathbf{z}} \times$ effectively multiplies times -1 a vector lying in the plane orthogonal to $\hat{\mathbf{z}}$ one obtains, for the incoming waves of the second term in (9.22):

$$P_{im}^{(e),(k)} = -jk_0\zeta_0 \begin{cases} j\beta_m^{(k)} \displaystyle\int_{S_k} \boldsymbol{\alpha}_i^{(e)} \cdot \boldsymbol{\Phi}_m^{(k)} dS & \text{for TE modes} \\ -j\omega\varepsilon \displaystyle\int_{S_k} \boldsymbol{\alpha}_i^{(e)} \cdot \boldsymbol{\Psi}_m^{(k)} dS & \text{for TM modes} \end{cases} \tag{9.26}$$

while for the third term in (9.22) the expression giving the integrals is still (9.26) but with the opposite sign due to the fact that waves are outgoing.

The local element equations can be written in a more compact matrix form:

$$\left[\left[E^{(e)}\right] - k_0^2\left[F^{(e)}\right]\right]\left[e^{(e)}\right] + \left[P^{(e),(k)}\right]\left[B^{(k)}\right] = \left[P^{(e),(k)}\right]\left[A^{(k)}\right] \tag{9.27}$$

By assembling all the local contributions of all the single elements of the mesh in a global matrix system of equations as described in the previous chapters the following full system is obtained:

$$\left[[E] - k_0^2[F]\right][e] + [P][B] = [I_H] \tag{9.28}$$

where the right-hand side term $[I_H] = \left[P^{(e),(k)}\right]\left[A^{(k)}\right]$ is fully known.

While the $\left[[E] - k_0^2[F]\right]$ matrix is assembled from the standard element matrices described in Chapter 7, the $[P]$ matrix takes into account magnetic field continuity across the port surfaces and vector $[I_H]$ represents the sources, as given by the chosen incident modes.

In a similar way it is possible to obtain a set of equations enforcing the continuity of the electric field starting from (9.20) and explicitating the integrals:

$$\begin{aligned} \sum_{j=1}^{N_e} e_j^{(e)} \int_{S_q} \bar{\mathbf{W}} \cdot \hat{\mathbf{z}}^{(q)} \times \boldsymbol{\alpha}_j^{(e)} \times \hat{\mathbf{z}}^{(q)} dS & \\ - \sum_{k=1}^{N}\sum_{m=1}^{M^{(k)}} B_m^{(k)} \int_{S_q} \bar{\mathbf{W}} \cdot \hat{\mathbf{z}}^{(q)} \times \mathbf{e}_m^{(k)} \times \hat{\mathbf{z}}^{(q)} dS & \\ = \sum_{k=1}^{N}\sum_{m=1}^{M^{(k)}} A_m^{(k)} \int_{S_k} \bar{\mathbf{W}} \cdot \hat{\mathbf{z}}^{(q)} \times \mathbf{e}_m^{(k)} \times \hat{\mathbf{z}}^{(q)} dS & \end{aligned} \tag{9.29}$$

As already stated, weights will be the eigenfunctions (9.9) and (9.10). In particular it is

$$\bar{\mathbf{W}} = \begin{cases} \frac{1}{jk_0\zeta_0}\mathbf{\Phi}_n^{(q)} & \text{weight deriving from TE mode } n \text{ at port } (q) \\ \mathbf{\Psi}_n^{(q)} & \text{weight deriving from TM mode } n \text{ at port } (q) \end{cases} \tag{9.30}$$

with mode index $n = 1, \ldots, M^{(q)}$ and port index $q = 1, 2, \ldots, N$:

Equation (9.29) can be recast in element matrix form by defining

$$C_{nj}^{(e),(q)} = \begin{cases} \frac{1}{jk_0\zeta_0}\int_{S_q} \mathbf{\Phi}_n^{(q)} \cdot \hat{\mathbf{z}}^{(q)} \times \boldsymbol{\alpha}_j^{(e)} \times \hat{\mathbf{z}}^{(q)}\, dS & \text{for TE-like weights} \\ \int_{S_q} \mathbf{\Psi}_n^{(q)} \cdot \hat{\mathbf{z}}^{(q)} \times \boldsymbol{\alpha}_j^{(e)} \times \hat{\mathbf{z}}^{(q)}\, dS & \text{for TM-like weights} \end{cases} \tag{9.31}$$

$$D_{nm}^{(q),(k)} = -\begin{cases} \frac{1}{jk_0\zeta_0}\int_{S_q} \mathbf{\Phi}_n^{(q)} \cdot \hat{\mathbf{z}}^{(q)} \times \mathbf{e}_m^{(k)} \times \hat{\mathbf{z}}^{(q)}\, dS & \\ \quad = \int_{S_q} \mathbf{\Phi}_n^{(q)} \cdot \mathbf{\Phi}_n^{(q)}\, dS & \text{for TE-like weights} \\ \int_{S_q} \mathbf{\Psi}_n^{(q)} \cdot \hat{\mathbf{z}}^{(q)} \times \mathbf{e}_m^{(k)} \times \hat{\mathbf{z}}^{(q)}\, dS & \\ \quad = j\beta_n^{(q)} \int_{S_q} \mathbf{\Psi}_n^{(q)} \cdot \mathbf{\Psi}_n^{(q)}\, dS & \text{for TM-like weights} \end{cases} \tag{9.32}$$

It is important to note that ports are not overlapping, that is $S_k \cap S_q = \emptyset$, hence integration between eigenmodes at separate ports are always null. Moreover, thanks to the orthogonality properties of waveguide eigenmodes, interaction integrals $D_{nm}^{(k),(k)}$ are null if $n \neq m$, whichever type of mode, TE or TM, is relevant for the indices. This leads to a diagonal $[D]$ matrix.

The matrix form of (9.29) is hence

$$\left[C^{(e),(q)}\right]\left[e^{(e)}\right] + \left[D^{(q),(k)}\right]\left[B^{(q)}\right] = -\left[D^{(q),(k)}\right]\left[A^{(q)}\right] \tag{9.33}$$

Assembling on all mesh elements and ports the global matrix equation becomes:

$$[C][E] + [D][B] = [I_E] \tag{9.34}$$

having defined $[I_E] = -[D][A]$. In (9.34) matrix $[C]$ is responsible for enforcing electric field continuity at the ports, while $[D]$ derives from modal weight-basis products.

In the end the global system to be solved is

$$\begin{bmatrix} [D] & [C] \\ [P] & [S] \end{bmatrix} \begin{bmatrix} [B] \\ [E] \end{bmatrix} = \begin{bmatrix} [I_E] \\ [I_H] \end{bmatrix} \tag{9.35}$$

where $[S] = [E] - k_0^2[F]$ for the sake of simplicity.

The number of independent equations and the number of unknowns of the problem is balanced, since weights are chosen equal to bases both within the finite element expansion and within the modal expansion at ports.

The Dirichlet conditions in (9.35) are of course satisfied if no bases allowing for a nonnull tangential component of the electric field is included in the expansion. This effectively requires to eliminate from the number of the degrees of freedom all the tetrahedron edges lying on a p.e.c. boundary. This can be either be enforced by requiring $e_i = 0$ or by eliminating from the global enumeration of edges the required edges. This second way leads to the reduction in size of the solving system, with lower memory occupation and CPU usage and will hence be followed.

9.2 Numerical Implementation

The algorithm described in the previous section has been implemented using first-order Nedelec tetrahedral elements as presented in Chapter 7 and, for the sake of simplicity, only homogeneous rectangular waveguides are considered as ports. As already stated, if general shape ports are present, the reader could implement FORTRAN routines to calculate eigenmodes at the two-dimensional ports, exploiting the code developed in Chapter 3.

In this chapter code for the computation of the global matrix is more complex than in the case of resonant cavities FEM analysis, because the global system to be solved is obtained assembling 4 blocks, as derived in (9.35). The assembly of matrix $[S]$, deriving from matrices $[E]$ and $[F]$ is essentially equivalent to the case treated in Chapter 8, for resonant cavities.

Calculation of matrices $[D]$, $[C]$, and $[P]$, taking into account respectively the interaction between different port modes and the interaction between port modes and inner field, will be outlined in the following.

As stated earlier the number of modes considered in the modal expansion at each port is necessarily finite. The finite set of modes which will be taken into account for the analysis, must be appropriately sorted at all ports, by their cutoff frequency value. The code developed here asks the user for two separate numbers at each port $M_H^{(k)}$ and $M_E^{(k)}$, stating how many TE and TM modes are to be considered, respectively. Modes are then computed and sorted and the required number of modes selected. Internally an appropriate array maps the single mode index m to the correct pair of indices i_m, j_m defining the mode, and to the mode type. This operation is repeated, in sequence, for all ports.

Using this conventional rule, it is possible to map the finite series of modes onto a single one-dimensional array.

Exploiting the field expression in (9.11) and (9.13) the integral in (9.32) can be computed. Entries $D_{mn}^{(q),(k)}$ are zero if $q \neq k$ since modes spans different ports. In case of TE modes m and n both at port k, for example, it is:

$$\begin{aligned} D_{mn}^{(k),(k)} &= -k_{y_m}^{(k)} k_{y_n}^{(k)} \int_0^{a^{(k)}} \int_0^{b^{(k)}} \cos\left(k_{xm}^{(k)} x^{(k)}\right) \cos\left(k_{xn}^{(k)} x^{(k)}\right) \\ &\quad \sin\left(k_{y_m}^{(k)} y^{(k)}\right) \sin\left(k_{y_n}^{(k)} y^{(k)}\right) dx^{(k)}\, dy^{(k)} + \\ &\quad - k_{xm}^{(k)} k_{xn}^{(k)} \int_0^{a^{(k)}} \int_0^{b^{(k)}} \sin\left(k_{xm}^{(k)} x^{(k)}\right) \sin\left(k_{xn}^{(k)} x^{(k)}\right) \\ &\quad \cos\left(k_{y_m}^{(k)} y^{(k)}\right) \cos\left(k_{y_n}^{(k)} y^{(k)}\right)\, dx^{(k)}\, dy^{(k)} \end{aligned} \tag{9.36}$$

Considering mode orthogonality all these integrals are zero except for $m = n$, that is:

$$D_{mn_H}^{(q),(k)} = -\begin{cases} \left[\left(k_{xm}^{(k)}\right)^2 + \left(k_{y_m}^{(k)}\right)^2\right] \dfrac{a^{(k)}}{2 - \delta_{0i_m}} \dfrac{b^{(k)}}{2 - \delta_{0j_m}} & \text{if } q = k \text{ and } m = n \\ 0 & \text{otherwise} \end{cases} \tag{9.37}$$

with δ_{ij} Kronecker's delta, which is 1 if $i = j$ and zero otherwise. Following a similar derivation, for TM modes one obtains:

$$D_{mn_E}^{(q),(k)} = \begin{cases} j\beta_m^{(k)} \left[\left(k_{xm}^{(k)}\right)^2 + \left(k_{y_m}^{(k)}\right)^2\right] \dfrac{a^{(k)}}{2} \dfrac{b^{(k)}}{2} & \text{if } q = k \text{ and } m = n \\ 0 & \text{otherwise} \end{cases} \tag{9.38}$$

and, of course, if a TE-like weight and a TM mode are considered at port k their integral is zero, again for mode orthogonality.

The evaluation of entries of matrix $[P]$ is a bit more complicated, because the integrals involved include both element basis functions and waveguide modes. Such integrals are better calculated via approximated numerical integration over triangular domains. Naming $\bar{\boldsymbol{\alpha}}_i^{(e)}$ the restriction of base $\boldsymbol{\alpha}_i^{(e)}$, to the face of (e) lying on port (k), explicitating its component as $\bar{\boldsymbol{\alpha}}_i^{(e)} = \bar{\alpha}_{i_x^{(k)}}^{(e)} \hat{\mathbf{x}}^{(k)} + \bar{\alpha}_{i_y^{(k)}}^{(e)} \hat{\mathbf{y}}^{(k)}$, and by using $\bar{\boldsymbol{\alpha}}_i^{(e)}$ in (9.26), for TE modes, one obtains:

$$\begin{aligned} P_{im}^{(e),(k)} &= j\beta_m^{(k)} k_{y_m}^{(k)} \int_0^{a^{(k)}} \int_0^{b^{(k)}} \bar{\alpha}_{i_x^{(k)}}^{(e)} \cos\left(k_{xm}^{(k)} x^{(k)}\right) \sin\left(k_{y_m}^{(k)} y^{(k)}\right)\, dx^{(k)} dy^{(k)} \\ &\quad - j\beta_m^{(k)} k_{xm}^{(k)} \int_0^{a^{(k)}} \int_0^{b^{(k)}} \bar{\alpha}_{i_y^{(k)}}^{(e)} \sin\left(k_{xm}^{(k)} x^{(k)}\right) \cos\left(k_{y_m}^{(k)} y^{(k)}\right)\, dx^{(k)} dy^{(k)} \end{aligned} \tag{9.39}$$

Similar expression can be derived in case of TM modes. It is worth reminding that in a vector tetrahedron element if a face is selected and the three edges pertaining to the face are considered, the associated bases exhibit only tangential components on the face itself, hence no $z^{(k)}$ component is present in $\bar{\boldsymbol{\alpha}}$.

A similar approach is valid also for the electric continuity matrix [C]. By first recognizing that $\cdot\hat{\mathbf{z}}^{(q)} \times \boldsymbol{\alpha}_j^{(e)} \times \hat{\mathbf{z}}^{(q)} = \bar{\boldsymbol{\alpha}}_j^{(e)}$ and by substituting (9.11) into (9.31) yields, for TE modes:

$$C_{nj}^{(e),(q)} = \frac{1}{jk_0\zeta_0}\int_0^{a^{(k)}}\int_0^{b^{(k)}} \bar{\alpha}_{i_x^{(k)}}^{(e)} \cos\left(k_{x_m}^{(k)} x^{(k)}\right) \sin\left(k_{y_m}^{(k)} y^{(k)}\right)\, dx^{(k)} dy^{(k)} - \frac{1}{jk_0\zeta_0}\int_0^{a^{(k)}}\int_0^{b^{(k)}} \bar{\alpha}_{i_y^{(k)}}^{(e)} \sin\left(k_{x_m}^{(k)} x^{(k)}\right) \cos\left(k_{y_m}^{(k)} y^{(k)}\right)\, dx^{(k)} dy^{(k)} \tag{9.40}$$

TM mode being similar.

It is evident that analytic computation of (9.39) and (9.40) is unfeasible. Hence in order to calculate the interaction matrix between internal and port fields for the continuity of the tangent magnetic field, it is necessary to resort to numerical integration.

Surface integrals must be computed on the triangular face, this can be done by approximating the integral with a finite summation. On a unitary triangle [Figure 9.2(a)] it is

$$\int_0^1\int_0^{1-v} f(\upsilon, v)\,d\upsilon\, dv \simeq \sum_{n=1}^{N} w_n f(\upsilon_n, v_n) \tag{9.41}$$

Points (υ_n, v_n) and weights w_n can be chosen so as to guarantee that the formula holds exactly when function f is a polynomial. In particular the formula is said of degree P if it is exact for all polynomials of degree no greater than P.

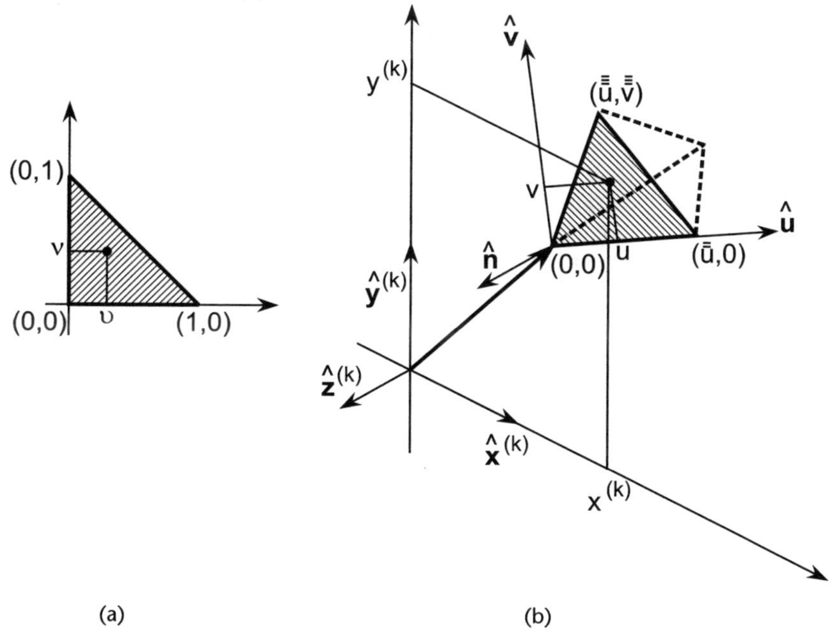

Figure 9.2 Unitary triangle (a), and its mapping (b) on a tetrahedron face on port (k).

In [12] formulas for integration over simplexes are given, and values for order up to $P = 5$ are presented for triangles. The number of points on which the summation must be done for $P = 5$ is $N = 7$ and Table 9.1 reports the points and coefficients for this order, which is the one used in the code.

Clearly a coordinate transformation from the space (υ, ν) to the space $(x^{(k)}, y^{(k)})$ is required. This is performed in two steps. First from (υ, ν) to a coordinate system (u, v) defined on the triangle so that vertex 1 is in $(0,0)$, vertex 2 in $(\bar{u}, 0)$, and vertex 3 in $(\bar{\bar{u}}, \bar{\bar{v}})$. This transformation maps $u = \bar{u}\upsilon + \bar{\bar{u}}\nu$ and $v = \bar{\bar{v}}\nu$ and its Jacobian is equal to twice the area of the triangle $|\mathbf{J}| = \bar{u}\bar{\bar{v}}$. The transformation from (u, v) to $(x^{(k)}, y^{(k)})$ can be handled implicitly since it is a mere rototranslation and hence its Jacobian is $|\mathbf{J}| = 1$. The integral can then be easily computed using $x^{(k)}$ and $y^{(k)}$ components of $\bar{\boldsymbol{\alpha}}_i^{(e)}$ and $\mathbf{e}_m^{(k)}$ while maintaining the point in the (u, v) space. In summary

$$\int_T^{(e)} f(x^{(k)}, y^{(k)}) dx^{(k)} dy^{(k)} \simeq \left[\sum_{n=1}^{N} w_n f(u_n, v_n) \right] \bar{u}\bar{\bar{v}} \tag{9.42}$$

being $u_n = \bar{u}\upsilon_n + \bar{\bar{u}}\nu_n$ and $v_n = \bar{\bar{v}}\nu_n$.

Since fields are already given on Cartesian components (9.11) and (9.12), the only issue is to obtain $\bar{\alpha}_{i_x^{(k)}}^{(e)}$ and $\bar{\alpha}_{i_y^{(k)}}^{(e)}$ which can be recovered easily by projecting its components on $\hat{\mathbf{x}}^{(k)}$ and $\hat{\mathbf{y}}^{(k)}$.

As a last note, mode normalization must be considered. Field expressions (9.11), (9.12), (9.13), and (9.14) do define orthogonal modes but the power associated with them is not unitary. For a correct implementation impinging wave amplitudes must be normalized and, consequently, outgoing wave amplitude must be denormalized. This being mandatory in case ports are of different sizes. The

Table 9.1 Points and Weights for Order $P = 5$ Over $N = 7$ Points for Integration on Triangles

Index n	*Point* (υ_n, ν_n)	*Weight* w_n
1	$\left(\frac{1}{3}, \frac{1}{3}\right)$	$\frac{9}{80}$
2	$\left(\frac{6-\sqrt{15}}{21}, \frac{6-\sqrt{15}}{21}\right)$	$\frac{155-\sqrt{15}}{2400}$
3	$\left(\frac{9+2\sqrt{15}}{21}, \frac{6-\sqrt{15}}{21}\right)$	$\frac{155-\sqrt{15}}{2400}$
4	$\left(\frac{6-\sqrt{15}}{21}, \frac{9+2\sqrt{15}}{21}\right)$	$\frac{155-\sqrt{15}}{2400}$
5	$\left(\frac{6+\sqrt{15}}{21}, \frac{6+\sqrt{15}}{21}\right)$	$\frac{155+\sqrt{15}}{2400}$
6	$\left(\frac{9-2\sqrt{15}}{21}, \frac{6+\sqrt{15}}{21}\right)$	$\frac{155+\sqrt{15}}{2400}$
7	$\left(\frac{6+\sqrt{15}}{21}, \frac{9-2\sqrt{15}}{21}\right)$	$\frac{155+\sqrt{15}}{2400}$

normalization/denormalization factor being the power associated to each single mode which is, for TE modes:

$$\begin{aligned} P_m^{(k)} &= \frac{1}{2}\Re\int_{S_k} \mathbf{E}\times\mathbf{H}^* dS = \int_0^{a^{(k)}}\int_0^{b^{(k)}} j\omega\mu\mathbf{\Phi}_m^{(k)} \times \left[\mp j\beta_m^{(k)}\bar{\mathbf{z}}\times\mathbf{\Phi}_m^{(k)}\right]^* \cdot \pm\bar{\mathbf{z}}ds = \\ &= \frac{\omega\mu\beta_m^{(k)}}{2}\left\{\left(k_{y_m}^{(k)}\right)^2\int_0^{a^{(k)}}\int_0^{b^{(k)}}\left(\cos\left(k_{x_m}^{(k)}x^{(k)}\right)\right)^2\left(\sin\left(k_{y_m}^{(k)}y^{(k)}\right)\right)^2 dx^{(k)}\,dy^{(k)}\right. \\ &\left.+\left(k_{x_m}^{(k)}\right)^2\int_0^{a^{(k)}}\int_0^{b^{(k)}}\left(\sin\left(k_{x_m}^{(k)}x^{(k)}\right)\right)^2\left(\cos\left(k_{y_m}^{(k)}y^{(k)}\right)\right)^2 dx^{(k)}\,dy^{(k)}\right\} \\ &= \frac{\omega\mu\beta_m^{(k)}}{2}\left[\left(k_{x_m}^{(k)}\right)^2+\left(k_{y_m}^{(k)}\right)^2\right]\frac{a^{(k)}}{2-\delta_{0i_m}}\frac{b^{(k)}}{2-\delta_{0j_m}} \end{aligned} \quad (9.43)$$

9.3 The Code WDEV

The `WDEV` code implemented for this chapter is by far the most complex of the book and it can be used to analyze waveguide devices with up to four rectangular ports, this limit can be easily augmented changing the pertinent parameters in the FORTRAN code. The code has been implemented to solve the vector Helmholtz equation inside the device domain in presence of anisotropic materials both of electric and magnetic type and possibly with finite conductivity, both of electric and magnetic type.

The number of FORTRAN routines which have to be compiled and linked together is about 50, to facilitate the task a `Makefile` is also provided, tailored on a `GNU make` and `GNU g77` compiler, but the reader can of course use any compiler. `WDEV` is designed to be a general software, although not comparable with commercial codes.

The execution flow is essentially similar to the `EHDEV` code presented in Chapter 4. The `Main` program calls sequentially four high-level routines:

1. `MainIn` : providing data acquisition for geometry, mesh, and simulation parameter;
2. `bBuild` : assembling the global matrices $[S]$, $[P]$, $[C]$, $[D]$ as well as the right-hand side vectors $[I_E]$ and $[I_H]$;
3. `bSolve` : implementing the solution of the linear algebra problem exploiting the block form of the system matrix;
4. `MainOut` : providing data output and writing set of output file storing internal fields and scattering parameters.

The program needs three separate input files, usually sharing the same name but presenting different extensions:

`*.in`: containing all the simulation information and the name of geometry files.

Table 9.2 Input File "*.in" Format Description

Input Data	*Data Type and Description*
HEADER	character*80
[MESH INPUT FILE]	A one-line tag of the MESH input file block
<filename>.fem	character*80 mesh file name
[MODE MATCHING FILE]	A one-line tag of the MM input file block
<filename>.fmm	character*80 mode-matching file name
[FREQUENCY]	A one-line tag of the frequency definition block
freq	1 Real frequency analysis
[SCALING]	A one-line tag of the scaling definition block
scx, scy, scz	3 Reals scaling factors along the 3 dimensions
[MATERIALS]	A one-line tag of the materials definition block
Ndiel	1 Integer number of materials
ND1,ND2,...	Ndiel labels label of materials
Material # = ND1	1 Integer defining the material
Relative Permittivity	
(Rxx,Ixx), (Rxy,Ixy), (Rxz,Ixz)	9 Complex values
(Ryx,Iyx), (Ryy,Iyy), (Ryz,Izz)	defining the permittivity
(Rzx,Izx), (Rzy,Izy), (Rzz,Izz)	tensor
Relative Permeability	
(Rxx,Ixx), (Rxy,Ixy), (Rxz,Ixz)	9 Complex values
(Ryx,Iyx), (Ryy,Iyy), (Ryz,Izz)	defining the permeability
(Rzx,Izx), (Rzy,Izy), (Rzz,Izz)	tensor
Electric Conductivity	
Rxx, Rxy, Rxz	9 Real values
Ryx, Ryy, Ryz	defining the electric
Rzx, Rzy, Rzz	conductivity
Magnetic Conductivity	
Rxx, Rxy, Rxz	9 Real values
Ryx, Ryy, Ryz	defining the magnetic
Rzx, Rzy, Rzz	conductivity
Relative Permittivities	
⋮	Repeat previous blocks as many times as there are materials
[MODE MATCHING DATA]	A one-line tag of the port definition block
Nports	1 Integer number of ports
K	1 Integer the port index
Origin = Ox, Oy, Oz	3 Real local coordinate origin point
Versor x = xx, xy, xz	3 Real local $x^{(1)}$ versor in global (x, y, z) coordinates
Versor y = yx, yy, yz	3 Real local $y^{(1)}$ versor in global (x, y, z) coordinates
modes = Nm_1	1 Integer number of modes at port 1
TE modes = Nm_te_1	1 Integer number of TE modes at port 1
TM modes = Nm_tm_1	1 Integer number of TM modes at port 1
(Ar1, Ai1)	1 Complex amplitude of impinging mode 1
⋮	Follows as many amplitude lines as there are modes

Table 9.2 *(continued)*

⋮	Repeat previous blocks as many times as there are ports
`[OUTPUTS]`	A one-line tag of the outputs definition block
`Interior Field [Y|N] =` flag_field	`1 character` if 'Y' inner field are computed and stored
`Dump System to file [Y|N] =` flag_dump	`1 character` if 'Y' matrices $[S]$, $[P]$ $[C]$ $[D]$ and vectors $[I_E]$ and $[I_H]$ are dumped to files

`*.fem`: three-dimensional mesh geometry file, providing all the data structures describing the mesh of the device under test.
`*.mm`: two-dimensional mesh geometry file of ports, containing the mesh information for each of the device port and its connections to the three-dimensional mesh.

The name of the file with extension `*.in` is requested interactively by the code, the names of the other two files are contained in this first file. The syntax of the file is reported in example and explained in Table 9.2. Please note that file reading occurs with FORTRAN `FORMAT` statements which are pretty strict.

Example 9.1

```
*******INPUT DATA FILE******
[MESH INPUT FILE]
pshift.fem
[MODE MATCHING FILE]
pshift.mm
[FREQUENCY]
9.000000                              ! Frequency is in GHz
[SCALING]
0.001, 0.001, 0.001                   ! Scaling along X,Y and Z.
[MATERIALS]
1                                     ! Number of different materials
1                                     ! Material Labels
Material # = 1
Relative Permittivity
( 1.00, 0.00), ( 0.00, 0.00), ( 0.00, 0.00)
( 0.00, 0.00), ( 1.00, 0.00), ( 0.00, 0.00)
( 0.00, 0.00), ( 0.00, 0.00), ( 1.00, 0.00)
Relative Permeability
( 1.00, 0.00), ( 0.00, 0.00), ( 0.00, 0.00)
( 0.00, 0.00), ( 1.00, 0.00), ( 0.00, 0.00)
( 0.00, 0.00), ( 0.00, 0.00), ( 1.00, 0.00)
Electrical Conductivity
0.00, 0.00, 0.00
0.00, 0.00, 0.00
0.00, 0.00, 0.00
Magnetic Conductivity
0.00, 0.00, 0.00
0.00, 0.00, 0.00
0.00, 0.00, 0.00
[MODE MATCHING DATA]
```

```
2                                                 ! Number of ports
1                                                 ! Data for Port 1
Origin    =   22.860,  0.000,  0.000
Versor u  =   -1.000,  0.000,  0.000
Versor v  =    0.000,  1.000,  0.000
dimensions = 22.860,  10.160
modes = 4
TE modes = 4
TM modes = 0
( 1.00, 0.00)                                     ! Amplitudes of incoming waves
( 0.00, 0.00)                                     ! First all TEs, then all TMs
( 0.00, 0.00)
( 0.00, 0.00)
2                                                 ! Data for port 2
Origin = 0.000, 0.000, 10.000
Versor u = 1.000, 0.000, 0.000
Versor v = 0.000, 1.000, 0.000
dimensions = 22.860, 10.160
modes = 4
TE modes = 4
TM modes = 0
( 0.00, 0.00)                                     ! Amplitudes of incoming waves
( 0.00, 0.00)                                     ! First all TEs, then all TMs
( 0.00, 0.00)
( 0.00, 0.00)
[OUTPUTS]
Interior Field [Y|N] = N
Dump System to file [Y|N] = N
```

The number of information that the user has to include into the input file is quite large, but necessary to describe a generic waveguide device.

The file contains frequency as a single value since, to maintain a lower code complexity, no loop in frequency is implemented.

Material description is quite lengthy, since permittivities and permeabilities must be described as complex tensors and conductivities, both electric and magnetic, must be given as real tensors.

The information about ports comprehends the local reference as well as dimensions, the number and type of modes that must be used during the computation is defined for each port separately. For each mode at each port an impinging amplitude can be provided.

At the end of the file some output settings provides output file control.

To better explain the input file format Example 9.1 provides the input file relative to the first example, the phase shifter, which will be considered in the following section.

The name of the two files containing mesh information and mode-matching auxiliary data are included into the input file. The file format for the mesh file is the same described in Chapter 7 and used for the `CAVITY` program and is hence not described here.

The mode matching file format is reported in Table 9.3. Essentially it provides an enumeration of the edges and faces of the tetrahedrons lying on the device ports. This file is automatically generated by the `plc2qf` application together with the mesh file if appropriate positive labels are defined.

Table 9.3 Mode-Matching File Format Syntax

Input Data	*Data Type and Description*
`Number of ports = NP`	`1 Integer` number of ports
`FACES AND RELATIVE EDGES`	A one-line tag of the face definition block
`Port: IP`	`1 Integer` port number, IP=1,. . . ,NP
`Number of faces = NF`	`1 Integer` number of face on port IP
`IF NE E1 E2 E3 IE`	`6 Integers`, face index IF, number of edges defining the face NE, local edge numbers E1, E2, E3 within the tetrahedron, tetrahedron index IE
⋮	
⋮	Previous block is repeated as many times as there are ports
`PORTS WITH RELATIVE EDGES`	A one-line tag of the edge definition block
`Port: IP`	`1 Integer` port number, IP=1,. . . , NP
`Number of edges = NED`	`1 Integer` number of edges on port IP
⋮	`NED` lines, each containing
`IEL IEG`	`2 Integers` edge local to the port (IEL) and global to the volume (IEG) number
⋮	
⋮	Previous block is repeated as many times as there are ports

The `*.mm` file stores information about faces, which is essential to the numerical computation of [*P*] and [*C*] matrices, and information about the global number of the edges lying on the port. The local face edge numbering to global volume edge numbering vector is provided to allow correct interpretation of [*P*] and [*C*] matrices, as it will be explained hereafter.

Example 9.2 shows some lines from the `*.mm` file of the first example of this chapter.

Example 9.2

```
Number of ports    =   2
FACES AND RELATIVE EDGES
Port:   1
Number of faces    =   49
    1    3    1    4    2    381
    2    3    5    6    4    165
   ...
   49    3    6    3    2    392
Port:   2
Number of faces    =    57
    1    3    6    3    2      5
    2    3    1    4    2    147
   ...
   57    3    6    3    2    265
PORTS WITH RELATIVE EDGES
Port:    1
Number of edges    =    84
    1    1
```

```
    2    2
    3    4
   ...
    84    640
Port:    2
Number of edges   =   97
    1    16
    2    17
    3    19
   ...
   97   683
```

After having loaded the input files, initialized all the data structures, stored mesh, labels, and port data, the program start assembling the global matrices, first renumbering edges so as to discard those on Dirichelet boundaries. Then $[S]$ matrix assembly takes place. This procedure is straightforward except for the anisotropic nature of the materials. The following stage builds $[P]$, $[C]$, and $[D]$ matrices and the right-hand side vectors $[I_E]$ and $[I_H]$.

Matrix $[S]$ is stored in band-storage mode, as described in Chapter 2. The reader might wish to experiment with sparse storage for it. Matrix $[D]$ is diagonal and small. It is stored as a full matrix since the waste of memory is anyway very reduced and this same memory will be reused later on as a workspace to hold data neccessary for the solution of the scattered mode amplitudes.

Matrices $[P]$ and $[C]$ are ideally very sparse, since they have as many rows and columns, respectively, as there are degrees of freedom in the volume but exhibits nonzero entries only for those edges which are on a port. To save memory these matrices are stored in a peculiar way, defining them as three-dimensional arrays. For example $[P]$ is array `P(MAXEDSP,MAXMODES,MAXPORTS)` the dimensions being `MAXEDSP`, the maximum number of edges lying on a port, which is subscript i in (9.25); `MAXMODES`, the maximum number of modes on a port, which is subscript m in (9.25); and `MAXPORTS` the maximum number of ports in the device which is superscript (k) in (9.25); superscript (e) being lost in the global matrix $[P]$.

This structure is optimal since the array `P` is indexed, in its first column, with the local number of the edge on the port, as defined in the `.mm` file. Hence the first index i is not the global edge number, but rather the local one.

In a similar way the $[C]$ matrix is stored as the three-dimensional array `C(MAXMODES,MAXEDSP,MAXPORTS)`.

This data storage structure is particularly suitable also for the solution of the block system. In fact, in almost all the analysis of microwave devices, the designer is interested in extracting the values of the generalized scattering matrix (GSM) at the ports of the device itself, rather than in computing the internal electric field.

If only the scattered wave amplitude are requested, then the same kind of block solver outlined in Chapter 4 can be used. By eliminating the internal field unknowns the system can be reduced yielding:

$$\left[[D] - [C]\,[S]^{-1}\,[P]\right][B] = [I_E] - [C]\,[S]^{-1}\,[I_H] \tag{9.44}$$

and only the columns which are relative, in the global edge numbering scheme, to an edge on a port correspond to entries in $[P]$ and in $[I_H]$, so only the columns of

$[S]^{-1}$ pertinent to these unknowns need to be computed. The gain in speed achieved is even more noticeable in three-dimensional than in two-and-a-half-dimensional since the ratio between the number of the columns of the inverse of $[S]$ to be computed and the total number of columns of $[S]$ is smaller in three-dimensional.

For what concerns outputs, if so requested the code dumps the whole linear system. This is quite a time-consuming operation since all matrices entries are dumped. The code, if so requested also writes the coefficient of the field expansion on the mesh edges. This might be useful in case some sort of visualization software is devised. In any case the format of these files is quite self-explanatory and will not be discussed here.

The main output of the code are five files. The first four, named `IncomingWaves_A_N.out`, `IncomingWaves_A_NN.out`, `OutgoingWaves_A_N.out` and `OutgoingWaves_A_NN.out` are relative to the normalized (`N`) and not normalized (`NN`) amplitudes of the incoming and outgoing waves at the various ports. Examples 9.3 and 9.4 presents the not-normalized amplitudes of incoming and outgoing wave, respectively, relative to a 10 mm long waveguide (see following section). It is apparent how the unitary incoming TE_{10} wave at port 1 generates a nearly unitary outgoing TE_{10} wave at port two, the other entries being due to numerical approximations.

Example 9.3

```
# Port, Mode, TX[m,n] Real, Imag, Amplitude, Phase
1  1  TE[ 1, 0]  ( 0.100000E+01  0.000000E+00)  0.10000000E+01  0.00000000E+00
1  2  TE[ 2, 0]  ( 0.000000E+00  0.000000E+00)  0.00000000E+00  0.00000000E+00
1  3  TE[ 0, 1]  ( 0.000000E+00  0.000000E+00)  0.00000000E+00  0.00000000E+00
1  4  TE[ 1, 1]  ( 0.000000E+00  0.000000E+00)  0.00000000E+00  0.00000000E+00
2  1  TE[ 1, 0]  ( 0.000000E+00  0.000000E+00)  0.00000000E+00  0.00000000E+00
2  2  TE[ 2, 0]  ( 0.000000E+00  0.000000E+00)  0.00000000E+00  0.00000000E+00
2  3  TE[ 0, 1]  ( 0.000000E+00  0.000000E+00)  0.00000000E+00  0.00000000E+00
2  4  TE[ 1, 1]  ( 0.000000E+00  0.000000E+00)  0.00000000E+00  0.00000000E+00
```

Example 9.4

```
# Port,  Mode,  TX[m,n]  Real,  Imag,  Amplitude,  Phase
1 1  TE[ 1, 0] (-0.626486E-02  -0.621997E-03)  0.62956563E-02  -0.30426335E+01
1 2  TE[ 2, 0] (-0.139838E-02   0.237090E-02)  0.27525653E-02   0.21036909E+01
1 3  TE[ 0, 1] ( 0.929192E-03  -0.124986E-02)  0.15574149E-02  -0.93150824E+00
1 4  TE[ 1, 1] (-0.249490E-03  -0.500022E-03)  0.55880868E-03  -0.20336099E+01
2 1  TE[ 1, 0] ( 0.261175E+00  -0.965271E+00)  0.99998039E+00  -0.13065513E+01
2 2  TE[ 2, 0] (-0.961318E-03  -0.201702E-02)  0.22343916E-02  -0.20155516E+01
2 3  TE[ 0, 1] ( 0.148671E-03   0.118855E-02)  0.11978121E-02   0.14463564E+01
2 4  TE[ 1, 1] (-0.761326E-03  -0.488521E-03)  0.90458273E-03  -0.25710948E+01
```

The fifth output file, `ScatteringMatrixColumn.out` is produced only if there is a single incident wave at the fundamental mode and consists in the ratio between $B_1^{(k)}$ and $A_1^{(l)}$, effectively providing one column of the scattering matrix of the device (Example 9.5).

Example 9.5

```
# Fed Port is 1
S_{1,1} = (-0.626486E-02 -0.621997E-03)  0.62956568E-02  -0.30426335E+01
S_{2,1} = ( 0.261175E+00 -0.965271E+00)  0.99998039E+00  -0.13065513E+01
```

9.4 Some Examples

In this section the code WDEV is run with a few examples to show its capabilities.

As first test case, to provide a simple validation of the code, the analysis of a fixed phase shifter similar to the one already analyzed in Chapter 4 is presented. The phase shifter is made of a section of rectangular waveguide with $a = 22.86$ mm and $b = 10.16$ mm, $l = 10$ mm-long; the geometry of this simple device is displayed in Figure 9.3.

The ideal scattering matrix of this device for only the fundamental mode TE_{10}, mode index $m = 1$, wavenumber β_1, being both ports equal superscript (k) is omitted, is

$$[S] = \begin{bmatrix} 0 & e^{j\beta_1 l} \\ e^{j\beta_1 l} & 0 \end{bmatrix} \tag{9.45}$$

Obviously ports are perfectly matched and outputs delayed with respect to inputs according to the mode wavenumber and device length. Phase shift $\phi = \beta_1 l$ itself changes as the wave frequency f changes as β_1 is a function of frequency:

$$\phi = \beta l = \frac{2\pi}{\lambda_{wg_1}} l = \frac{2\pi f}{c_0} l \sqrt{1 - \left(\frac{c_0}{2af}\right)^2} \tag{9.46}$$

where c_0 is the speed of light.

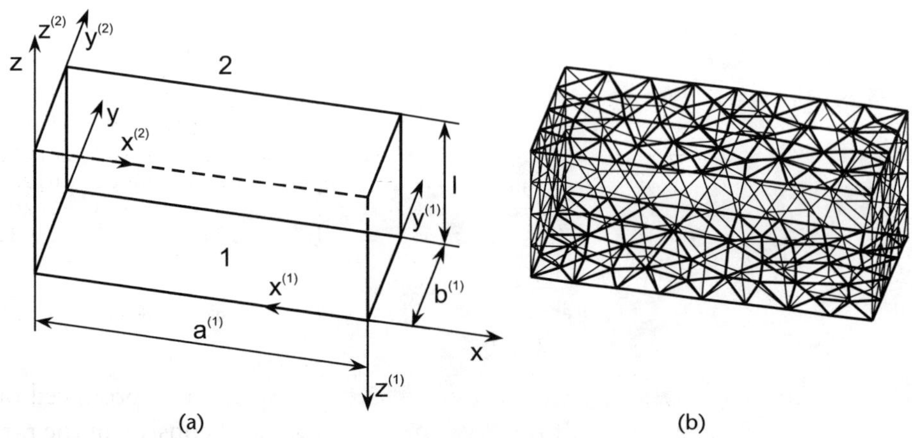

Figure 9.3 Fixed phase shifter: geometry (a), ports are equal, length is $l = 10$ mm; mesh (b) only tetrahedron faces on boundaries are drawn, ports are drawn with thicker lines.

The device has been analyzed as a function of frequency and at increasing number of elements within the mesh. The input file used for this test case is `pshift_#.in` and can be found in the CD-ROM, # being the frequency in GHz defined in the file. A set of directories `pshift_#`, with # the number of degrees of freedom in the mesh holds files `pshift.poly` `pshift.fem`, `pshif.mm`, as well as the output files generated by the code. The reader is encouraged to thoughtfully analyze the `*.in` file especially for what concerns the definition of the port local reference, which is a tricky issue.

The phase of the S_{21} scattering parameters has been extracted from the output data and reported in Table 9.4, where the analytically evaluated phase shift is also reported.

Figure 9.4 shows the data graphically, on a larger number of different meshes (779, 1011, 1438, 1680, 2164, and 3254 degrees of freedom). It is apparent how accuracy increases as the mesh is refined.

As second test case the same waveguide H-plane bend presented in Chapter 4 is considered. Waveguide dimensions are $a = 18.35$ and $b = 9.175$, the chosen bend is the mitered one, with mitering parameter $p = 3$ mm. The enclosed CD-ROM holds all the input files `bend_#.in` relative to the various frequency points #, plus two directories `bend_1580` and `bend_3596` holding all the `bend.poly`, `bend.fem`, `bend.mm` and output files for the pertinent number of degrees of freedom. The analysis of the device is performed expanding the electric field by 4-TE modes at each port. The device structure and boundary mesh is displayed in Figure 9.5.

The analysis results are compared into the graphs of Figures 9.6 and 9.7, in which are respectively displayed the magnitude and the phase of the scattering parameters

As expected there is little difference between the two simulation on the frequency band, being the larger mesh simulation more accurate. Furthermore, comparison with results presented in Chapter 4 crossvalidate the implementations.

At last, as a truly three-dimensional example, the analysis of a four-port hybrid-T junction is considered. The geometry of such a microwave device is depicted in Figure 9.8 and is realized joining 4-rectangular waveguides with $a = 18.35$ and $b = 9.175$, three of them on the H-plane and the fourth joining the other three from the direction normal to the H-plane.

This is a very simple device providing an acceptable isolation between ports 1 and 2, while all the other ports are coupled. By inserting suitable reactive elements within the device the hybrid-T junction can be transformed to the well-known magic-T junction [13]. Referring to Figure 9.8 port 1 is also known as sum port,

Table 9.4 Waveguide Fixed Phase Shift Convergence Varying the Mesh Size and Analysis Frequency

Freq.	*Analyt.*	*FEM(779 DoF)*		*FEM(1438 DoF)*		*FEM(3254 DoF)*	
GHz	*$\Delta\phi$[deg]*	*$\Delta\phi$[deg]*	*Error*	*$\Delta\phi$[deg]*	*Error*	*$\Delta\phi$[deg]*	*Error*
9.00	−74.03	−74.86	1.12%	−74.43	0.54%	−74.20	0.23%
10.0	−90.66	−91.68	1.13%	−91.18	0.57%	−90.87	0.23%
11.0	−106.06	−107.32	1.20%	−106.73	0.63%	−106.31	0.24%

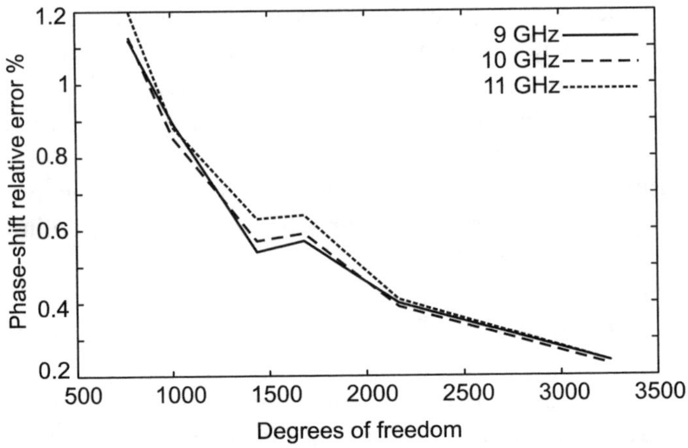

Figure 9.4 Phase shift relative error convergence with increasing mesh size.

since waves coming from ports 3 and 4 exit port 1 in phase; port 2 is called difference port, since waves coming from port 3 and 4 exit port 4 out of phase. Ports 3 and 4 can be named input ports and, due to the device symmetry, exhibit the same behavior. Unless other reactive components are included within the domain of the junction none of the device ports is matched, ports 3 and 4 are coupled, and the power splitting between sum and difference ports when a single input port is fed is not even. The generic scattering matrix of this device is:

$$\begin{bmatrix} B_1 \\ B_2 \\ B_3 \\ B_4 \end{bmatrix} = \begin{bmatrix} S_{11} & S_{12} & S_{13} & S_{14} \\ S_{21} & S_{22} & S_{23} & S_{24} \\ S_{31} & S_{32} & S_{33} & S_{34} \\ S_{41} & S_{42} & S_{43} & S_{44} \end{bmatrix} \begin{bmatrix} A_1 \\ A_2 \\ A_3 \\ A_4 \end{bmatrix} \tag{9.47}$$

The first consideration is that, due to the reciprocity of the device, its scattering matrix must be symmetric and, hence $S_{ij} = S_{ji}$. It is then evident, from Figure 9.8,

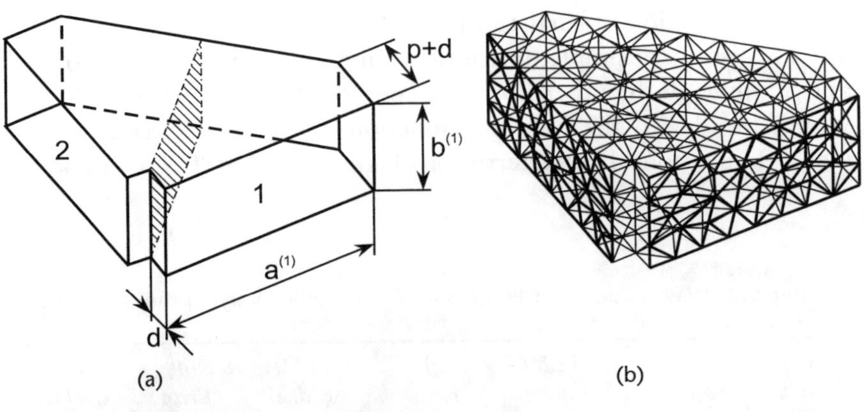

Figure 9.5 H-plane rectangular waveguide mitered bend: geometry (a), hatched area highlights symmetry plane, ports are equal, $d = 2$ mm, $p = 3$ mm; mesh (b) only tetrahedron faces on boundaries are drawn, ports are drawn with thicker lines.

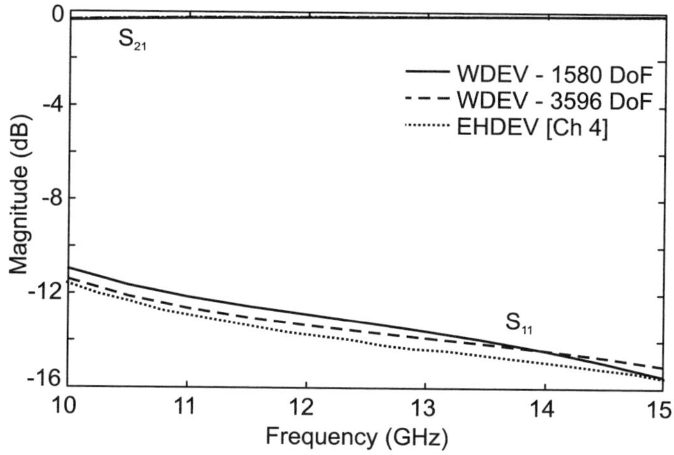

Figure 9.6 H-plane rectangular waveguide mitered bend-scattering parameters magnitude as a function of frequency. Curves represent two separate analyses on two different meshes as well as, for comparison, the results obtained in Chapter 4.

that the device is symmetric if viewed from the port 1 while it is antisymmetric if viewed from the port 2. For this reason in the scattering matrix the coupling coefficients S_{31} and S_{41} are equal and in the following will be named C_s, while the coupling coefficients, S_{32} and S_{42} must be equal magnitude and opposite in phase: $S_{32} = -S_{42} = C_d$. Furthermore, again due to symmetry, $S_{33} = S_{44} = R_i$. In summary:

$$\begin{bmatrix} S_{11} & S_{12} & S_{13} & S_{14} \\ S_{21} & S_{22} & S_{23} & S_{24} \\ S_{31} & S_{32} & S_{33} & S_{34} \\ S_{41} & S_{42} & S_{43} & S_{44} \end{bmatrix} = \begin{bmatrix} R_s & I_{sd} & C_s & C_s \\ I_{sd} & R_d & C_d & -C_d \\ C_s & C_d & R_i & I_{ii} \\ C_s & -C_d & I_{ii} & R_i \end{bmatrix} \tag{9.48}$$

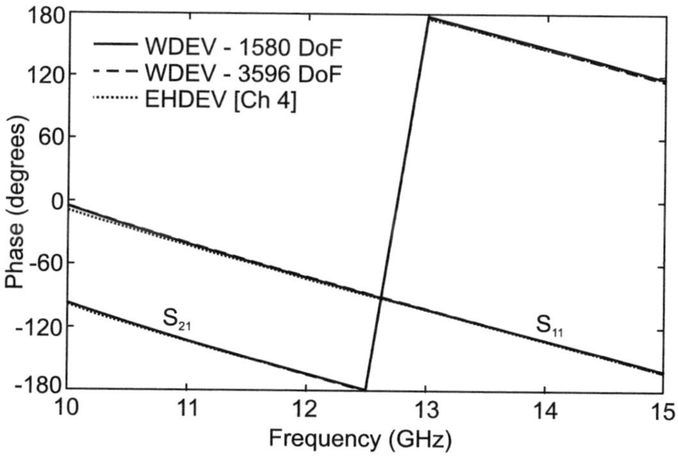

Figure 9.7 H-plane rectangular waveguide mitered bend-scattering parameters phase as a function of frequency. Curves represent two separate analyses on two different meshes as well as, for comparison, the results obtained in Chapter 4.

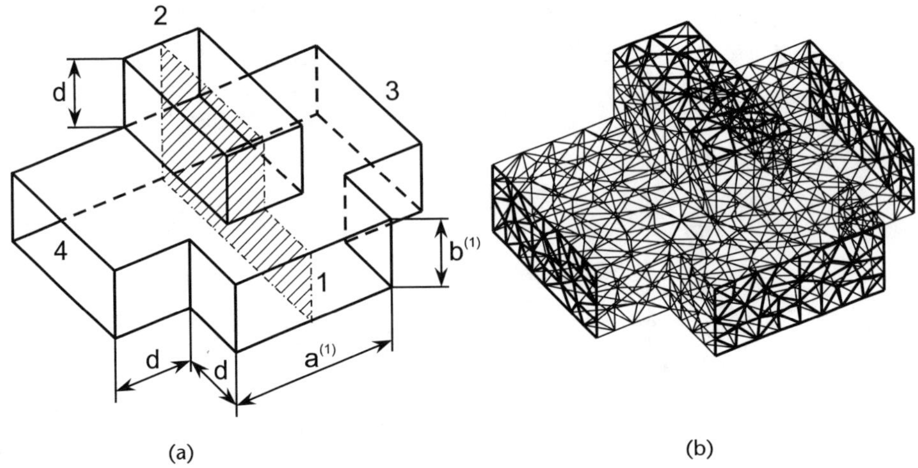

Figure 9.8 Hybrid-T junction: geometry (a) junction is symmetric with respect of the hatched plane, all ports are equal and $d = 10$ mm; mesh (b) only tetrahedron faces on boundaries are drawn, ports are drawn with thicker lines.

Generally speaking, R indicates a reflection coefficient, C indicates coupling between ports, which is a desired feature, while I indicates isolation between ports, isolated ports should ideally have no coupling ($I = 0$). In a magic-T junction, $R_s = R_d = R_i = 0$, $C_s = C_d = 1/\sqrt{2}$, and $I_{sd} = I_{ii} = 0$.

The qualitative consideration discussed above can be tested using the `WDEV` program. the simulation has been carried out with the files in the `HYBRID` directory on the CD-ROM. The several input files are named `#hybrid_##.#.in` where the first # is the fed port, while the ##.# group indicates frequency mesh information and mode-matching data are stored in `hybrid.fem` and `hybrid.mm` files respectively.

Scattering parameters have been extracted from the output data files ad plotted separately in different graphs. Figure 9.9 shows the reflection parameters at the

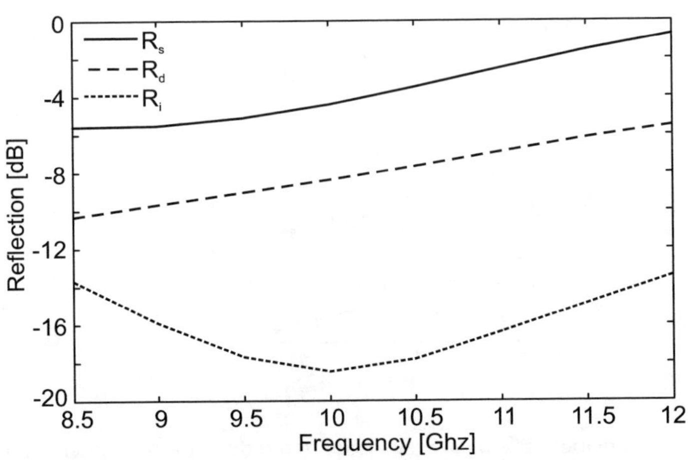

Figure 9.9 Magnitude of the reflection coefficients for the 4-port hybrid-T junction.

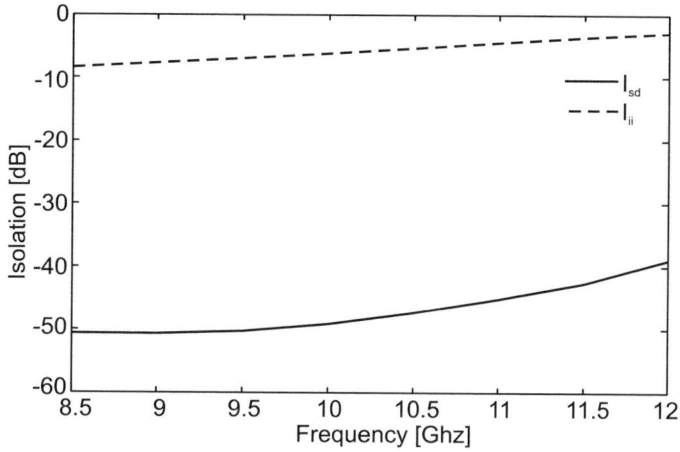

Figure 9.10 Isolation between sum-difference ports (I_{sd}), and between the two input ports (I_{ii}).

device ports, it can be easily recognized how matching is quite bad, particularly for the sum, R_s, and difference, R_d, ports. By analyzing the isolation graph in Figure 9.10 it is evident that the sum and difference ports are well decoupled, since the fundamental TE_{10} mode feeding the sum port has a structure which is related with the TE_{01} mode in port 2. Such a TE_{01} is obviously not propagating in the unimodal band of the waveguide. The nonzero value of such isolation is mainly due to the fact that even if the device is symmetric the mesh is not. On the contrary the two output ports 3 and 4 are badly isolated, but this was expected since no obstruction is present on the port 3 to 4 line of sight.

The magnitude and phase of the coupling coefficients C_s and C_d are reported in Figure 9.11 showing how the two coefficients relative phase is almost exactly 0 and 180°, respectively. Finally Figure 9.12 is relative to the usage of the hybrid-T as a comparator. Both ports 3, and 4 are fed with an incidents wave of amplitude $A_1 = \frac{1}{\sqrt{2}}$, for a total incoming power of 1W. Most of the power exits from port

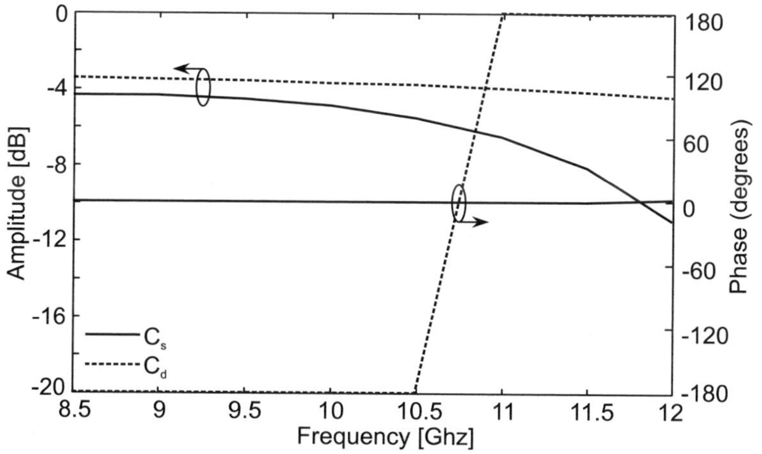

Figure 9.11 Coupling (magnitude and phase) between output and sum ports (C_s), and between output and difference ports (C_d).

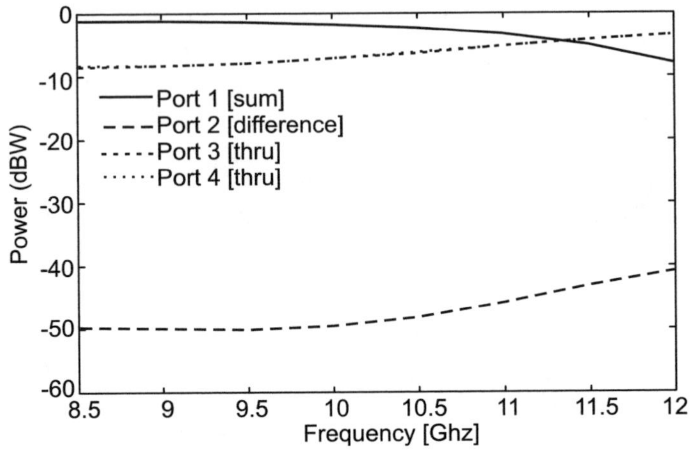

Figure 9.12 Output power at the four ports when ports 3 ad 4 are fed, in phase, with 0.5W each.

1, while output power from port 2 is nearly zero. A nonnegligible power outputs from ports 3 and 4 since these two are neither matched nor decoupled, as it would be the case in a magic-T.

9.5 Disc Content

The software described here is contained in the subdirectory WDEV of the directory CODES3D in the enclosed CD-ROM. The program files are too numerous to be listed here, the main program being stored in MAIN.F. The code is self-contained and the directory also contain a GNU format Makefile for quick compilation.

The three subdirectories PSHIFT, BEND, and HYBRID, contain all the input and mesh files of the examples discussed in the previous section. A more complete description of each file is contained in the README file in the directory WDEV itself.

References

[1] Picon, O., "Three-Dimensional Finite-Element Formulation for Deterministic Waveguide Problems," *Microwave and Optical Technology Letters*, Vol. 1(5), pp. 170–172, July 1988.

[2] Ise, K., Inoue, K., and Koshiba, M., "Three-Dimensional Finite-Element Solution of Dielectric Scattering Obstacles in a Rectangular Waveguide," *IEEE Transactions on Microwave Theory and Techniques*, Vol. 38(9), pp. 1352–1359, Sept. 1990.

[3] Dyczij-Edlinger, R., et al., "A Deterministic Approach to the Analysis of Three-Dimensional Waveguide Configurations by Finite Elements and Mode Matching," *IEEE Transactions on Magnetics*, Vol. 28(2), pp. 1235–1238, Mar. 1992.

[4] Valor, L. and Zapata, J., "Efficient Finite Element Analysis of Waveguides with Lossy Inhomogeneous Anisotropic Materials Characterized by Arbitrary Permittivity and Permeability Tensors," *IEEE Transactions on Microwave Theory and Techniques*, Vol. 43, pp. 2452–2459, 1995.

[5] Rubio, J., Arroyo, J., and Zapata, J., "Analysis of Passive Microwave Circuits by Using a Hybrid Two-Dimensional and Three-Dimensional Finite-Element Mode-Matching Method," *IEEE Transactions on Microwave Theory and Techniques*, Vol. 47, pp. 1746–1749, 1999.

[6] Rubio, J., Arroyo, J., and Zapata, J., "SFELP-An Efficient Methodology for Microwave Circuit Analysis," *IEEE Transactions on Microwave Theory and Techniques*, Vol. 49(3), pp. 509–516, Mar. 2001.

[7] Martini, E., Pelosi, G., and Selleri, S., "A Hybrid Finite-Element Modal-Expansion Method With a New Type of Curvilinear Mapping for the Analysis of Microwave Passive Devices," *IEEE Transactions on Microwave Theory and Techniques*, Vol. 51(6), pp. 1712–1717, June 2003.

[8] Ilić, M. M., Ilić, A. Ž., and Notaroš, B. M., "Higher Order Large-Domain FEM Modeling of 3-D Multiport Waveguide Structures With Arbitrary Discontinuities," *IEEE Transactions on Microwave Theory and Techniques*, Vol. 52(6), pp. 1608–1614, June 2004.

[9] Skolnik, M. I., *Radar Handbook*, 2nd edition, New York: McGraw-Hill, 1990.

[10] Easwaran, J., et al., "Computer-Aided Design of Planar Waveguide Monopulse Comparator for Low-Height Airborne Antennas," In *Radar 97*, pp. 160, 522–525, Oct. 14-16, 1997.

[11] Giunta, G., et al., "Tolerance Manufacturing Effects in Crossfeed Monopulse Radars in Presence of Rough Sea Scattering," In *IEEE Antennas and Propagation International Symposium*, pp. 160, 4817–4820, June 9-15, 2007.

[12] Hammer, P. C., Marlowe, O, J., and Stroud, A.H., "Numerical Evaluation Over Simplexes and Cones," *Mathematical Tables and Other Aids to Computation*, Vol. 10(55), pp. 130–137, July 1956.

[13] Collin, R. E., *"Foundations for Microwave Engineering,"* New York: McGraw-Hill, 1992.

PART III

To Probe Further

CHAPTER 10

Selected Bibliography

This chapter reviews the history and present state of finite element methods as applied to practical electromagnetic field propagation problems. Many papers have been published on this topic since the first edition of this book and the exponential increase of available computational power, as predicted by Moore's law, has substantially expanded the use of finite element methods to solve microwave engineering problems. The bibliography presented here is basically an updated version of a survey originally published on *IEEE Antennas and Propagation Magazine* [1] by the authors. This updated version is aimed at organizing and expanding the concepts described in the previous chapters providing annotations and evaluative comments on the works considered, where the interested reader can find valuable additional materials on finite elements methods. Due to the great rate of scientific publications on FEM and space limits, it has not been possible to include all the valuable existing works on FEM, so many others may be found in electronic databases of scientific literature.

This bibliography is organized in three sections. In the first section, the available books on the subject are reviewed. Section 10.2 gives an overview of papers in journals and transactions, considering separately

1. Those addressing fundamental issues such as element definitions and error estimation;
2. Those dealing with guided wave propagation;
3. Those treating scattering and antenna problems.

The third section discusses advanced topics and FE techniques. It describes the latest developments and applications of the FEM for wave propagation problems and might also provide insights on future research trends and topics.

10.1 Books

For the student or novice researcher, a few good textbooks can furnish initial guidance. They should be followed by a critical bibliography or two and access to the relevant articles in peer-reviewed journals. This section gives a brief overview of some available books that may serve as an initial set of research tools.

Courses on computational electromagnetics for undergraduate students, where they exist, rarely treat the finite element method at length. However, if accompanied by suitable laboratory exercises, the method can be introduced quickly and effectively even in basic courses. It can be useful, for example, when the rigorous analysis of a wave guiding structure requires unfamiliar mathematical tools such as Bessel functions, or involves oversimplification of the model, as in the usual treatment of the finline. Such early use of the finite element method requires basic textbooks. Their number is unfortunately still small, but sufficient.

There are some general texts on numerical methods for applied electromagnetism in which the finite element method is treated, though usually introduced through static or quasistatic potentials [2–6]. The literature on finite elements in general is extensive; unfortunately, much of it is concerned with structural engineering so it uses physical concepts and mathematical notations foreign to microwave engineers. Among the books presently available, that by Oden and Reddy [7] is easily readable in some parts, fortunately the most important ones! The text by Strang and Fix [8], recently reprinted, provides a good description of the basic theory and can be suggested for a first reading. However, it is not extensive enough to satisfy the demanding reader.

The first book completely committed to finite elements in electromagnetics is that by Silvester and Ferrari [9], whose much extended third edition contains not only the quasi-static potential development but also some example of three-dimensional vectorial element applications. The book covers step by step the mathematical development of FEM starting from simple static field problems and leading the reader to applications of FEM to more complex propagation problems, three-dimensional problems, and to problems involving nonlinear materials.

The text by Davidson [10], although not fully dedicated to FEM, includes a clear treatment of issues in numerical methods for electromagnetics and provides the reader with clear comparison between different numerical techniques. The examples included are of practical interest and numerical results are thoroughly analyzed to verify accuracy and to point out possible pitfalls in setting up the models for simulation.

A couple of more advanced books are those by Jin [11] and by Volakis, Chatterjee, and Kempel [12]. Both texts offer a detailed treatment of three-dimensional nodal and vector shape functions on different kind of elements and the implementation of finite element techniques for open and closed three-dimensional problems. Hybrid techniques such as finite element-boundary integral and fast integral methods are also presented in both books. Implementation of finite element in time domain is also presented in [11].

An anthology of the fundamental works in the development of finite elements, with particular emphasis on open problems, is that by Silvester and Pelosi [13]. It will be particularly useful for engineers already familiar with the fundamentals of finite elements, for it focuses on surveying the available methods for wave propagation problems rather than on detailed description of their implementation. This monograph, sponsored by the IEEE Antennas and Propagation Society, contains a selected and heavily annotated bibliography of about 200 articles dated up to 1994. A survey of applications and development in finite element software for microwave engineering, with particular emphasis to implementation and including the characteristics of popular commercial software packages, can be found in the book edited by Itoh, Pelosi, and Silvester [14]. The more advanced topic of self-adaptive mesh generation, which is extremely important to reduce discretization errors, is thoroughly covered in the book by Salazar-Palma, Sarkar, Garcia-Castillo, and Roy [15].

10.2 Scientific Literature

Among the journals dealing with finite element electromagnetics, *IEEE Transactions on Magnetics* has assumed a leading role by publishing the largest number of research and application papers on finite elements. Indeed, so far as finite element methods are concerned, the material published in this journal far transcends the magnetic components and materials implied by its title. In 1990–1997, *IEEE Transactions on Antennas and Propagation* and *IEEE Transactions on Microwave Theory and Techniques* averaged one article per issue on finite elements, while *IEEE Transactions on Magnetics* published more than 100 papers per year. The average number of papers published in the following decade 1998–2008 decreased, but is still above 70 per year. What is significantly increased in the last decade is the number of papers using finite element-based commercial software as a benchmark for other numerical techniques, or as a design tool, proving the maturity level reached by this numerical technique. Other useful resources for this subject include *IEE Proceedings-H* and *COMPEL*. Numerical aspects of finite elements, as apart from electromagnetics, are considered in the *International Journal for Numerical Methods in Engineering*. Much of the remaining material is spread over a wide range of scientific journals, some belonging to disciplines other than electrical engineering and hence using notations different from those electrical engineers are familiar with.

10.2.1 Fundamental Issues

Since its first description given by Courant [16] and its first application to the classic electromagnetic problem of guided propagation developed by Silvester [17], vast improvements have enhanced the accuracy and efficiency of FEM. Higher order nodal elements were introduced both in two- and three-dimensions and the pertinent local matrices expressed in terms of tabulated universal matrices [18–20]. The geometrical modeling capabilities of the method were greatly improved, without sacrificing simplicity, by introducing curved *isoparametric* elements [21].

More recently, many research efforts have been aimed at the development of two- and three-dimensional vectorial elements of higher order than those described in Chapters 2 and 7 and employed in Chapters 3, 4, 8, and 9 for waveguide analysis. On the basis of the pioneering works of Nedelec [22, 23] higher order vectorial elements have been developed for simplexes (triangles and tetrahedra) [24–27] as well as Cartesian elements [28–31]. These are also called orthospectral ("spectrally correct") mixed-order elements [32–34], or "curl-conforming" elements [35–37], because they employ different order vectorial basis functions to approximate the electromagnetic field in orthogonal directions. That allows ensuring tangential continuity between adjacent elements without imposing total continuity. Edge elements were initially used to avoid the spurious modes that can arise in numerical solution of Maxwell's equations. This problem was reported early in the development of finite-element applications to electromagnetic problems [38, 39], and its origin and possible remedies have been a topic of considerable interest among researchers in computational electromagnetics [40–44]. Their usage has rapidly gained popularity not only because they avoid spurious modes, but also because they allow enforcing boundary conditions in a very straightforward manner, regardless the geometric complexity of the problem being analyzed [45, 46].

Because of the flexibility and ease of use of edge-elements, the general formulation proposed in [37] to generate higher order vectorial elements, on tetrahedra has been extended to prism, pyramidal, and hexaedral elements in [47], [48], and [49], respectively. However, the accuracy of FE solutions provided by higher order elements is limited by the accuracy of the mesh, that is by how accurately the discretized domain actually models the real domain that, in many practical cases, includes objects with curved faces.

A systematic procedure for the construction of higher order and curvilinear vector finite elements was outlined in [50] and applied in [51, 52] to solve a resonant cavity and a guided propagation problem using conventional interpolating polynomial to implement the coordinate transformation from linear to curvilinear elements. These type of curvilinear or isoparametric elements, although very flexible, are somewhat limited and more general transformations are required for an exact representation of arbitrary geometries. A different kind of mapping based on nonuniform rational B-spline (NURBS) was proposed for two-dimensional FEM with quadrilateral elements in [53]. In [54, 55] the use of curved triangular and tetrahedral elements defined by a rational Bézier mapping is demonstrated and applied to the solution of resonant cavity modes and to guided propagation problems. The papers show how Bézier mapping provides a remarkable increase in accuracy and flexibility over the conventional polynomial mapping without a significant increase in computational burden.

The accuracy of finite element solutions can at times be increased by employing special shape functions with singular behavior near edges or corners, which closely approximate the electromagnetic fields there [56–61].

Even using higher order and curvilinear elements, the accuracy of the numerical solution can only be guaranteed by implementing a procedure for the a posteriori estimation of the error and the automatic mesh refinement [62, 63]. The mesh adaptive refinement can be executed following two different strategies: h-refinement and p-refinement. The h-refinement strategy increments the number of elements in the areas of the domain where the estimated error is highest [64]; the p-refinement strategy increases the order of the shape functions in those elements where the estimated error is highest [65, 66]. Hierarchical elements [67–71], whose shape functions are polynomial of increasing order and not all of the same order, are particularly suited for the implementation of p-refinement strategies since they naturally allow to use different order polynomial on adjacent elements [65, 72, 73]. Hierarchical elements appear to be particularly convenient in conjunction with curvilinear elements [74] when the geometry to be analyzed includes curved surfaces.

Some studies have also been carried on to evaluate numerical dispersion in finite element methods for both nodal and edge elements [75–80], with the aim to reducing it by a posteriori assessment in terms of higher order shape functions [81].

10.2.2 Microwave Circuits and Devices

The success of finite elements in electromagnetics can be attributed largely to their great versatility and flexibility, which allow geometrically complex structures containing inhomogeneous, anisotropic, or even nonlinear materials to be treated. In addition, finite element methods exhibit the rather pleasing characteristics of computational economy, which arises from the sparsity of their coefficient matrices. However, they have suffered from the difficulty of not being directly applicable to open problems, at least not in their basic forms, for they do not naturally account for the radiation condition. This may explain the long delay in the application of finite elements to radiation problems, in contrast to open problems of static and diffusive fields whose treatment appeared in the literature at an early date. Several hundreds of papers had been published by the end of the 1970s on finite-element applications, but articles on problems involving the radiation condition began to appear only in the early 1980s, and full three-dimensional scattering problems have been successfully solved only in the early 1990s.

Finite elements as applied to guided wave propagation initially tended to concentrate on the transverse problem subclass. Here the classical process of variable separation is used to eliminate the propagation direction, reducing the problem to that of solving a two-dimensional field problem in the plane transverse to the propagation direction, as much as it has been done in Chapter 3. Typical examples include determining the modal structure of metallic homogeneous and inhomogeneous waveguides of arbitrary cross-section (including optical guides) and finding the equivalent circuit parameters of inhomogeneous multiconductor lines (microstrip, coplanar waveguide, finline, etc.). Transverse problems pertaining to open structures, where the field is not restricted to a bounded geometric region enclosed by metallic walls but slowly decays with distance from the wave guiding structure, require special treatment. The techniques able to deal with this case can be classified into four groups [82]:

1. Ad hoc methods [83, 84];
2. Use of infinite elements [85–87];
3. Recursive condensation or ballooning [88–93];
4. Geometric transformation [94–96].

In each technique, the infinitely extending region where the field is defined is subdivided in two parts: an interior region, where the field is computed by means of the finite element method, and an exterior one, whose influence on the interior region must be taken into account while a detailed field solution in the exterior itself is of no interest.

The earliest finite-element applications to rectilinear propagation problems were concerned with finding the modal structure of arbitrarily complex closed metallic waveguides. A formulation employing the longitudinal components of the fields was initially used to analyze the loaded hollow waveguide [17, 38], also with magnetic materials [97], optical fibers [98–101], and lossy waveguides [102, 103]. This formulation is quite efficient since it use only two components of the electromagnetic field inside the waveguide, but suffers from the presence of spurious solutions. To overcome this drawback, three different methods have been proposed starting in the mid 1980s: the *penalty function* technique [104, 105], which does not completely eliminate the spurious modes but shifts their cutoff frequency outside the frequency range of interest; the field reformulation technique [106–112], which explicitly enforces solenoidality of the magnetic field as a separate condition; and finally, the use of vectorial or "curl-conforming" elements [113], which guarantee a correct representation of the null-space of the curl operator [35–37, 44], thus avoiding the appearance of spurious modes. A review and comparison of the above mentioned techniques can be found in [114–116].

Such techniques have been used to characterize waveguides loaded with both lossless ferrite, axially magnetized [117, 118] as well as magnetized in arbitrary direction [119], and lossy ferrite with longitudinal or transverse magnetization [120]. The case of closed metallic waveguides loaded with lossy bianisotropic [121] and lossy anisotropic [122] materials has been also investigated using higher order elements and model order reduction techniques [123, 124] (see Section 10.3 for references on MOR techniques). Finite elements methods have been widely exploited to attack the problem of characterization of propagation in planar structures as lossless shielded microstrips [125, 126], finlines [127], striplines [128], as well as dielectric waveguides [129] and open microstrips [130]. Formulations able to account for losses in planar structures due to conductors and/or dielectric materials have been reported in [131–133], and some efforts have been devoted to the analysis of multilayer and multiconductor transmission lines [134–137].

Dielectric waveguides comprised of lossy and/or anisotropic lossy [138], as well as anisotropic nonlinear [139] material have also been treated. A formulation able to deal with dielectric waveguides with inhomogeneous anisotropic materials characterized by arbitrary permittivity and permeability tensors has been presented in [140].

Finding the modal spectrum of a cavity poses problems similar to those encountered in the characterization of wave guiding structures. A two-dimensional approach can be used for axisymmetric cavities [141, 142], while a three-dimensional formulation is required for loaded cavities [143, 144] or when the effect of the coupled network have to be considered [145].

Discontinuities in guiding structures, such as posts, irises, junctions, devices, bends, and corners, require analytic techniques able to account for the infinitely extending part of the guiding structure not discretized with finite elements, in which the fields does not decay but does propagate. The techniques proposed make use of finite elements in conjunction with either integral equations [146–149], or modal expansion [150–154], or recursive condensation [155], or absorbing boundary conditions [156, 157]. These latter are special kinds of boundary condition able to absorb almost completely waves that impinge on the boundary. They were initially developed to absorb waves propagating in free space as they occur in scattering or antenna problems; for this reason, papers related to derivation and application will be more thoroughly considered in the next section.

Depending on the type of discontinuity and guiding structure, the problem may be analyzed using either two- or three-dimensional formulations. A simplified and efficient two-dimensional treatment may be given, for instance, to axisymmetric coaxial waveguide discontinuities [158], to homogeneous and inhomogeneous E-

and H-plane rectangular waveguide junctions or discontinuities [148–153], and to dielectric slab waveguide discontinuities [156, 159]. A full three-dimensional finite element method has been applied to the analysis of high-frequency interconnects [160], shielded microstrip discontinuities in conjunction with a recursive technique [161], or with a cavity resonance procedure [162], as well as to waveguide filters and components [154, 163, 164]. Unshielded configurations have been dealt with by using absorbing boundary conditions to terminate the finite element mesh [165].

10.2.3 Radial Propagation

In the following, the term radial propagation will denote those propagation problems in which wavefronts change amplitude and energy density as the wave progresses in a radial direction, and the wave vectors at various space points are not all parallel. Such propagation is usually associated with the radiation from an antenna or scattering from an obstacle.

As already mentioned in Chapter 5, the central difficulty encountered in applying finite elements to practical radial propagation problems remains that of treating an unbounded region subject to a radiation condition. The latter can be introduced into the finite-element solution scheme by converting the original open problem into an equivalent interior problem with appropriate boundary conditions at a fictitious surface Γ. Such surface delimits the computational domain Ω and should enclose all the scatterers and sources (Figure 5.1). The appropriate boundary conditions may be of two kinds:

- Nonlocal, that is, the relationship between tangential electric and magnetic field at a point P on the boundary Γ (Figure 5.1) involves the values of the field on the whole boundary;
- Local, that is, the relationship between tangential electric and magnetic field valid at the point P is expressed in terms of the field values at the neighboring points of P.

Boundary conditions of the first kind are derived by expressing the electromagnetic field in the exterior region in terms of either a modal expansion, or as an integral representation employing the appropriate Green's function. On the other hand, local boundary conditions, often referred to as absorbing boundary conditions (ABC), may be derived either analytically or numerically, and those analytically derived may be based on asymptotic expansion of the field or on the perfectly matched layer concept.

Finite element methods that make use of nonlocal boundary conditions are often referred to as hybrid methods or exact methods because the boundary conditions are derived from exact expressions of the field in the region exterior to the fictitious surface Γ. They have been briefly reviewed at the beginning of Chapter 5, and are mentioned here again for completeness. They include the unimoment [166–169], the transfinite element method [170–173], and the bymoment method [174, 175], all of which rely on a modal expansion in the region outside Γ. Finite element-boundary integral (FE-BI) method [85, 176–182], the hybrid FEM/method of moments [183–185], and the so-called *field feedback formulation* [186, 187], also belong to the first group, but employ an integral equation representation of the field exterior to the fictitious surface Γ. The analysis of scattering problems and antenna radiation including the effects of the feeding structure may require to express the field outside the region Γ analyzed with FEM using both integral equation and modal expansion [188, 189].

Caution must be exercised when formulating hybrid methods since they are subject to the interior resonance problem associated with the integral equations employed to terminate the mesh [190–195].

An efficient version of the finite element-boundary integral method that makes use of the fast multipole method (FMM) to reduce the storage and computational requirement of the boundary integral in FE-BI, has been proposed [196, 197], showing significant improvements in the boundary-integral portion of the algorithm. Further improvements of the FE-BI technique include using higher order basis functions [198], an a posteriori error estimation technique [199, 200], and the development of fast and robust preconditioners for the solution of the resulting linear system of equations [201]. An overview of hybrid finite-element methodologies suitable to deal with scattering and radiation problems can be found in [202]. Although somewhat dated, the survey still is a good starting point for further investigation on the subject.

The boundary conditions derived from hybrid methods give rise to full matrices because integral formulation or modal expansions express action at a distance. On the other hand, they are based on exact representations of the field outside the fictitious surface Γ, so that Γ itself may be placed arbitrarily close to the radiating system or to the scatterer. The loss of sparsity is thus compensated by at least some decrease in the matrix size. Local boundary conditions or absorbing boundary conditions, were developed later, but have quickly gained popularity and the attention of many researchers because of their feature of maintaining the sparsity of the FEM linear system of equations. As a matter of fact, at any point P on the surface Γ they are expressed in terms of the electromagnetic field at P and its neighboring points only; they do not involve points distant from P. Because they are local, they produce sparse coefficient matrices, unlike the hybrid methods, and this fact renders them very convenient for numerical computation.

Absorbing boundary conditions were originally developed either assuming the simple case of a plane wave perpendicularly impinging on a planar interface [203], or using the asymptotic expansion of outward-traveling electromagnetic fields at a circular fictitious surface Γ [204, 205]. They have been subsequently improved to include the cases of skew incidence [206, 207], or fictitious surfaces Γ of elliptic type [208]. Several different types of absorbing boundary conditions have been proposed starting in the early 1990s, from the simplest case of conditions suitable for two-dimensional scalar problems [208–215] to the more involved situation of a three-dimensional problem with vectorial unknowns [216–224]. However, since all these kinds of absorbing boundary conditions have been derived by expressing the outward-traveling field in an asymptotic series, which is strictly valid only in the far field, they must be applied at some distance from any radiating or scattering object, and this restriction poses a limit on their efficiency in terms of minimization of the geometrical domain of computation. To overcome this limit, numerical procedures to create a relationship between the tangential field components at the fictitious surface Γ have been also proposed [147, 225–236]. These do not rely on the assumption of far-field behavior of the outward-traveling fields, hence may be enforced at a surface arbitrarily close to the antenna or scatterer. However, they do have an additional computational cost to numerically evaluate the coefficients of the relation among tangential field components at Γ.

A third group of local boundary conditions for truncating the finite-element mesh has also been developed starting in the early 1990s when Jin, Volakis, and Liepa [237] proposed the usage of an absorbing material backed by perfectly conducting walls to enclose the computational domain. This engineering approach closely resembles a measurement setup for radiation and scattering, where the device or scatterer under test is located into an anechoic chamber with walls covered by absorbers. In their work, Jin, Volakis, and Liepa employed an absorbing material designed to absorb TE and TM waves impinging the absorber at a wide range of angles. A similar idea was developed by Berenger [238], who analytically derived the characteristic parameters of an ideal absorbing material able to absorb arbitrarily polarized plane waves for any incidence angle and frequency. In its original formulation, this concept, also called perfectly matched layer (PML), although very successful in absorbing outward-traveling waves, has heavy computational overheads and gives rise to a physically unrealizable absorber. Consequently, it must be used with some caution since the experimenter's physical intuition no longer provides a reliable guide. Moreover, Berenger conceived the PML concept in the framework of a two-dimensional finite difference time domain (FDTD) scheme, hence further work has been done to apply the concept to three-dimensional problems [239] and to reformulate it in the framework of finite elements in frequency domain [240, 241]. The idea of Berenger has been refined by Sacks et al. [242] who analytically derived, in the frequency domain, the physical parameters of an anisotropic material (perfectly matched anisotropic absorber, PMA) able to absorb arbitrarily polarized incident waves at any frequency and incidence angle. The performances provided by the PMA are superior to those obtainable by using analytical ABC or numerical absorbing boundary conditions [243], and the PMA concept is gaining popularity because of its easy implementation, as seen also in the example given in Chapter 6. Nonplanar, cylindrical, and spherical PMAs have also been investigated [244–247].

The use of finite elements for solving practical radial propagation problems has been extensively studied and fairly complex problems have recently been analyzed. Scattering from complex bodies of revolution [248–251], from three-dimensional multidielectric and multiconducting bodies [252, 253], and from a cavity-backed antenna on a circular cylinder [254] have been analyzed by means of hybrid finite element-boundary integral techniques, while vector absorbing boundary conditions have been employed for the analysis of scattering from homogeneous and inhomogeneous dielectric objects, as well as from inhomogeneous anisotropic or lossy scatterers [220, 223, 255]. Efficient far-field transforms have also been proposed for the computation of echo-width [256].

Considerable efforts have been devoted to the solution of electromagnetic scattering from periodic structures. The large amount of published material in this field is probably due to both the wide range of practical applications of this kind of devices and to the relatively simple formulation of the problem. Structures with periodicity along one or two directions have been analyzed using different hybrid techniques: finite elements combined with the method of moments (MoM) [257], finite elements in combination with boundary integrals [258, 259], and transfinite elements [260–263]. Simplified analyses have been proposed for periodic structures with a simple geometry [264, 265], and numerically derived boundary conditions have also been applied [234].

As for antenna problems, finite element solutions have been obtained for axisymmetric radiators [188, 266–270], taking into account also the effects of radomes [266], cavity-backed antennas of arbitrary shape and conformal antennas [271–275], edge slots in rectangular waveguide considering also the wall thickness [276], and, in the hypothesis of an infinitely extending structure, also phased array antennas [277].

10.3 Advanced Topics and Methodologies in Finite Elements

The efficient solution of practical electromagnetic problems by finite element methods still poses many challenges. One of the main problems concerns electromagnetic scattering from objects whose dimensions are large with respect to the wavelength. Schemes to truncate the finite element mesh by hybrid methods or absorbing boundary conditions can both prove unsatisfactory for this case because of the high number of unknowns in the linear system of equations obtained. Even though modern computational resources may be sufficient to solve the problem, such schemes solve for the near field, while the far-field is of interest in scattering problems. More convenient approaches should apply the finite elements only in those regions which cannot be otherwise treated, while high-frequency techniques ought to be used to represent the electromagnetic field elsewhere. This approach has been initially investigated in [278–281], and subsequently refined in [282–287].

Another promising way to reduce the number of unknowns in the finite element solution of electromagnetic problem consists in modeling the envelope behavior of the field rather than its phase variations [288–290]. A speed up in CPU time can be also achieved by using wavelets as basis functions to diagonalize the finite element coefficient matrix [291, 292].

For problems involving complex geometries or geometries with fine details in localized areas, classic adaptive refinement procedures can quickly lead to a great number of unknowns. This problem has been recently addressed with the introduction of domain decomposition methods (DDM) [293] that subdivide the space of the original problem in a set of subdomains. Each subdomain is treated separately and possibly with different numerical techniques and the solution in the original large domain is sought by enforcing appropriate boundary conditions at each subdomain boundary. The original problem can be solved either iteratively, enforcing boundary conditions between adjacent domain till convergence is reached, or enforcing boundary conditions at all subdomain interfaces and solving the resulting global system of linear equations. Domain decomposition methods can be broadly classified as nonoverlapping methods and overlapping methods. The most efficient and popular nonoverlapping technique used in computational electromagnetics is the finite element tearing and interconnecting (FE-TI) [294–299]. The Schwarz technique [300] is among the most widely used overlapping methods and among these, the nonconforming techniques such as the mortar DDM [301] present particular advantages because they do not require exact alignment of unknowns at subdomain boundaries. This is of particular interest for the final element solution of periodic structures since it eliminates constraints on mesh at periodic boundaries [302]. DDMs not only accelerate significantly the solution time for large electromagnetic problems, they are also conducive to algorithms well suited for implementation on parallel architecture machines [303–307] that nowadays are relatively low cost and easily accessible.

Another important practical consideration affecting the computation time of EM simulator in general, is the frequency range of interest. Design of modern multigigabit-per-second communication or storage systems requires the FEM analysis of interconnects over a broad frequency range, from DC to tens of Gigahertz. To increase efficiency and speed, MOR techniques have been developed [308]. The fundamental idea behind these techniques is to model the real, complex, linear system over a broad frequency range through a simpler model with a transfer function equivalent to that of the original system. The transfer function of a complex EM system is comprised of an infinite number of poles and zeros. The MOR techniques develop a transfer function with a finite number of dominant poles and zeros, that best approximate the behavior of the original system in the frequency range of interest. There are many MOR techniques that can be applied to implement fast frequency sweep using a FEM analysis, the most popular are the asymptotic waveform evaluation (AWE) [309], the Padé via Lanczos (PVL) [310, 311], and the adaptive Lanczos-Padé sweep (ALPS) [312]. The common feature of these techniques is that the EM system transfer function is computed over the frequency range of interest by FEM solution of the system on a limited number of frequencies. MOR techniques have been extended and applied to solution of both closed and open domain problems using either hybrid FEM methods or absorbing boundary conditions [313–319].

The development of MOR techniques and consequent implementation of fast-frequency sweeps has enabled designers to apply full-wave electromagnetic design and optimization to practical problems. Optimization of linear microwave circuit using FEM as the analysis tools has been proposed in several papers [320–323], and has actually become a mature procedure and part of the design flow for practical problems as soon as fast frequency sweep technology [311, 312, 324, 325] has been implemented in commercial FEM-based software tools.

Theory and application of finite element in time domain [326] have greatly advanced since the first edition of this book was published. The change from frequency to time domain is not very difficult from the programming point of view and the method does not suffer from the step gridding approximations present in the finite difference time domain technique. Different aspects of finite element time domain implementations, using either nodal or vectorial elements, were considered in [327–333], and studies of numerical dispersion have been carried out using both triangular [334] and quadrilateral edge elements [77].

Although the finite difference time domain (FDTD) remains the most widely used technique for time domain analysis, the finite element time domain has the typical advantages of finite elements of using an unstructured grid that allows to accurately discretize complex geometries. Using edge elements in time domain, continuity condition of tangential field at interface is automatically enforced. FE-TD also maintains the possibility to apply p-refinement or h-refinement, and using unconditionally stable implementations [335, 336], a grid refinement does not force a decrease in the time step. The FDTD has a much simpler implementation but does not allow this kind of flexibility for mesh refinement: spatial and time steps are linked through the Courant condition to achieve stability, the finer the spatial grid, smaller the time step. Although the FDTD is intrinsically faster than FE-TD due to its explicit iterative algorithm, the use of unstructured grids and higher order basis functions may lead to an FE-TD problem with a smaller number of unknowns than those required by the FDTD, for complex geometries and same level of solution accuracy [337]. For this reason, FE-TD technique started to emerge and gain popularity as efficient meshers and higher order hierarchical elements have been introduced.

Several applications of FE-TD have been reported, concerning scattering problems [338], planar transmission lines characterization [339, 340], and axisymmetric as well as aperture radiators [341, 342]. Analysis of propagation in optical waveguides [343] and dispersive media [344] have been reported and several papers have been also dedicated to the application of FETD to the analysis of scattering from periodic structures [345–348].

Current research on FE-TD techniques aims to the hybridization of FE-TD with other techniques like FDTD, to achieve speedup in computer time due to the use of FDTD and overcome the staircasing approximation entailed in the FDTD [349–351], and with other integral equation methods to deal with scattering and antenna problems [352]. Development of accurate absorbing boundary conditions in time domain for closed [337, 353, 354] and open problems [355, 356], as well as use of DDM and parallel algorithms are also of interest, and application using up to 10-billions elements have been reported [357].

Another current research effort is the development of FEM schemes to introduce lumped components into the finite element analysis of microwave circuits. An example of introduction of linear passive elements can be found in [358–361], while introduction of nonlinear elements have been proposed by Ping, Wang, and Itoh in the framework of a FE-TD formulation [336, 362]. Alternatively, circuit simulation with full-wave accuracy level of packages and interconnect can be achieved by creating a reduced order model of the interconnect transfer function, obtained through electromagnetic simulations, and running the circuit simulation in SPICE or equivalent circuit simulator [363]. Moreover, as clock rates increase and rise/fall time of high-speed digital signals decrease, there is an ever-increasing need for full-wave simulation tools also for on-chip and off-chip interconnects. Given the extremely complex geometry and size of this kind of problems, especially for on-chip interconnects and power delivery networks, DDM [364] and other special formulations have been developed to efficiently simulate interconnects in multilayer structures. These formulations work best by using vector basis functions defined on prism elements [365–367]. Given the nature of high-speed digital signals, with spectral components from DC to multigigahertz frequencies, there is also the need for finite element formulations working on the full frequency spectrum, down to DC [368].

Last but not least, other fields of application and research for FEM techniques are material characterization [369–371] and inverse scattering [372,373].

References

[1] Coccioli, R., et al., "Finite Element Methods in Microwaves: a Selected Bibliography," *IEEE Antennas and Propagation Magazine*, Vol. 38, No. 6; pp. 34–48, 1996.

[2] Hoole, S. R. H., *Computer-Aided Analysis and Design of Electromagnetic Devices*, New York: Elsevier, 1989.

[3] Sabonnadiere, J. C. and Coulomb, J. L., *Finite Element Methods in CAD: Electrical and Magnetic Fields*, S. Salon, translator, New York: Springer-Verlag, 1987.

[4] Binns, K. J., Lawrenson, P. J., and Trowbridge, C. W., *The Analytical and Numerical Solution of Electric and Magnetic Field*, Chichester, UK: John Wiley & Sons, 1992.

[5] Steele, C.W., *Numerical Computation of Electric and Magnetic Fields*, New York: Van Nostrand Reinhold, 1987.

[6] Itoh, T., *Numerical Techniques for Microwave and Millimeter-Wave Passive Structures*, New York: John Wiley & Sons, 1989.

[7] Oden, J. T. and Reddy, J. N., *An Introduction to the Mathematical Theory of Finite Elements*, New York: John Wiley & Sons, 1976.

[8] Strang, G. and Fix, G. J., *An Analysis of the Finite Element Method*, Englewood Cliffs, NJ: Prentice-Hall, 1973.

[9] Silvester, P. P. and Ferrari, R. L., *Finite Elements for Electrical Engineers*, 3rd edition, Cambridge, UK: Cambridge University Press, 1996.

[10] Davidson, D. B., *Computational Electromagnetics for RF and Microwave Engineering*, Cambridge, UK: Cambridge University Press, 2005.

[11] Jin, J. M., *The Finite Element Method in Electromagnetics*, New York: John Wiley & Sons, 2002.

[12] Volakis, J. L., Kempel, L. C., Chatterjee, A., *Finite Element Method for Electromagnetics*, IEEE Press, 1998.

[13] Silvester, P. P. and Pelosi, G., *Finite Elements for Wave Electromagnetics*, New York: IEEE Press, 1994.

[14] Itoh, T., Pelosi, G., and Silvester, P. P., *Finite Element Software for Microwave Engineering*, New York: John Wiley & Sons, 1996.

[15] Salazar-Palma, M., *Self-Adaptive Finite-Element Electromagnetic Modeling*, London, UK: Artech House, 1998.

[16] Courant, R. L., "Variational Methods for the Solution of Problems of Equilibrium and Vibration," *Bulletin of the American Mathematical Society*, Vol. 5, pp. 1–23, 1943.

[17] Silvester, P. P., "Finite-Element Solution of Homogeneous Waveguide Problems," *Alta Frequenza*, Vol. 38, pp. 313–317, 1969.

[18] Silvester, P., "High-Order Polynomial Triangular Finite Elements for Potential Problems," *International Journal for Engineering Science*, Vol. 7, pp. 849–861, 1969.

[19] Silvester, P. P., "Universal Finite Element Matrices for Tetrahedra," *International Journal for Numerical Methods in Engineering*, Vol. 18, pp. 1055–1061, 1982.

[20] Silvester, P. P., "Permutation Rules for Simplex Finite Elements," *International Journal for Numerical Methods in Engineering*, Vol. 18, pp. 1245–1248, 1982.

[21] Ergatoudis, I. B., Irons, M., and Zienkiewicz, O. C., "Curved Isoparametric 'Quadrilateral' Elements for Finite Element Analysis," *International Journal for Solids and Structures*, Vol. 4, pp. 31–42, 1968.

[22] Nedelec, J. C., "Mixed Finite Elements in R^3," *Numerische Mathematik*, Vol. 35, pp. 315–341, 1980.

[23] Nedelec, J. C., "A New Family of Mixed Finite Elements in R^3," *Numerische Mathematik*, Vol. 50, pp. 57–81, 1986.

[24] Cendes, Z. J., "Vector Finite Elements for Electromagnetic Field Computation," *IEEE Transactions on Magnetics*, Vol. 27, pp. 3958–3966, 1991.

[25] Tanner, D. R. and Peterson, A. F., "Vector Expansion Functions for the Numerical Solution of Maxwell's Equations," *Microwave and Optical Technology Letters*, Vol. 2, pp. 331–334, 1989.

[26] Lee, J. F., Sun, D. K., and Cendes, Z. J., "Full-Wave Analysis of Dielectric Waveguides Using Tangential Vector Finite Elements," *IEEE Transactions on Microwave Theory and Techniques*, Vol. MTT-39, pp. 1262–1271, 1991.

[27] Webb, J. P. and Forghani, B., "Hierarchal Scalar and Vector Tetrahedra," *IEEE Transactions on Magnetics*, Vol. 29, pp. 1495–1498, 1993.

[28] Mur, G. and de Hoop, A. T., "A Finite-Element Method for Computing Three-Dimensional Electromagnetic Fields in Inhomogeneous Media," *IEEE Transactions on Magnetics*, Vol. MAG-21, pp. 2188–2191, 1985.

[29] van Welij, J. S., "Calculation of Eddy Currents in Terms of H on Hexahedra," *IEEE Transactions on Magnetics*, Vol. MAG-21, pp. 2239–2241, 1985.

[30] Kameari, A., "Calculation of Transient Three-Dimensional Eddy Current Using Edge-Elements," *IEEE Transactions on Magnetics*, Vol. 26, pp. 466–469, 1990.

[31] Dillon, B. M., Liu, P. T. S., and Webb, J. P., "Spurious Modes in Quadrilateral and Triangular Edge Elements," *COMPEL—The International Journal for Computation and Mathematics in Electrical and Electronic Engineering*, Vol. 13, Suppl. A, pp. 311–316, 1994.

[32] Crowley, C. W., Silvester, P. P., and Hurwitz Jr., H., "Covariant Projection Elements for Three-Dimensional Vector Field Problems," *IEEE Transactions on Magnetics*, Vol. 24, pp. 397–400, 1988.

[33] Crowley, C. W., *Mixed Order Covariant Projection Vector Fields*, PhD thesis, McGill University, Montreal, 1988.

[34] Hirayama, K., et al., "Vector Finite Element Method with Mixed-Interpolation-Type Triangular-Prism Element for Waveguide Discontinuities," *IEEE Transactions on Microwave Theory and Techniques*, Vol. 42, pp. 2311–2316, 1994.

[35] Peterson, A. F., "Vector Finite Element Formulation for Scattering from Two-Dimensional Heterogeneous Bodies," *IEEE Transactions on Antennas and Propagation*, Vol. 43, pp. 357–365, 1994.

[36] Savage, J. S. and Peterson, A. F., "Higher-Order Vector Finite Elements for Tetrahedral Cells," *IEEE Transactions on Microwave Theory and Techniques*, Vol. 44(6), pp. 874–879, June 1996.

[37] Graglia, R. D., Wilton, D. R., and Peterson, A. F., "Higher Order Interpolatory Vector Bases for Computational Electromagnetics," *IEEE Transactions on Antennas and Propagation*, Vol. 45, pp. 329–332, 1997.

[38] Cendes, Z. J. and Silvester, P. P., "Numerical Solution of Dielectric Loaded Waveguides: I-Finite Element Analysis," *IEEE Transactions on Microwave Theory and Techniques*, Vol. MTT-18, pp. 1124–1131, 1970.

[39] Corr, D. G. and Davies, J. B., "Computer Analysis of the Fundamental and Higher Order Modes in Single and Coupled Microstrip," *IEEE Transactions on Microwave Theory and Techniques*, Vol. MTT-20, pp. 669–678, 1972.

[40] Lynch, D. R. and Paulsen, K. D., "Origin of Vector Parasites in Numerical Maxwell Solutions," *IEEE Transactions on Microwave Theory and Techniques*, Vol. 39, pp. 383–394, 1991.

[41] Paulsen, K. D. and Lynch, D. R., "Elimination of Vector Parasites in Finite Element Maxwell Solutions," *IEEE Transactions on Microwave Theory and Techniques*, Vol. 39, pp. 395–404, 1991.

[42] Lynch, D. R., Paulsen, K. D., and Boyse, W. E., "Synthesis of Vector Parasites in Finite Element Maxwell Solutions," *IEEE Transactions on Microwave Theory and Techniques*, Vol. 41, pp. 1439–1448, 1993.

[43] Schroeder, W. and Wolff, I., "The Origin of Spurious Modes in Numerical Solutions of Electromagnetic Field Eigenvalue Problems," *IEEE Transactions on Microwave Theory and Techniques*, Vol. 42, pp. 644–653, 1994.

[44] Sun, D., et al., "Spurious Modes in Finit-Element Methods," *IEEE Antennas and Propagation Magazine*, Vol. 37, No. 5, pp. 12–24, 1995.

[45] Mur, G., "The Finite-Element Modeling of Three-Dimensional Electromagnetic Fields Using Edge and Nodal Elements," *IEEE Transactions on Antennas and Propagation*, Vol. 41, pp. 948–953, 1993.

[46] Boyse, W. E. and Seidl, A. A., "A Hybrid Finite Element Method for Three-Dimensional Scattering Using Nodal and Edge Elements," *IEEE Transactions on Antennas and Propagation*, Vol. 42, pp. 1436–1442, 1994.

[47] Graglia, R. D., "Higher Order Interpolatory Vector Bases on Prism Elements," *IEEE Transactions on Antennas and Propagation*, Vol. 46(3), pp. 442–450, March 1998.

[48] Graglia, R. D. and Gheroma, I.-L., "Higher Order Interpolatory Vector Bases on Pyramidal Elements," *IEEE Transactions on Antennas and Propagation*, Vol. 47(5), pp. 775–782, May 1999.

[49] Rieben, R. N., White, D. A., and Rodrigue, G. H., "Improved Conditioning of Finite Element Matrices Using New High-Order Interpolatory Bases," *IEEE Transactions on Antennas and Propagation*, Vol. 52(10), pp. 2675–2683, Oct. 2004.

[50] Wang, J. S. and Ida, N., "Curvilinear and Higher Order Edge Finite Elements in Electromagnetic Field Computation," *IEEE Transactions on Magnetics*, Vol. 29(2), pp. 1491–1494, 1993.

[51] Webb, J. P. and Miniowitz, R., "Analysis of Three-Dimensional Microwave Resonators Using Covariant-Projection Elements," *IEEE Transactions on Microwave Theory and Techniques*, Vol. 29(11), pp. 1895–1899, Nov. 1991.

[52] Loo, S. L. and Silvester, P. P., "Finite Element Analysis of Inductive Strips in Unilateral Finlines," *IEEE Transactions on Microwave Theory and Techniques*, Vol. 41, pp. 298–304, 1993.

[53] Selleri, S., "A New Class of NURBS-Based Hybrid Hyperparametric Elements," In *1998 Actes J. Int. Nice sur les Antennes*, Nice, France, pp. 160, 165–168, Nov. 1998–1999.

[54] Martini, E. and Selleri, S., "An Innovative Class of Curvilinear Tetrahedral Elements," *Electronics Letters*, Vol. 37, pp. 557–558, 2001.

[55] Martini, E., Pelosi, G., and Selleri, S., "A Hybrid Finite-Element-Modal-Expansion Method with a New Type of Curvilinear Mapping for the Analysis of Microwave Passive Devices," *IEEE Transactions on Microwave Theory and Techniques*, Vol. 51(6), pp. 1712–1717, June 2003.

[56] Webb, J. P., "Finite Element Analysis of Dispersion in Waveguides with Sharp Metal Edges," *IEEE Transactions on Microwave Theory and Techniques*, Vol. 36, pp. 1819–1824, 1988.

[57] Gil, J. M. and Zapata, J., "Efficient Singular Element for Finite Element Analysis of Quasi-TEM Trasmission Lines and Waveguides with Sharp Metal Edges," *IEEE Transactions on Microwave Theory and Techniques*, Vol. 42, pp. 92–98, 1994.

[58] Gil, J. M. and Zapata, J., "A New Scalar Transition Finite Element for Accurate Analysis of Waveguides with Field Singularities," *IEEE Transactions on Microwave Theory and Techniques*, Vol. 43, pp. 1978–1982, 1995.

[59] Graglia, R. D. and Lombardi, G., "Singular Higher Order Complete Vector Bases for Finite Methods," *IEEE Transactions on Antennas and Propagation*, Vol. 52(7), pp. 1672–1685, July 2004.

[60] Sun, D.-K., Vardapetyan, L., and Cendes, Z., "Dimensional Curl-Conforming Singular Elements for FEM Solutions of Dielectric Waveguide Structures," *IEEE Transactions on Microwave Theory and Techniques*, Vol. 53(3), pp. 984–992, March 2005.

[61] Masoni, J., Pelosi, G., and Selleri, S., "Substitutive Divergent Bases for FEM Modeling of Field Singularities Near a Wedge," *Microwave and Optical Technology Letters*, Vol. 44(4), pp. 327–328, 20 Feb. 2005.

[62] Salazar-Palma, M. and Garcia-Castillo, L. E., "Self-Adaptive Procedures for Waveguiding Strucure Analysis," In *Finite Element Software for Microwave Engineering*, Itoh, T., Pelosi, G., and Silvester, P. P., (editors), New York: John Wiley & Sons, 1996.

[63] Salazar-Palma, M. and Recio-Pelaez, J. M., "A Convergence Study of a Self-Adaptive Mesh Algorithm," In *Software for Electrical Engineering Analysis and Design*, Silvester, P. P., (editor), Southampton, UK: Computational Mechanics Publications, 1996.

[64] Diaz-Morcillo, A. B., Balbastre, J. V., and Nuno, L., "New Recovery Error Indicator for Adaptive Finite-Element Analysis on Waveguiding Structures," *IEEE Transactions on Microwave Theory and Techniques*, Vol. 51(5), pp. 1467–1475, May 2003.

[65] Ilic, M. M., Ilic, A. Z., and Notaros, B. M., "Efficient Large-Domain Two-Dimensional FEM Solution of Arbitrary Waveguides Using P-Refinement on Generalized Quadrilaterals, *IEEE Transactions on Microwave Theory and Techniques*, Vol. 53(4), pp. 1377–1383, April 2005.

[66] Choi, C. T. M. and Webb, J. P., "P-Adaptive Finite Element Analysis of Unbounded Electromagnetic Wave Problems," *IEE Proceedings, Part H*, Vol. 148(6), pp. 388–392, Dec. 2001.

[67] Rossow, M. P. and Katz, I. N., "Hierarchal Finite Elements and Precomputed Arrays," *International Journal for Numerical Methods in Engineering*, Vol. 12, pp. 977–999, 1978.

[68] Zienkiewicz, O. C., de S.R. Gago, J. P., Kelly, D. W., "The Hierarchical Concept in Finite Element Analysis," *Computers and Structures*, Vol. 16, pp. 53–65, 1983.

[69] Sears, G. J., "Hierarchical Quadrilateral Elements in Field Analysis," *IEEE Transactions on Magnetics*, Vol. 24, pp. 404–407, 1988.

[70] Webb, J. P. and McFee, S., "The Use of Hierarchical Triangles an Finite-Element Analysis of Microwave and Optical Devices," *IEEE Transactions on Magnetics*, Vol. 27, pp. 4040–4043, 1991.

[71] Webb, J. P. and Abouchacra, R., "Hierarchal Triangular Elements Using Orthogonal Polynomials," *International Journal for Numerical Methods in Engineering*, Vol. 38, pp. 245–257, 1995.

[72] McFee, S. and Webb, J. P., "Adaptive Finite Elements Analysis of Microwave and Optical Devices Using Hierarchal Triangles," *IEEE Transactions on Magnetics*, Vol. 28, pp. 1708–1711, 1992.

[73] Golias, N. A., Papagiannakis, A. G., and Tsiboukis, T. D., "Efficient Mode Analysis with Edge Elements and Three-Dimensional Adaptive Refinements," *IEEE Transactions on Microwave Theory and Techniques*, Vol. MTT-42, pp. 99–107, 1994.

[74] Marais, N. and Davidson, D. B., "Numerical Evaluation of Hierarchical Vector Finite Elements on Curvilinear Domains in Two-Dimensional," *IEEE Transactions on Antennas and Propagation*, Vol. 54(2), pp. 734–738, Feb. 2006.

[75] Peterson, A. F. and Baca, R. J., "Error in the Finite Element Discretization of the Scalar Helmholtz Equation Over Electrically Large Regions," *IEEE Microwave and Guided Waves Letters*, Vol. 1, pp. 219–222, 1991.

[76] Lee, R. and Cangellaris, A. C., "A Study of Discretization Error in the Finite Element Approximation of Wave Solutions," *IEEE Transactions on Antennas and Propagation*, Vol. AP-40, pp. 542–549, 1992.

[77] Warren, G. S. and Scott, W. R., "An Investigation of Numerical Dispersion in the Vector Finite-Element Method Using Quadrilateral Elements," *IEEE Transactions on Antennas and Propagation*, Vol. AP-42, pp. 1502–1508, 1994.

[78] Scott, W. R, "Errors Due to Spatial Discretization and Numerical Precision in the Finite-Element Method," *IEEE Transactions on Antennas and Propagation*, Vol. AP-42, pp. 1565–1570, 1994.

[79] Warren, G. S. and Scott, W. R., "Numerical Dispersion in the Finite-Element Method Using Triangular Edge Elements," *Microwave and Optical Technology Letters*, Vol. 9, pp. 315–319, 1995.

[80] Warren, G. S. and Scott, W. R., "Numerical Dispersion of Higher Order Nodal Elements in the Finite-Element Method," *IEEE Transactions on Antennas and Propagation*, Vol. 44, pp. 317–320, 1996.

[81] Pekel, U. and Lee, R., "An *a Posteriori* Error Reduction Scheme for the Three-Dimensional Finite Element Solution of Maxwell's Equation," *IEEE Transactions on Microwave Theory and Techniques*, Vol. AP-43, pp. 421–427, 1995.

[82] Emson, C. R. I., "Methods for the Solution of Open-Boundary Electromagnetic Field Problems," *IEE Proceedings, Part A*, Vol. 135, pp. 151–158, 1988.

[83] McDougall, M. J. and Webb, J. P., "Infinite Elements for the Analysis of Open Dielectric Waveguides," *IEEE Transactions on Microwave Theory and Techniques*, Vol. MTT-37, pp. 1724–1731, 1989.

[84] Pantic, Z. and Mittra, R., "Quasi-TEM Analysis of Microwave Transmission Lines by the Finite-Element Method," *IEEE Transactions on Microwave Theory and Techniques*, Vol. MTT-34, pp. 1096–1103, 1986.

[85] Silvester, P. and Hsieh, M. S., "Finite-Element Solution of Two-Dimensional Exterior-Field Problems," *IEE Proceedings*, Vol. 118, pp. 1743–1747, 1971.

[86] Kagawa, Y., Murai, T., and Kitagami, S., "On the Compatibility of Finite Element-Boundary Element Coupling in Field Problems," *COMPEL—The International Journal for Computation and Mathematics in Electrical and Electronic Engineering*, Vol. 1, pp. 197–217, 1982.

[87] Onuki, T., "Hybrid Finite Element and Boundary Element Method Applied to Electromagnetic Problems," *IEEE Transactions on Magnetics*, Vol. 26, pp. 582–587, 1990.

[88] Silvester, P. P., et al., "Exterior Finite Elements for Two-Dimensional Field Problems with Open Boundaries," *IEE Proceedings*, Vol. 124, pp. 1267–1270, 1977.

[89] Dasgupta, G., "Computation of Exterior Potential Fields by Infinite Substructuring," *Computer Methods in Applied Mechanics and Engineering*, Vol. 46, pp. 295–305, 1984.

[90] Lowther, D. A., Rajanathan, C. B., and Silvester, P. P., "A Finite-Element Technique for Solving Two-Dimensional Open Boundary Problems," *IEEE Transactions on Magnetics*, Vol. MAG-14, pp. 467–469, 1978.

[91] Chiang, K. S., "Finite-Element Analysis of Optical Fibres with Iterative Treatment of the Infinite Two-Dimensional Space," *Optical and Quantum Electronics*, Vol. 17, pp. 381–391, 1985.

[92] Hurwitz, H. and Wong, Jr., "Infinitesimal Scaling a New Procedure for Modeling Exterior Field Problems," *IEEE Transactions on Magnetics*, Vol. MAG-20, pp. 1918–1923, 1984.

[93] Hurwitz, H. and Wong, Jr., "The Surface Projection Approach to Vector Field Calculations," *Mathematical and Computer Modelling*, Vol. 11, pp. 313–316, 1988.

[94] Imhoff, J. F., et al., "An Original Solution for Unbounded Electromagnetic Two-Dimensional and Three-Dimensional-Problems Throughout the Finite-Element Method," *IEEE Transactions on Magnetics*, Vol. 26, pp. 1659–1661, 1990.

[95] Chang, C. N., Wong, Y. C., and Chen, C. H., "Full-Wave Analysis of Coplanar Waveguides by Variational Conformal Mapping Technique," *IEEE Transactions on Microwave Theory and Techniques*, Vol. 38, pp. 1339–1344, 1990.

[96] Stochniol, A., "A General Transformation for Open Boundary Finite-Element Method for Electromagnetic Problems," *IEEE Transactions on Magnetics*, Vol. 28, pp. 1679–1681, 1992.

[97] Gibson, A. A. P. and Helszajn, J., "Finite Element Solution of Longitudinally Magnetized Elliptical Gyromagnetic Waveguides," *IEEE Transactions on Microwave Theory and Techniques*, Vol. 28, pp. 999–1005, 1992.

[98] Yeh, C., et al., "Single-Mode Optical Waveguides," *Applied Optics*, Vol. 18, pp. 1490–1504, 1979.

[99] Oyamada, K. and Okoshi, T., "Two-Dimensional Finite-Element Calculation of Propagation Characteristics of Axially Nonsymmetrical Optical Fibers," *Radio Science*, Vol. 17, pp. 109–116, 1982.

[100] Rahman, B. M. A. and Davies, J. B., "Finite-Element Analysis of Optical and Microwave Waveguide Problems," *IEEE Transactions on Microwave Theory and Techniques*, Vol. MTT-32, pp. 20–28, 1984.

[101] Vandenbulcke, P., and Lagasse, P. E., "Eigenmode Analysis of Anisotropic Optical Fibers or Integrated Optical Waveguides," *Electronics Letters*, Vol. 12, pp. 120–122, 1976.

[102] McAulay, A. D., "Variational Finite-Element Solution for Dissipative Waveguides and Transportation Application," *IEEE Transactions on Microwave Theory and Techniques*, Vol. MTT-25, pp. 382–392, 1977.

[103] Aubourg, M., et al., "Finite Element Analysis of Lossy Waveguides Application to Microstrip Lines on Semiconductor Substrate," *IEEE Transactions on Microwave Theory and Techniques*, Vol. MTT-31, pp. 326–331, 1983.

[104] Rahman, B. M. A. and Davies, J. B., "Penalty Function Improvement of Waveguide Solution by Finite Elements," *IEEE Transactions on Microwave Theory and Techniques*, Vol. MTT-32, pp. 922–928, 1984.

[105] Young, T. P., "Design of Integrated Optical Circuits Using Finite Elements," *IEE Proceedings, Part A*, Vol. 135, pp. 135–144, 1988.

[106] Hayata, K., et al., "Vectorial Finite-Element Method Without Any Spurious Solutions for Dielectric Wave-Guiding Problems Using Transverse Magnetic-Field Component," *IEEE Transactions on Microwave Theory and Techniques*, Vol. MTT-34, pp. 1120–1124, 1986.

[107] Angkaew, T., Matsuhara, M., and Kumagai, N., "Finite-Element Analysis of Waveguide Modes: A Novel Approach that Eliminates Spurious Modes," *IEEE Transactions on Microwave Theory and Technique*, Vol. 35, pp. 117–123, 1987.

[108] Hayata, K., Eguchi, M., and Koshiba, M., "Finite Element Formulation for Guided-Wave Problems Using Transverse Electric Field Component," *IEEE Transactions on Microwave Theory and Techniques*, Vol. 37, pp. 256–258, 1989.

[109] Chew, W. C. and Nasir, M. A., "A Variational Analysis of Anisotropic, Inhomogeneous Dielectric Waveguides," *IEEE Transactions on Microwave Theory and Techniques*, Vol. MTT-37, pp. 661–668, 1989.

[110] Fernandez, F. A. and Lu, Y., "A Variational Finite-Element Formulation for Dielectric Waveguides in Terms of Transverse Magnetic Fields," *IEEE Transactions on Magnetics*, Vol. 27, pp. 3864–3867, 1991.

[111] Matsuhara, M., Yunoki, H., and Maruta, A., "Analysis of Open-Type Waveguides by the Vector Finite-Elements Analysis," *IEEE Microwave and Guided Wave Letters*, Vol. 1, pp. 376–378, 1991.

[112] Koshiba, M. and Inoue, K., "Simple and Efficient Finite-Element Analysis of Microwave and Optical Waveguides," *IEEE Transactions on Microwave Theory and Techniques*, Vol. 40, pp. 371–377, 1992.

[113] Sachdev, S. and Cendes, Z. J., "Combined Tangential-Normal Vector Elements for Computing Electric and Magnetic Fields," *IEEE Transactions on Magnetics*, Vol. 29, pp. 1456–1459, 1993.

[114] Rahman, B. M. A., Fernandez, F. A., and Davies, J. B., "Review of Finite-Element Methods for Microwave and Optical Waveguides," *Proceedings of the IEEE*, Vol. 79, pp. 1442–1448, 1991.

[115] Davies, J. B., "Finite-Element Analysis of Waveguides and Cavities: A Review," *IEEE Transactions on Magnetics*, Vol. 29, pp. 1578–1583, 1993.

[116] Dillon, B. M. and Webb, J. P., "A Comparison of Formulations for the Vector Finite Element Analysis of Waveguides," *IEEE Transactions on Microwave Theory and Techniques*, Vol. 42, pp. 308–316, 1994.

[117] Dillon, B. M., Gibson, A. A. P., and Webb, J. P., "Cutoff and Phase Constants of Partially Filled Axially Magnetized, Gyromagnetic Waveguides Using Finite Elements," *IEEE Transactions on Microwave Theory and Techniques*, Vol. 41, pp. 803–808, 1993.

[118] Dillon, B. M. and Gibson, A. A. P., "Applied Field/Frequency Dependency of Propagation in Axially Magnetized Ferrite Waveguides," *IEEE Transactions on Microwave Theory and Techniques*, Vol. 41, pp. 2092–2098, 1993.

[119] Zhou, L. and Davis, L. E., "Finite-Element Method with Edge Elements for Waveguides Loaded with Ferrite Magnetized in Arbitrary Direction," *IEEE Transactions on Microwave Theory and Techniques*, Vol. 44, pp. 809–815, 1996.

[120] Zhou, L. and Davis, L. E., "Calculation of Phase and Attenuation Constants for Waveguides Loaded with Lossy Ferrite with Longitudinal or Transverse Magnetisation," *IEE Proceedings Microwaves, Antennas and Propagation*, Vol. 144, pp. 111–117, 1997.

[121] Valor, L. and Zapata, J., "An Efficient Finite-Element Formulation to Analyze Waveguides with Lossy Inhomogeneous Bianisotropic Materials," *IEEE Transactions on Microwave Theory and Techniques*, Vol. 44, pp. 291–296, 1996.

[122] Nuno, L., Balbastre, J. V., and Castane, H., "Analysis of General Lossy Inhomogeneous and Anisotropic Waveguides by the Finite-Element Method (FEM) Using Edge Elements," *IEEE Transactions on Microwave Theory and Techniques*, Vol. 45, pp. 446–449, 1997.

[123] Savi, P., Gheorma, I. L., and Graglia, R. D., "Full-Wave High-Order FEM Model for Lossy Anisotropic Waveguides," *IEEE Transactions on Microwave Theory and Techniques*, Vol. 50(2), pp. 495–500, Feb. 2002.

[124] Bertazzi, F., et al., "A Fast Reduced-Order Model for the Full-Wave FEM Analysis of Lossy Inhomogeneous Anisotropic Waveguides," *IEEE Transactions on Microwave Theory and Techniques*, Vol. 50(9), pp. 2108–2114, Sept. 2002.

[125] Slade, G. W. and Webb, K. J., "Computation of Characteristic Impedance for Multiple Microstrip Transmission Lines Using a Vector Finite Element Method," *IEEE Transactions on Microwave Theory and Techniques*, Vol. 40, pp. 34–40, 1992.

[126] Alam, M. S., et al., "A Vector Finite-Element Analysis of Complex Modes in Shielded Microstrip Lines," *Microwave and Optical Technology Letters*, Vol. 6, pp. 873–875, 1993.

[127] Alam, M. S., et al., "Finite-Element Analysis of Propagating, Evanescent, and Complex Modes in Finlines," *IEE Proceedings, Part H*, Vol. 141, pp. 65–69, 1994.

[128] Tsitsos, S., Gibson, A. A. P., and McCormick, A. H. I., "Higher Order Modes in Coupled Striplines: Prediction and Measurements," *IEEE Transactions on Microwave Theory and Techniques*, Vol. 42, pp. 2071–2077, 1994.

[129] Koshiba, M., Maruyama, S., and Hirayama, K., "A Vector Finite-Element Method with the High-Order Mixed-Interpolation-Type Triangular Elements for Optical Waveguiding Problems," *Journal of Lightwave Technology*, Vol. 12, pp. 495–502, 1994.

[130] Crain, B. R. and Peterson, A. F., "Analysis of Propagation on Open Microstrip Lines Using Mixed-Order Covariant Projection Vector Finite Elements," *International Journal of Microwave and Millimeter-Wave Computer-Aided Engineering*, Vol. 5, pp. 59–67, 1995.

[131] Alam, M. S., et al., "Analysis of Shielded Microstrip Lines with Arbitrary Metallization Cross-Section Using a Vector Finite-Element Method," *IEEE Transactions on Microwave Theory and Techniques*, Vol. 42, pp. 2112–2117, 1994.

[132] Lee, J. F., "Finite Element Analysis of Lossy Dielectric Waveguides," *IEEE Transactions on Microwave Theory and Techniques*, Vol. 42, pp. 1025–1031, 1994.

[133] Alam, M. S., et al., "Analysis of Lossy Planar Transmission Lines by Using a Vector Finite-Element Method," *IEEE Transactions on Microwave Theory and Techniques*, Vol. 43, pp. 2466–2471, 1995.

[134] Alam, M. S., et al., "Hybrid-Mode Analysis of Multilayered and Multiconductor Transmission Lines," *IEEE Transactions on Microwave Theory and Techniques*, Vol. 45, pp. 205–211, 1997.

[135] Zhao, Y. and Wang, Y.-Y., "A New Finite-Element Solution for Parameter Extraction of Multilayer and Multiconductor Interconnects," *IEEE Microwave and Guided Wave Letters*, Vol. 7, pp. 156–158, 1997.

[136] Wu, K. L., Ding, Y., and Fang, D. G., "A Parameter Extraction Technique for MM/FEM Analysis of Multilayered RF Passives Using TRL Calibration," *IEEE Microwave and Wireless Components Letters*, Vol. 14(9), pp. 452–454, Sept. 2004.

[137] Bertazzi, F., Ghione, G., and Goano, M., "Efficient Quasi-TEM Frequency-Dependent Analysis of Lossy Multiconductor Lines Through a Fast Reduced-Order FEM Model," *IEEE Transactions on Microwave Theory and Techniques*, Vol. 51(9), pp. 2029–2035, Sept. 2003.

[138] Lu, Y. and Fernandez, F. A., "An Efficient Finite Element Solution of Inhomogeneous Anisotropic and Lossy Dielectric Waveguide," *IEEE Transactions on Microwave Theory and Techniques*, Vol. 41, pp. 1215–1223, 1993.

[139] Selleri, S. and Zoboli, M., "An Improved Finite-Element Method Formulation for the Analysis of nonlinear Anisotropic Dielectric Waveguides," *IEEE Transactions on Microwave Theory and Techniques*, Vol. 43, pp. 887–892, 1995.

[140] Valor, L. and Zapata, J., "Efficient Finite-Element Analysis of Waveguides with Lossy Inhomogeneous Anisotropic Materials Characterized by Arbitrary Permittivity and Permeability Tensors," *IEEE Transactions on Microwave Theory and Techniques*, Vol. 43, pp. 2452–2459, 1995.

[141] Kooi, P. S., "Finite-Element Analysis of Shielded Cylindrical Dielectric Resonator, *IEE Proceedings, Part H*, Vol. 136, pp. 7–16, 1985.

[142] Lee, J. F., Wilkins, G. M., and Mittra, R., "Finite-Element Analysis of Axysimmetric Cavity Resonator Using a Hybrid Edge Element Technique," *IEEE Transactions on Microwave Theory and Techniques*, Vol. 41, pp. 1981–1987, 1993.

[143] Chatterjee, A., Jin, J. M., and Volakis, J. L., "Computation of Cavity Resonances Using Edge-Based Finite Elements," *IEEE Transactions on Microwave Theory and Techniques*, Vol. 40, pp. 2106–2108, 1992.

[144] Krupka, J., et al., "Study of Whispering Gallery Modes in Anysotropic Single-Crystal Dielectric Resonators," *IEEE Transactions on Microwave Theory and Techniques*, Vol. 42, pp. 56–61, 1994.

[145] Cousty, J. P., et al., "Finite Elements for Microwave Device Simulation: Application to Microwave Dielectric Resonator Filters," *IEEE Transactions on Microwave Theory and Techniques*, Vol. 40, pp. 925–932, 1992.

[146] Hirayama, K. and Koshiba, M., "Numerical Analysis of Arbitrarily Shaped Discontinuities Between Planar Dielectric Waveguides with Different Thicknesses," *IEEE Transactions on Microwave Theory and Techniques*, Vol. 38, pp. 260–264, 1990.

[147] Ramahi, O. M. and Mittra, R., "A Surface Integral Equation Method for the Finite Element Solution of Waveguide Discontinuity Problems," *IEEE Transactions on Microwave Theory and Techniques*, Vol. 39, pp. 604–608, 1991.

[148] Kiyoshi, I. and Masanori, K., "Numerical Analysis of H-Plane Waveguide Junctions by Combination of Finite and Boundary Elements," *IEEE Transactions on Microwave Theory and Techniques*, Vol. 36, pp. 1343–1351, 1988.

[149] Wu, K. L., et al., "Waveguide Discontinuity Analysis with a Coupled Finite-Boundary Element Method," *IEEE Transactions on Microwave Theory and Techniques*, Vol. 37, pp. 993–998, 1989.

[150] Webb, J. P. and Parihar, S., "Finite-Element Analysis of H–Plane Rectangular Waveguide Problems," *IEE Proceedings, Part H*, Vol. 133, pp. 91–94, 1986.

[151] Lee, J. F. and Cendes, Z. J., "The Transfinite Element Method for Modeling MMIC Devices," *IEEE Transactions on Microwave Theory and Techniques*, Vol. 36, pp. 1639–1649, 1988.

[152] Kanellopoulos, V. N. and Webb, J. P., "A Complete E-Plane Analysis of Waveguide Junctions Using the Finite-Element Method," *IEEE Transactions on Microwave Theory and Techniques*, Vol. 38, pp. 290–295, 1990.

[153] Coccioli, R., et al., "Design of Matched Bends in Rectangular Waveguides by FEM," *International Journal of Microwave and Millimeter-Wave Computer-Aided Engineering*, Vol. 6, pp. 391–398, 1996.

[154] Arndt, F., et al., "Automated Design of Waveguide Components Using Hybrid Mode-Matching/Numerical EM Building-Blocks in Optimization-Oriented CAD Frameworks-State of the Art and Recent Advances," *IEEE Transactions on Microwave Theory and Techniques*, Vol. 45, pp. 747–760, 1997.

[155] Foo, S. L. and Silvester, P. P., "Boundary-Marching Method for Discontinuity Analysis in Waveguides of Arbitrary Cross Section," *IEEE Transactions on Microwave Theory and Techniques*, Vol. 40, pp. 1889–1893, 1992.

[156] Webb, J. P., "Absorbing Boundary Conditions for the Finite-Element Analysis of Planar Devices," *IEEE Transactions on Microwave Theory and Techniques*, Vol. 38, pp. 1328–1332, 1990.

[157] Wilkins, G. M., Lee, J. F., and Mittra, R., "Numerical Modeling of Axisymmetric Coaxial Waveguide Discontinuities," *IEEE Transactions on Microwave Theory and Techniques*, Vol. 39, pp. 1323–1328, 1991.

[158] Zhu, N. Y. and Landstorfer, F. M., "An Efficient FEM Formulation for Rotationally Symmetric Coaxial Waveguides," *IEEE Transactions on Microwave Theory and Techniques*, Vol. 43, pp. 410–415, 1995.

[159] Hirayama, K. and Koshiba, M., "Analysis of Discontinuities in Asymmetric Dielectric Slab Waveguide by Combination of Finite and Boundary Elements," *IEEE Transactions on Microwave Theory and Techniques*, Vol. 40, pp. 686–691, 1992.

[160] Yook, J. G., Dib, N. I., and Katehi, L. P. B., "Characterization of High Frequency Interconnects Using Finite Difference Time Domain and Finite-Element Methods," *IEEE Transactions on Microwave Theory and Techniques*, Vol. 42, pp. 1727–1736, 1994.

[161] Foo, S. L. and Silvester, P. P., "Finite Element Analysis of Inductive Strips in Unilateral Finlines," *IEEE Transactions on Microwave Theory and Techniques*, Vol. 41, pp. 298–304, 1993.

[162] Wang, J. S. and Mittra, R., "A Finite Element Cavity Resonance Method for Waveguide and Microstrip Line Discontinuity Problems," *IEEE Transactions on Microwave Theory and Techniques*, Vol. 42, pp. 433–440, 1994.

[163] Montejo-Garai, J. R. and Zapata, J., "Full-Wave Design and Realization of Multicoupled Dual-Mode Circular Waveguide Filters," *IEEE Transactions on Microwave Theory and Techniques*, Vol. 43, pp. 1290–1297, 1995.

[164] Brauer, J. R. and Lizalek, G. C., "Microwave Filter Analysis Using a New Three-Dimensional Finite-Element Modal Frequency Method," *IEEE Transactions on Microwave Theory and Techniques*, Vol. 45, pp. 810–818, 1997.

[165] Wang, J. S. and Mittra, R., "Finite Element Analysis of MMIC Structures and Electronic Packages Using Absorbing Boundary Conditions," *IEEE Transactions on Microwave Theory and Techniques*, Vol. 42, pp. 441–449, 1994.

[166] Mei, K. K., "Unimoment Method of Solving Antenna and Scattering Problems," *IEEE Transactions on Antennas and Propagation*, Vol. AP-22, pp. 760–766, 1974.

[167] Chang, S. K. and Mei, K. K., "Application of the Unimoment Method to Electromagnetic Scattering of Dielectric Cylinders," *IEEE Transactions on Antennas and Propagation*, Vol. AP-24, pp. 35–42, 1976.

[168] Mei, K. K., "Unimoment Method for Electromagnetic Wave Scattering," *Journal of Electromagnetic Waves and Applications*, Vol. 1, pp. 201–222, 1987.

[169] Jin, J.-M. and Lu, N., "The Unimoment Method Applied to Elliptical Boundaries," *IEEE Transactions on Antennas and Propagation*, Vol. 45, pp. 564–566, 1997.

[170] Jeng, S. K. and Chen, C. H., "On Variational Electromagnetics: Theory and Application," *IEEE Transactions on Antennas and Propagation*, Vol. AP-32, pp. 902–907, 1984.

[171] Lee, J. F. and Cendes, Z. J., "Transfinite Elements: A Highly Efficient Procedure for Modeling Open-Field Problems," *Journal of Applied Physics*, Vol. 61, pp. 3913–3915, 1987.

[172] Graglia, R. D., Freni, A., and Pelosi, G., "A Finite-Element Approach to the Electromagnetic Interaction with Rotating Penetrable Cylinder of Arbitrary Cross-Section," *IEEE Transactions on Antennas and Propagation*, Vol. 41, pp. 635–650, 1993.

[173] Pelosi, G., Freni, A., and Coccioli, R., "A Hybrid Technique for Analyzing the Scattering from Periodic Structures," *IEE Proceedings, Part H*, Vol. 140, pp. 65–70, 1993.

[174] Cangellaris, A. C. and Lee, R., "The Bymoment Method for Two-Dimensional Electromagnetic Scattering," *IEEE Transactions on Antennas and Propagation*, Vol. 38, pp. 1429–1437, 1990.

[175] Cangellaris, A. C. and Lee, R., "Finite-Element Analysis of Electromagnetic Scattering from Inhomogeneous Cylinders at Oblique Incidence," *IEEE Transactions on Antennas and Propagation*, Vol. 39, pp. 645–650, 1991.

[176] McDonald, B. H. and Wexler, A., "Finite-Element Solution of Unbounded Field Problems," *IEEE Transactions on Microwave Theory and Techniques*, Vol. MTT-20, pp. 841–847, 1972.

[177] Jin, J. M. and Liepa, V. V., "Application of Hybrid Finite-Element Method to Electromagnetic Scattering from Coated Cylinders," *IEEE Transactions on Antennas and Propagation*, Vol. 36, pp. 50–54, 1988.

[178] Jin, J. M. and Liepa, V. V., "A Note on Hybrid Finite-Element Method for Solving Scattering Problems," *IEEE Transactions on Antennas and Propagation*, Vol. 36, pp. 1486–1490, 1988.

[179] Collins, J. D., Volakis, J. L., and Jin, J. M., "A Combined Finite-Element Boundary Integral Formulation for Solution of Two-Dimensional Scattering Problems Via CG-FFT," *IEEE Transactions on Antennas and Propagation*, Vol. 38, pp. 1852–1858, 1990.

[180] Jin, J. M. and Volakis, J. L., "Electromagnetic Scattering by and Transmission Through a Three-Dimensional Slot in a Thick Conducting Plane," *IEEE Transactions on Antennas and Propagation*, Vol. 39, pp. 543–550, 1991.

[181] Jin, J. M. and Volakis, J. L., "A Hybrid Finite-Element Method for Scattering and Radiation by Microstrip Patch Antennas and Arrays Residing in a Cavity," *IEEE Transactions on Antennas and Propagation*, Vol. 39, pp. 1598–1604, 1991.

[182] Cwik, T, "Coupling Finite Element and Integral Equation Solutions Using Decoupled Boundary Meshes," *IEEE Transactions on Antennas and Propagation*, Vol. 40, pp. 1496–1504, 1992.

[183] Yuan, X., Lynch, D. R., and Strohbehn, J. W., "Coupling of Finite Element and Moment Methods for Electromagnetics Scattering from Inhomogeneous Objects," *IEEE Transactions on Antennas and Propagation*, Vol. 38, pp. 386–393, 1990.

[184] Yuan, X., "Three-Dimensional Electromagnetic Scattering from Inhomogeneous Objects by the Hybrid Moment and Finite-Element Method," *IEEE Transactions on Microwave Theory and Techniques*, Vol. 38, pp. 1053–1058, 1990.

[185] Gedney, S. and Mittra, R., "Analysis of the Electromagnetic Scattering by Thick Gratings Using a Combined FEM/MM Solution," *IEEE Transactions on Antennas and Propagation*, Vol. 39, pp. 1605–1614, 1991.

[186] Morgan, M. A. and Welch, B. E., "The Field Feedback Formulation for Electromagnetic Scattering Computations," *IEEE Transactions on Antennas and Propagation*, Vol. AP-34, pp. 1377–1382, 1986.

[187] Morgan, M. A. and Welch, T. B., "Field Feedback Computation of Scattering by Two-Dimensional Penetrable Objects," *IEEE Transactions on Antennas and Propagation*, Vol. 40, pp. 445–450, 1992.

[188] Zuffada, C., Cwik, T., and Jamnejad, V., "Modeling Radiation with an Efficient Hybrid Finite-Element Integral-Equation Waveguide Mode-Matching Technique," *IEEE Transactions on Antennas and Propagation*, Vol. 45, pp. 34–39, 1997.

[189] Barka, A., Soudais, P., and Volpert, D., "Scattering from Three-Dimensional Cavities with a Plug and Play Numerical Scheme Combining IE, PDE, and Modal Techniques," *IEEE Transactions on Antennas and Propagation*, Vol. 48(5), pp. 704–712, May 2000.

[190] Peterson, A. F., "The 'Interior Resonance' Problem Associated with Surface Integral Equations of Electromagnetics: Numerical Consequences and a Survey of Remedies," *Electromagnetics*, Vol. 10, pp. 293–312, 1990.

[191] Murphy, W. D., Rokhlin, V., and Vassiliou, M. S., "Solving Electromagnetic Scattering Problems at Resonance Frequencies," *Journal of Applied Physics*, Vol. 67, pp. 6061–6065, 1990.

[192] Collins, J. D., Jin, J. M., and Volakis, J. L., "Eliminating Interior Resonances in Finite-Element Boundary Integral Methods for Scattering," *IEEE Transactions on Antennas and Propagation*, Vol. 40, pp. 1583–1585, 1992.

[193] Pearson, L. W., et al., "Inward-Looking and Outward-Looking Formulations for Scattering from Penetrable Objects," *IEEE Transactions on Antennas and Propagation*, Vol. 40, pp. 714–720, 1992.

[194] Hoppe, D. J., Epp, L. W., and Lee, J. F., "Hybrid Symmetric FEM/MOM Formulation Applied to Scattering by Inhomogeneous Bodies of Revolution," *IEEE Transactions on Antennas and Propagation*, Vol. 42, pp. 798–805, 1994.

[195] Ji, Y., Wang, H., and Hubing, T. H., "A Numerical Investigation of Interior Resonances in the Hybrid FEM/MoM Method," *IEEE Transactions on Antennas and Propagation*, Vol. 51(2), pp. 347–349, Feb. 2003.

[196] Lu, N. and Jin, J.-M., "Application of Fast Multipole Method to Finite-Element Boundary-Integral Solution of Scattering Problems," *IEEE Transactions on Antennas and Propagation*, Vol. 44, pp. 781–786, 1996.

[197] Bindiganavale, S. S. and Volakis, J. L., "A Hybrid FE-FMM Technique for Electromagnetic Scattering," *IEEE Transactions on Antennas and Propagation*, Vol. 45, pp. 180–181, 1997.

[198] Liu, J. and Jin, J. M., "A Novel Hybridization of Higher Order Finite Element and Boundary Integral Methods for Electromagnetic Scattering and Radiation Problems," *IEEE Transactions on Antennas and Propagation*, Vol. 49(12), pp. 1794–1806, Dec. 2001.

[199] Botha, M. M. and Davidson, D. B., "An Explicit a Posteriori Error Indicator for Electromagnetic, Finite Element-Boundary Integral Analysis," *IEEE Transactions on Antennas and Propagation*, Vol. 53(11), pp. 3717–3725, Nov. 2005.

[200] Botha, M. M. and Davidson, D. B., "The Implicit, Element Residual Method for a Posteriori Error Estimation in FE-BI Analysis," *IEEE Transactions on Antennas and Propagation*, Vol. 54(1), pp. 255–258, Jan. 2006.

[201] Liu, J. and Jin, J. M., "A Highly Effective Preconditioner for Solving the Finite Element-Boundary Integral Matrix Equation of Three-Dimensional Scattering," *IEEE Transactions on Antennas and Propagation*, Vol. 50(9), pp. 1212–1221, Sept. 2002.

[202] Volakis, J. L., Ozdemir, T., and Gong, J., "Hybrid Finite-Element Methodologies for Antennas and Scattering," *IEEE Transactions on Antennas and Propagation*, Vol. 45, pp. 493–507, 1997.

[203] Engquist, B. and Majda, A., "Absorbing Boundary Conditions for the Numerical Simulation of Waves," *Mathematics of Computation*, Vol. 31, pp. 629–651, 1977.

[204] Bayliss, A. and Turkel, E., "Radiation Boundary Conditions for Wave-Like Equations," *Communications on Pure and Applied Mathematics*, Vol. 33, pp. 707–725, 1980.

[205] Bayliss, A., Gunzburger, M., and Turkel, E., "Boundary Conditions for the Numerical Solution of Elliptic Equations in Exterior Regions," *SIAM Journal of Applied Mathematics*, Vol. 42, pp. 430–451, 1982.

[206] Hingdon, R. L., "Numerical Absorbing Boundary Conditions for the Wave Equation," *Mathematics of Computation*, Vol. 49, pp. 65–90, 1987.

[207] Trefethen, L. N. and Halpern, L., "Well-Posedness of One-Way Wave Equations and Absorbing Boundary Conditions," *Mathematics of Computation*, Vol. 47, pp. 421–435, 1986.

[208] Lee, C. F., et al., "Absorbing Boundary Conditions on Circular and Elliptical Boundaries," *Journal of Electromagnetic Waves and Applications*, Vol. 4, pp. 945–962, 1990.

[209] Mur, G., "Absorbing Boundary Conditions for the Finite-Difference Approximation of the Time-Domain Electromagnetic-Field Equation," *IEEE Transactions on Electromagnetic Compatibility*, Vol. EMC-23, pp. 377–382, 1981.

[210] Moore, T. G., et al., "Theory and Application of Radiation Boundary Operators," *IEEE Transactions on Antennas and Propagation*, Vol. 36, pp. 1797–1812, 1988.

[211] Peterson, A. F. and Castillo, S. P., "A Frequency-Domain Differential Equation Formulation for Electromagnetic Scattering from Inhomogeneous Cylinders," *IEEE Transactions on Antennas and Propagation*, Vol. 37, pp. 601–607, 1989.

[212] Ramahi, O. and Mittra, R., "Finite-Element Analysis of Dielectric Scatterers Using the Absorbing Boundary Conditions," *IEEE Transactions on Magnetics*, Vol. 25, pp. 3043–3045, 1989.

[213] Mittra, R. and Ramahi, O., "Absorbing Boundary Conditions for the Direct Solution of Partial Differential Equations Arising in Electromagnetic Scattering Problems," In *Finite Element and Finite Difference Methods in Electromagnetic Scattering*, Morgan, M., editor, Vol. 2, New York: Elsevier, 1990.

[214] D'Angelo, J. and Mayergoyz, I. D., "Finite-Element Methods for the Solution of RF Radiation and Scattering Problems," *Electromagnetics*, Vol. 10, pp. 177–199, 1990.

[215] Meade, D. B., et al., "Comparison of Local Radiation Boundary Conditions for the Scalar Helmholtz Equation with General Boundary Shapes," *IEEE Transactions on Antennas and Propagation*, Vol. 43, pp. 6–10, 1995.

[216] Peterson, A. F., "Absorbing Boundary Conditions for the Vector Wave Equation," *Microwave and Optical Technology Letters*, Vol. 1, pp. 62–64, 1988.

[217] Webb, J. P. and Kanellopoulos, V. N., "Absorbing Boundary Conditions for the Finite-Element Solution of the Vector Wave Equation," *Microwave and Optical Technology Letters*, Vol. 2, pp. 370–372, 1989.

[218] Kanellopoulos, V. N. and Webb, J. P., "A Numerical Study of Vector-Absorbing Boundary Conditions for the Finite Element Solution of Maxwell's Equations," *IEEE Microwave and Guided Wave Letters*, Vol. 1, pp. 325–327, 1991.

[219] Peterson, A. F., "Accuracy of Three-Dimensional Radiation Boundary Conditions for Use with the Vector Helmholtz Equation," *IEEE Transactions on Antennas and Propagation*, Vol. 40, pp. 351–355, 1992.

[220] Chatterjee, A. and Volakis, J. L., "Conformal Absorbing Boundary Conditions for the Vector Wave Equation," *Microwave and Optical Technology Letters*, Vol. 6, pp. 886–889, 1993.

[221] Stupfel, B., "Absorbing Boundary Conditions on Arbitrary Boundaries for the Scalar and Vector Wave Equations," *IEEE Transactions on Antennas and Propagation*, Vol. 42, pp. 773–780, 1994.

[222] Weimin, W. and Balanis, C. A., "Vector One-Way Wave Absorbing Boundary Conditions for FEM Applications," *IEEE Transactions on Antennas and Propagation*, Vol. 42, pp. 872–878, 1994.

[223] Kanellopoulos, V. N. and Webb, J. P., "The Importance of the Surface Divergence Term in the Finite Element-Vector Absorbing Boundary Condition Method," *IEEE Transactions on Microwave Theory and Techniques*, Vol. 43, pp. 2168–2169, 1995.

[224] Botha, M. M. and Davidson, D. B., "Rigorous, Auxiliary Variable-Based Implementation of a Second-Order ABC for the Vector FEM," *IEEE Transactions on Antennas and Propagation*, Vol. 54(11), pp. 3499–3504, Nov. 2006.

[225] Fan, D. and Wang, Y., "Global Numerical Boundary Condition-Based PDE Solution Technique for Open-Region Electromagnetic Field Problems," *IEEE Transactions on Antennas and Propagation*, Vol. 41, pp. 253–260, 1993.

[226] Zhaoqing, C., et al., "Measured Equation of Invariance: A New Concept in Field Computations," *IEEE Transactions on Antennas and Propagation*, Vol. 42, pp. 320–328, 1994.

[227] Jevtic, J. O. and Lee, R., "Theoretical and Numerical Analysis of the Measured Equation of Invariance," *IEEE Transactions on Antennas and Propagation*, Vol. 42, pp. 1097–1105, 1994.

[228] Mei, K. K., et al., "Comments on 'A Theoretical and Numerical Analysis of the Measured Equation of Invariance,'" *IEEE Transactions on Antennas and Propagation*, Vol. 43, pp. 1168–1171, 1995.

[229] Jevtic, J. O. and Lee, R., "An Analytical Characterization of the Error in the Measured Equation of Invariance," *IEEE Transactions on Antennas and Propagation*, Vol. 43, pp. 1109–1115, 1995.

[230] Wright, D. B. and Cangellaris, A. C., "Finite-Element Gid Truncation Schemes Based on the Measured Equation of Invariance," *Radio Science*, Vol. 29, pp. 907–921, 1994.

[231] Boag, A., et al., "Numerical Absorbing Boundary Condition for Finite-Difference and Finite-Element Analysis of Open Structures," *Microwave and Optical Technology Letters*, Vol. 7, pp. 395–398, 1994.

[232] Boag, A., Boag, A., and Mittra, R., "Numerical Absorbing Boundary Condition for Edge-Based Finite-Element Analysis," *Microwave and Optical Technology Letters*, Vol. 7, pp. 733–737, 1994.

[233] Li, Y. and Cendes, Z. J., "High-Accuracy Absorbing Boundary Condition," *IEEE Transactions on Magnetics*, Vol. 31, pt. 1, pp. 1524–1529, 1995.

[234] Boag, A. and Mittra, R., "Numerical Absorbing Boundary Condition for Three-Dimensional Edge-Based Finite-Element Analysis of Very Low-Frequency Fields," *Microwave and Optical Technology Letters*, Vol. 9, pp. 22–27, 1995.

[235] Stupfel, B. and Mittra, R., "A Theoretical Study of Numerical Absorbing Boundary Conditions," *IEEE Transactions on Antennas and Propagation*, Vol. 43, pp. 478–487, 1995.

[236] Stupfel, B. and Mittra, R., "Numerical Absorbing Boundary Conditions for the Scalar and Vector Wave Equations," *IEEE Transactions on Antennas and Propagation*, Vol. 44, pp. 1015–1022, 1996.

[237] Jin, J. M., Volakis, J. L., and Liepa, V. V., "Fictitious Absorber for Truncating Finite Element Meshes in Scattering," *IEE Proceedings, Part H*, Vol. 139, pp. 472–476, 1992.

[238] Berenger, J. P., "A Perfectly Matched Layer for the Absorption of Electromagnetic Waves," *Journal of Computational Physics*, Vol. 114, pp. 185–200, 1994.

[239] Katz, D. S., Thiele, E. T., and Taflove, A., "Validation and Extension to Three-Dimensions of the Berenger PML Absorbing Boundary Condition for FD-TD Meshes," *IEEE Microwave and Guided Wave Letters*, Vol. 4, pp. 268–270, 1994.

[240] Mittra, R. and Pekel, U., "New Look at the Perfectly Matched Layer (PML) Concept for the Reflectionless Absorption of Electromagnetic Waves," *IEEE Microwave and Guided Wave Letters*, Vol. 5, pp. 84–86, 1995.

[241] Pekel, U. and Mittra, R., "Finite-Element-Method Frequency-Domain Application of the Perfectly Matched Layer (PML) Concept," *Microwave and Optical Technology Letters*, Vol. 9, pp. 84–86, 1995.

[242] Sacks, Z. S., et al., "A Perfectly Matched Anisotropic Absorber for Use as an Absorbing Boundary Condition," *IEEE Transactions on Antennas and Propagation*, Vol. 43, pp. 1460–1463, 1995.

[243] Kingsland, D. M., et al., "Performance of an Anisotropic Artificial Absorber for Truncating Finite-Element Meshes," *IEEE Transactions on Antennas and Propagation*, Vol. 44, pp. 975–982, 1996.

[244] Kuzuoglu, M. and Mittra, R., "Investigation of Nonplanar Perfectly Matched Absorbers for Finite-Element Mesh Truncation," *IEEE Transactions on Antennas and Propagation*, Vol. 5, pp. 474–486, 1997.

[245] Teixeira, F. L., and Chew, W. C., "Systematic Derivation of Anisotropic PML Absorbing Media in Cylindrical and Spherical Coordinates," *IEEE Microwave and Guided Wave Letters*, Vol. 7(11), pp. 371–373, Nov. 1997.

[246] Ozgun, O. and Kuzuoglu, M., "Non-Maxwellian Locally Conformal PML Absorbers for Finite-Element Mesh Truncation," *IEEE Transactions on Antennas and Propagation*, Vol. 55(3), pp. 931–937, Mar. 2007.

[247] Davidson, D. B. and Botha, M. M., "Evaluation of a Spherical PML for Vector FEM Applications," *IEEE Transactions on Antennas and Propagation*, Vol. 55(2), pp. 494–498, Feb. 2007.

[248] Khebir, A., D'Angelo, J., and Joseph, J., "A New Finite-Element Formulation for RF Scattering from Complex Bodies of Revolution," *IEEE Transactions on Antennas and Propagation*, Vol. 41, pp. 534–541, 1993.

[249] Greenwood, A. D. and Jin, J. M., "A Novel Efficient Algorithm for Scattering from a Complex BOR Using Mixed Finite Elements and Cylindrical PML," *IEEE Transactions on Antennas and Propagation*, Vol. 47(4), pp. 620–629, Apr. 1999.

[250] Jin, J. M., "A Highly Robust and Versatile Finite Element-Boundary Integral Hybrid Code for Scattering by BOR Objects," *IEEE Transactions on Antennas and Propagation*, Vol. 53(7), pp. 2274–2281, July 2005.

[251] Dunn, E. A., et al., Numerical Simulation of BOR Scattering and Radiation Using a Higher Order FEM," *IEEE Transactions on Antennas and Propagation*, Vol. 54(3), pp. 945–952, Mar. 2006.

[252] Soudais, P., "Iterative Solution of a Three-Dimensional Scattering Problem from Arbitrary Shaped Multidielectric and Multiconducting Bodies," *IEEE Transactions on Antennas and Propagation*, Vol. 42, pp. 954–959, 1994.

[253] Cwik, T., Zuffada, C., and Jamnejad, V., "Modeling Three-Dimensional Scatterers Using a Coupled Finite Element-Integral Equation Formulation," *IEEE Transactions on Antennas and Propagation*, Vol. 44, pp. 453–459, 1996.

[254] Kempel, L. C. and Volakis, J. L., "Scattering by a Cavity-Backed Antenna on a Circular Cylinder," *IEEE Transactions on Antennas and Propagation*, Vol. 42, pp. 1268–1279, 1994.

[255] Sun, W. and Balanis, C. A., "Edge-Based FEM Solution of Scattering from Inhomogeneous and Anisotropic Objects," *IEEE Transactions on Antennas and Propagation*, Vol. 42, pp. 627–632, 1994.

[256] Baucke, R. C., D'Angelo, J., and Crabtree, G. D., "Far-Field Transforms for Finite-Element Frequency Domain Solutions," *IEEE Transactions on Antennas and Propagation*, Vol. 41, pp. 511–514, 1993.

[257] Gedney, S. D., Lee, J. F., and Mittra, R., "A Combined FEM/MoM Approach to Analyze the Plane Wave Diffraction by Arbitrary Gratings," *IEEE Transactions on Microwave Theory and Techniques*, Vol. 40, pp. 363–370, 1992.

[258] Jin, J. M. and Volakis, J. L., "Scattering and Radiation Analysis of Three-Dimensional Cavity Arrays Via a Hybrid Finite Element Method," *IEEE Transactions on Antennas and Propagation*, Vol. 41, pp. 1580–1586, 1993.

[259] Lucas, E. W. and Fontana, T. P., "A Three-Dimensional Hybrid Finite Element/Boundary Element Method for the Unified Radiation and Scattering Analysis of General Infinite Periodic Array," *IEEE Transactions on Antennas and Propagation*, Vol. 43, pp. 145–154, 1995.

[260] Calamia, M., et al., "Finite-Element Analysis of Electromagnetic Scattering from Periodic Structures at Oblique Incidence," *Electronics Letters*, Vol. 29, pp. 847–848, 1993.

[261] Cecchini, R., Coccioli, R., and Pelosi, G., "PERIODIC^3: A Software Package for the Analysis of Artificially Anisotropic Surfaces," *IEEE Antennas and Propagation Magazine*, Vol. 37, pp. 83–86, 1995.

[262] Mias, C., Webb, J. P., and Ferrari, R. L., "Finite-Element Analysis of Electromagnetic Plane Wave Scattering from an Infinite Doubly Periodic Structure," *COMPEL—The International Journal for Computation and Mathematics in Electrical and Electronic Engineering*, Vol. 13, Supplement A, pp. 393–398, 1994.

[263] Bardi, I., et al., "Plane Wave Scattering from Frequency-Selective Surfaces by the Finite-Element Method," *IEEE Transactions on Magnetics*, Vol. 38(2), pp. 641–644, Mar. 2002.

[264] Chung, S. J. and Chen, J. L., "A Modified Finite-Element Method for Analysis of Finite Periodic Structures," *IEEE Transactions on Microwave Theory and Techniques*, Vol. 42, pp. 1561–1566, 1994.

[265] Pinello, W. P., Lee, R., and Cangellaris, A. C., "Finite-Element Modeling of Electromagnetic Wave Interactions with Periodic Dielectric Structures," *IEEE Transactions on Microwave Theory and Techniques*, Vol. 42, pp. 2294–2301, 1994.

[266] Gordon, R. K. and Mittra, R., "Finite Element Analylsis of Axisymmetric Radomes," *IEEE Transactions on Antennas and Propagation*, Vol. 41, pp. 975–981, 1993.

[267] Wang, Y. and Fan, D., "Accurate Global Solutions of EM Boundary-Value Problems for Coaxial Radiators," *IEEE Transactions on Antennas and Propagation*, Vol. 42, pp. 767–770, 1994.

[268] Chinn, G. C., Epp, L. W., and Hoppe, D. J., "A Hybrid Finite-Element Method for Axisymmetric Waveguide-Fed Horns," *IEEE Transactions on Antennas and Propagation*, Vol. 44, pp. 280–285, 1996.

[269] Gentili, G. G., et al., "High-Order FEM Mode Matching Analysis of Circular Horns With Rotationally Symmetric Dielectrics," *IEEE Transactions on Antennas and Propagation*, Vol. 55(10), pp. 2915–2918, Oct. 2007.

[270] Greenwood, A. D. and Jin, J. M., "Finite-Element Analysis of Complex Axisymmetric Radiating Structures," *IEEE Transactions on Antennas and Propagation*, Vol. 47(8), pp. 1260–1266, Aug. 1999.

[271] Gong, J., et al., "A Hybrid Finite Element-Boundary Integral Method for the Analysis of Cavity-Backed Antennas of Arbitrary Shape," *IEEE Transactions on Antennas and Propagation*, Vol. 42, pp. 1233–1242, 1994.

[272] Ozdemir, T. and Volakis, J. L., "Triangular Prisms for Edge-Based Vector Finite-Element Analysis of Conformal Antennas," *IEEE Transactions on Antennas and Propagation*, Vol. 45, pp. 788–797, 1997.

[273] Gonzalez de Aza, M. A., Zapata, J., and Encinar, J. A., "Broad-Band Cavity-Backed and Capacitively Probe-Fed Microstrip Patch Arrays," *IEEE Transactions on Antennas and Propagation*, Vol. 48(5), pp. 784–789, May 2000.

[274] Vouvakis, M. N., et al., "Ferrite-Loaded Cavity-Backed Antennas Including Nonuniform and Nonlinear Magnetization Effects," *IEEE Transactions on Antennas and Propagation*, Vol. 51(5), pp. 1000–1010, May 2003.

[275] Vouvakis, M. N., et al., "Multilayer Effects on Cavity-Backed Slot Antennas," *IEEE Transactions on Antennas and Propagation*, Vol. 52(3), pp. 880–887, Mar. 2004.

[276] Jan, C.-G., et al., "Analysis of Edge Slots in Rectangular Waveguide with Finite Waveguide Wall Thickness," *IEEE Transactions on Antennas and Propagation*, Vol. 44, pp. 1120–1126, 1996.

[277] McGrath, D. T., and Pyati, V. P., "Phased Array Antenna Analysis with the Hybrid Finite-Element Method," *IEEE Transactions on Antennas and Propagation*, Vol. 42, pp. 1625–1630, 1994.

[278] Calamia, M., et al., "A Hybrid FEM/UTD Technique for the Analysis of the Scattering from a Cavity on the Face of a Perfectly Conducting Wedge," *COMPEL—The International Journal for Computation and Mathematics in Electrical and Electronic Engineering*, Vol. 13, Supplement A, pp. 229–235, 1994.

[279] Pelosi, G., et al., "Scattering from a Wedge with Cavity-Backed Apertures in its Faces and Related Configurations: TE case," *IEE Proceedings*, Part H, Vol. 142, pp. 183–188, 1995.

[280] Pelosi, G., Coccioli, R., and Silvester, P. P., "Finite Elements for Open-Boundary Problems: Algorithms and Prospects," In *1995 International Symposium on Electromagnetic Theory*, St. Petersburg, Russia, May 23-26, 1995.

[281] Ross, D. C., Volakis, J. L., and Anastassiu, H. T., "Hybrid Finite Element-Modal Analysis of Jet-Engine Inlet Scattering," *IEEE Transactions on Antennas and Propagation*, Vol. 43, pp. 277–285, 1995.

[282] Jin, J. M., Ni, S. S., and Lee, S. W., "Hybridization of SBR and FEM for Scattering by Large Bodies with Gaps and Cracks," *IEEE Transactions on Antennas and Propagation*, Vol. 43, pp. 1130–1139, 1995.

[283] Reddy, C. J., et al., "Radiation Characteristics of Cavity Backed Aperture Antennas in Finite Ground Plane Using the Hybrid FEM/MoM Technique and Geometrical Theory of Diffraction," *IEEE Transactions on Antennas and Propagation*, Vol. 44, pp. 1327–1333, 1996.

[284] Ozdemir, T., et al., "A Hybridization of Finite-Element and High-Frequency Methods for Pattern Prediction for Antennas on Aircraft Structures," *IEEE Antennas and Propagation Magazine*, Vol. 38, pp. 28–38, 1996.

[285] Pelosi, G., Coccioli, R., and Manara, G., "An FEM-Based Numerical Diffraction Coefficient for Irregular Wedge Configurations," *IEEE Transactions on Antennas and Propagation*, Vol. 45, pp. 563–564, 1997.

[286] Chiniard, R., Barka, A., and Pascal, O., "Hybrid FEM/Floquet Modes/PO Technique for Multi-Incidence RCS Prediction of Array Antennas," *IEEE Transactions on Antennas and Propagation*, Vol. 56(6), pp. 1679–1686, June 2008.

[287] Fernandez-Recio, R., "Fully Coupled Hybrid FEM-UTD Method Using NURBS for the Analysis of Radiation Problems," *IEEE Transactions on Antennas and Propagation*, Vol. 56(3), pp. 774–783, Mar. 2008.

[288] Astley, R. J., Macaulay, G. J., and Coyette, J. P., "Mapped Wave Envelope Elements for Acoustical Radiation and Scattering," *Journal of Sound and Vibration*, Vol. 170, pp. 97–118, 1994.

[289] Bettess, P. and Chadwick, E., "Wave Envelope Examples for Progressive Waves," *International Journal for Numerical Methods in Engineering*, Vol. 38, pp. 2487–2508, 1995.

[290] Choi, C. M. T. and Webb, J. P., "The Wave-Envelope Method and Absorbing Boundary Conditions," *IEEE Transactions on Magnetics*, Vol. 33, pp. 1420–1423, 1997.

[291] Sarkar, T. K., et al., "Utilization of Wavelet Concepts in Finite Elements for an Efficient Solution of Maxwell's Equations," *Radio Science*, Vol. 29, pp. 965–977, 1994.

[292] Garcia-Castillo, L. E. and Salazar-Palma, M., "Efficient Solution of the Differential Form of Maxwell's Equations in Rectangular Regions," *IEEE Transactions on Microwave Theory and Techniques*, Vol. 43, pp. 647–654, 1995.

[293] Chan, T. F., et al., *Domain Decomposition Methods*. Philadelphia, PA: SIAM, 1989.

[294] Stupfel, B., "A Fast-Domain Decomposition Method for the Solution of Electromagnetic Scattering by Large Objects," *IEEE Transactions on Antennas and Propagation*, Vol. 44, pp. 1375–1385, 1996.

[295] Stupfel, B. and Mognot, M., "A Domain Decomposition Method for the Vector Wave Equation," *IEEE Transactions on Antennas and Propagation*, Vol. 48(5), pp. 653–660, May 2000.

[296] Li, Y. and Jin, J. M., "A Vector Dual-Primal Finite Element Tearing and Interconnecting Method for Solving 3-D Large-Scale Electromagnetic Problems," *IEEE Transactions on Antennas and Propagation*, Vol. 54(10), pp. 3000–3009, Oct. 2006.

[297] Li, Y. and Jin, J. M., "A New Dual-Primal Domain Decomposition Approach for Finite Element Simulation of 3-D Large-Scale Electromagnetic Problems," *IEEE Transactions on Antennas and Propagation*, Vol. 55(10), pp. 2803–2810, Oct. 2007.

[298] Zhao, K., et al., "A Domain Decomposition Method With Nonconformal Meshes for Finite Periodic and Semi-Periodic Structures," *IEEE Transactions on Antennas and Propagation*, Vol. 55(9), pp. 2559–2570, Sept. 2007.

[299] Li, Y. and Jin, J. M., "Implementation of the Second-Order ABC in the FETI-DPEM Method for 3D EM Problems," *IEEE Transactions on Antennas and Propagation*, Vol. 56(8), pp. 2765–2769, Aug. 2008.

[300] Gander, M. J., "Optimized Schwarz Method for Helmholtz Problems," In *13th Int. Conf. Domain Decomposition*, pp. 160, 245–252, 2001.

[301] Bernardi, C., Maday, Y., and Patera, A., "A New Nonconforming Approach to Domain Decomposition: The Mortar Element Method," In *Nonlinear Partial Differential Equations and Their Applications*, Brezis, H. and Lions, J., editors, New York: Pitman, 1994.

[302] Vouvakis, M. N., Cendes, Z. J., and Lee, J. F., "A FEM Domain Decomposition Method for Photonic and Electromagnetic Band Gap Structures," *IEEE Transactions on Antennas and Propagation*, Vol. 54(2), pp. 721–733, Feb. 2006.

[303] Hamandi, L., Lee, R. , and Ozguner, F., "Review of Domain-Decomposition Methods for the Implementation of FEM on Massively Parallel Computers," *IEEE Antennas and Propagation Magazine*, Vol. 37, pp. 93–98, 1995.

[304] Choi-Grogan, Y. S., et al., "Sequential and Parallel Implementations of the Partitioning Finite-Element Method," *IEEE Transactions on Antennas and Propagation*, Vol. 44, pp. 1609–1616, 1996.

[305] Cwik, T., Katz, D. S., and Patterson, J., "Scalable Solutions to Integral-Equation and Finite-Element Simulations," *IEEE Transactions on Antennas and Propagation*, Vol. 45, pp. 544–555, 1997.

[306] Wolfe, C. T., Navsariwala, U., and Gedney, S. D., "A Parallel Finite-Element Tearing and Interconnecting Algorithm for Solution of the Vector Wave Equation with PML Absorbing Medium," *IEEE Transactions on Antennas and Propagation*, Vol. 48(2), pp. 278–284, Feb. 2000.

[307] Guarnieri, G., et al., "A Domain Decomposition Technique for Finite Element Based Parametric Sweep and Tolerance Analyses of Microwave Passive Devices,"*International*

Journal of RF and Microwave Computer-Aided Engineering, Vol. 19(3), pp. 328–337, May 2009.

[308] Cangellaris, A. C., et al., "Electromagnetic Model Order Reduction for System-Level Modeling," *IEEE Transactions on Microwave Theory and Techniques*, Vol. 47(6), pp. 840–850, June 1999.

[309] Chiprout, E. and Nakhla, M. S., *Asymptotic Waveform Evaluation and Moment Matching for Interconnect Analysis*. Norwell, MA: Kluwer, 1994.

[310] Celik, M. and Cangellaris, A. C., "Simulation of Dispersive Multiconductor Transmission Lines by Padé Approximation Via the Lanczos Process," *IEEE Transactions on Antennas and Propagation*, Vol. 44(12), pp. 2525–2535, Dec. 1996.

[311] Bracken, J. E., Sun, D. K., and Cendes, Z. J., "S-Domain Methods for Simultaneous Time and Frequency Characterization of Electromagnetic Devices," *IEEE Transactions on Microwave Theory and Techniques*, Vol. 46(9), pp. 1277–1290, Sept. 1998.

[312] Sun, D. K., et al., "ALPS-A New Fast Frequency-Sweep Procedure for Microwave Devices," *IEEE Transactions on Microwave Theory and Techniques*, Vol. 49(2), pp. 398–402, Feb. 2001.

[313] Rubio, J., Arroyo, J., and Zapata, J., "SFELP-An Efficient Methodology for Microwave Circuit Analysis," *IEEE Transactions on Microwave Theory and Techniques*, Vol. 49(3), pp. 509–516, Mar. 2001.

[314] Rubio, J., Gonzalez, M. A., and Zapata, J., "Generalized Scattering Matrix Analysis of a Class of Finite Arrays of Coupled Antennas by Using Three-Dimensional FEM and Spherical Mode Expansion," *IEEE Transactions on Antennas and Propagation*, Vol. 53(3), pp. 1133–1144, Mar. 2005.

[315] Schultschik, A., Farle, O., and Dyczij-Edlinger, R., "A Model Order Reduction Method for the Finite-Element Simulation of Inhomogeneous Waveguides," *IEEE Transactions on Magnetics*, Vol. 44(6); pp. 1394–1397, June 2008.

[316] Slone, R. D., Lee, R., and Lee, J. F., "Well-Conditioned Asymptotic Waveform Evaluation for Finite Elements," *IEEE Transactions on Antennas and Propagation*, Vol. 51(9), pp. 2442–2447, Sept. 2003.

[317] Slone, R. D., Lee, R., and Lee, J. F., "Multipoint Galerkin Asymptotic Waveform Evaluation for Model Order Reduction of Frequency Domain FEM Electromagnetic Radiation Problems," *IEEE Transactions on Antennas and Propagation*, Vol. 49(10), pp. 1504–1513, Oct. 2001.

[318] Slone, R. D., Lee, J. F., and Lee, R., "Automating Multipoint Galerkin AWE for a FEM Fast Frequency Sweep," *IEEE Transactions on Magnetics*, Vol. 38(3), pp. 637–640, Mar. 2002.

[319] Slone, R. D., Lee, J. F., and Lee, R., "Well-Conditioned Asymptotic Waveform Evaluation for Finite Elements," *IEEE Transactions on Antennas and Propagation*, Vol. 51(9), pp. 2442–2447, Sept. 2003.

[320] Lee, H. B., Jung, H. K., and Hahn, S. Y., "Shape Optimization of H-Plane Waveguide Tee Junction Using Edge Finite Element Method," *IEEE Transactions on Magnetics*, Vol. 31, pp. 1928–1931, 1995.

[321] Kim, J., et al., "Optimal Design Technique for Waveguide Device," *IEEE Transactions on Magnetics*, Vol. 32, pp. 1250–1253, 1996.

[322] Lee, H.-B. and Itoh, T., "A Systematic Optimum Design of Waveguide-to-Microstrip Transition," *IEEE Transactions on Microwave Theory and Techniques*, Vol. 45, pp. 803–809, 1997.

[323] Coccioli, R., Pelosi, R., and Selleri, S., "Optimization of Bends in Rectangular Waveguide by a Finite Element-Genetic Algorithm Procedure," *Microwave and Optical Technology Letters*, Vol. 16(5), pp. 287–290, Dec. 1997.

[324] Kolbehdari, M. A., et al., "Simultaneous Time and Frequency Domain Solutions of EM Problems Using Finite Element and CFH Techniques," *IEEE Transactions on Microwave Theory and Techniques*, Vol. 44, pp. 1526–1534, 1996.

[325] Polstyanko, S. V., Dyezij-Edlinger, R., and Lee, J.-F., "Fast Frequency Sweep Technique for the Efficient Analysis of Dielectric Waveguides," *IEEE Transactions on Microwave Theory and Techniques*, Vol. 45, pp. 1118–1126, 1997.

[326] Lee, J.-F., Lee, R., and Cangellaris, A., Time-Domain Finite-Element Methods," *IEEE Transactions on Antennas and Propagation*, Vol. 45, pp. 430–442, 1997.

[327] Cangellaris, A. C., Lin, C. C., and Mei, K. K., "Point-Matched Time Domain Finite-Element Methods for Electromagnetic Radiation and Scattering," *IEEE Transactions on Antennas and Propagation*, Vol. AP-35, pp. 1160–1173, 1987.

[328] Lynch, D. R. and Paulsen, K. D., "Time-Domain Integration of the Maxwell's Equations on Finite Elements," *IEEE Transactions on Antennas and Propagation*, Vol. 38, pp. 1933–1942, 1990.

[329] Mur, G., "The Finite-Element Modeling of Three-Dimensional Time-Domain Electromagnetic Fields in Strongly Inhomogeneous Media," *IEEE Transactions on Magnetics*, Vol. 28, pp. 1130–1133, 1992.

[330] Lee, J. F., "WETD-A Finite Element Time-Domain Approach for Solving Maxwell's Equations," *IEEE Microwave and Guided Wave Letters*, Vol. 4, pp. 11–13, 1994.

[331] Mahadevan, K., Mittra, R., and Vaidya, P. M., "Use of Whitney's Edge and Face Elements for Efficient Finite-Element Time Domain Solution of Maxwell's Equations," *Journal of Electromagnetic Waves and Applications*, Vol. 8, pp. 1173–1191, 1994.

[332] Wong, M. F., Picon, O., and Hanna, V. F., "Finite-Element Method Based on Whitney Forms to Solve Maxwell's Equations in the Time Domain," *IEEE Transactions on Magnetics*, Vol. 31, pp. 1618–1621, 1995.

[333] Lee, J. F. and Sacks, Z., "Whitney Elements Time Domain (WETD) Methods," *IEEE Transactions on Magnetics*, Vol. 31(3), pp. 1325–1329, Mar. 1995.

[334] Monorchio, A., et al., "A Dispersion Analysis for the Finite-Element Method in Time Domain with Triangular Edge Elements," *IEEE Antennas and Wireless Propagation Letters*, Vol. 1(1), pp. 207–210, Jan. 2002.

[335] Gedney, S. D., "Unconditionally Stable Finite Element Time-Domain Solution of the Vector Wave Equation," *IEEE Microwave and Guided Wave Letters*, Vol. 5, pp. 332–334, 1995.

[336] Ping, H. S., Wang, Y., and Itoh, T., "An Unconditionally Stable Extended (USE) Finite-Element Time-Domain Solution of Active Nonlinear Microwave Circuits Using Perfectly Matched Layers," *IEEE Transactions on Microwave Theory and Techniques*, Vol. 50(10), pp. 2226–2232, Oct. 2002.

[337] Sun, D. K., Lee, J. F., and Cendes, Z. J., "The Transfinite-Element Time Domain Method," *IEEE Transactions on Microwave Theory and Techniques*, Vol. 51(10), pp. 2097–2105, Oct. 2003.

[338] Mahadevan, K. and Mittra, R., "Radar Cross-Section Computation of Inhomogeneous Scatterers Using Edge-Based Finite Element Methods in Frequency and Time Domains," *Radio Science*, Vol. 28, pp. 1181–1193, 1993.

[339] Konrad, A., Rahim, R., and Hanna, R, "Time Domain Finite Element Analysis of Transients in Multiconductor Systems," *COMPEL—The International Journal for Computation and Mathematics in Electrical and Electronic Engineering*, Vol. 13, Supplement A, pp. 399–404, 1994.

[340] Zgainski, F. X., Meunier, G., and Konrad, A., "Analysis of Microwave Planar Circuits MIC2D: A User-Friendly Software Based on a Time Domain Finite Element Method," *IEEE Transactions on Magnetics*, Vol. 31, pp. 1650–1653, 1995.

[341] Choi, K., et al., "Time Domain Finite Element Analysis of High-Power Microwave Aperture Antennas," *IEEE Transactions on Magnetics*, Vol. 31, pp. 1622–1625, 1995.

[342] Ali, H. O. and Costache, G., "Finite-Element Time-Domain Analysis of Axysimmetrical Radiators," *IEEE Transactions on Antennas and Propagation*, Vol. 42, pp. 272–275, 1994.

[343] Obayya, S. S. A., "Efficient Finite-Element-Based Time-Domain Beam Propagation Analysis of Optical Integrated Circuits," *IEEE Journal of Quantum Electronics*, Vol. 40(5), pp. 591–595, May 2004.

[344] Jiao, D. and Jin, J. M., "Time-Domain Finite-Element Modeling of Dispersive Media," *IEEE Microwave and Wireless Letters*, Vol. 11(5), pp. 220–222, May 2001.

[345] Petersson, L. E. R. and Jin, J. M., "A Three-Dimensional Time-Domain Finite-Element Formulation for Periodic Structures," *IEEE Transactions on Antennas and Propagation*, Vol. 54(1), pp. 12–19, Jan. 2006.

[346] Petersson, L. E. R. and Jin, J. M., "Analysis of Periodic Structures Via a Time-Domain Finite-Element Formulation with a Floquet ABC," *IEEE Transactions on Antennas and Propagation*, Vol. 54(3), pp. 933–944, Mar. 2006.

[347] Cai, Y. and Mias, C., "Fast Finite Element Time Domain-Floquet Modal Absorbing Boundary Condition Modelling of Periodic Structures Using Recursive Convolution," *IEEE Transactions on Antennas and Propagation*, Vol. 55(9), pp. 2550–2558, Sept. 2007.

[348] Riley, D. J. and Jin, J. M., "Finite-Element Time-Domain Analysis of Electrically and Magnetically Dispersive Periodic Structures," *IEEE Transactions on Antennas and Propagation*, Vol. 56(11), pp. 3501–3509, Nov. 2008.

[349] Wu, R.-B. and Itoh, T., "Hybrid Finite-Difference Time-Domain Modeling of Curved Surfaces Using Tetrahedral Edge Elements," *IEEE Transactions on Antennas and Propagation*, Vol. 45, pp. 1302–1309, 1997.

[350] Yioultsis, T. V., et al., "A Fully Explicit Whitney Element-Time Domain Scheme with Higher Order Vector Finite-Elements for Three-Dimensional High Frequency Problems," *IEEE Transactions on Magnetics*, Vol. 34(9), pp. 3288–3291, Sept. 1998.

[351] Feliziani, M. and Maradei, F., "Mixed Finite-Difference/Whitney-Elements Time Domain (FD/WE-TD) Method," *IEEE Transactions on Magnetics*, Vol. 34(5), pp. 3222–3227, May 1998.

[352] Monorchio, A., et al., "A Hybrid Time-Domain Technique that Combines the Finite Element, Finite Difference and Method of Moment Techniques to Solve Complex Electromagnetic Problems," *IEEE Transactions on Antennas and Propagation*, Vol. 52(10), pp. 2666–2674, Oct. 2004.

[353] Loh, T. H. and Mias, C., "Implementation of an Exact Modal Absorbing Boundary Termination Condition for the Application of the Finite-Element Time-Domain Technique to Discontinuity Problems in Closed Homogeneous Waveguides," *IEEE Transactions on Microwave Theory and Techniques*, Vol. 52(3), pp. 882–888, Mar. 2004.

[354] Lou, Z. and Jin, J. M., "An Accurate Waveguide Port Boundary Condition for the Time-Domain Finite-Element Method," *IEEE Transactions on Microwave Theory and Techniques*, Vol. 53(3), pp. 3014–3023, Mar. 2005.

[355] Bermani, E., Caorsi, S., and Raffetto, M., "Performance of a Causal Anisotropic Absorber for the Truncation of Finite Element Meshes in Time Domain Electromagnetic Unbounded Problems," *IEEE Transactions on Antennas and Propagation*, Vol. 51(9), pp. 2523–2529, Sept. 2003.

[356] Rylander, T. and Jin, J.M., "Perfectly Matched Layer in Three-Dimensions for the Time-Domain Finite-Element Method Applied to Radiation Problems," *IEEE Transactions on Antennas and Propagation*, Vol. 53(4), pp. 1489–1499, Apr. 2005.

[357] Stowell, M. L., Fasenfest, B. J., and White, D. A., "Investigation of Radar Propagation in Buildings: A 10-Billion Element Cartesian-Mesh FETD Simulation," *IEEE Transactions on Antennas and Propagation*, Vol. 56(8), pp. 2241–2250, Aug. 2008.

[358] Guillouard, K., et al., "A New Global Finite Element Analysis of Microwave Circuits Including Lumped Elements," *IEEE Transactions on Microwave Theory and Techniques*, Vol. 4, pp. 2587–2594, 1996.

[359] Feliziani, M. and Maradei, F., "Modeling of Electromagnetic Fields and Electrical Circuits with Lumped and Distributed Elements by the WETD Method," *IEEE Transactions on Magnetics*, Vol. 35(3), pp. 1666–1669, May 1999.

[360] Feliziani, M. and Maradei, F., "Circuit-Oriented FEM: Solution of Circuit-Field Coupled Problems by Circuit Equations," *IEEE Transactions on Magnetics*, Vol. 38(2), pp. 965–968, Mar. 2002.

[361] Zhou, P., et al., "A General Cosimulation Approach for Coupled Field-Circuit Problems," *IEEE Transactions on Magnetics*, Vol. 42(4), pp. 1051–1054, Apr. 2006.

[362] Essakhi, B., Pichon, L., and Akoun, G., "Fast Analysis of a Broad-Band Microwave Rectenna Using 3-D FEM and Pad Approximation," *IEEE Transactions on Magnetics*, Vol. 43(4), pp. 1309–1312, Apr. 2007.

[363] Anderson, B., et al., "Full-Wave Analysis in SPICE Via Model-Order Reduction," *IEEE Transactions on Microwave Theory and Techniques*, Vol. 52(9), pp. 2314–2320, Sept. 2004.

[364] Wu, H. and Cangellaris, A. C., "A Finite-Element Domain-Decomposition Methodology for Electromagnetic Modeling of Multilayer High-Speed Interconnects," *IEEE Transactions on Advanced Packaging*, Vol. 31(2), pp. 339–350, May 2008.

[365] Gan, H. and Jiao, D., "A Time-Domain Layered Finite Element Reduction Recovery (LAFE-RR) Method for High-Frequency VLSI Design," *IEEE Transactions on Antennas and Propagation*, Vol. 55(12), pp. 3620–3629, Dec. 2007.

[366] Gan, H. and Jiao, D., "A Recovery Algorithm of Linear Complexity in the Time-Domain Layered Finite Element Reduction Recovery (LAFE-RR) Method for Large-Scale Electromagnetic Analysis of High-Speed ICs," *IEEE Transactions on Advanced Packaging*, Vol. 31(3), pp. 612–618, Aug. 2008.

[367] Jiao, D., Zhu, J., and Chakravarty, S., "A Fast Frequency-Domain Eigenvalue-Based Approach to Full-Wave Modeling of Large-Scale Three-Dimensional On-Chip Interconnect Structures," *IEEE Transactions on Advanced Packaging*, Vol. 31(4), pp. 890–899, Nov. 2008.

[368] Zhu, J. and Jiao, D., "A Unified Finite-Element Solution From Zero Frequency to Microwave Frequencies for Full-Wave Modeling of Large-Scale Three-Dimensional On-Chip Interconnect Structures," *IEEE Transactions on Advanced Packaging*, Vol. 31(4), pp. 873–881, Nov. 2008.

[369] Aregba, D. and Mazo-Merceur, G., "Modeling Multiport Using a Three-Dimensional Coupled Analytical/Finite Element-Method Application to Microwave Characterization of Material," *IEEE Transactions on Microwave Theory and Techniques*, Vol. 42, pp. 590–594, 1994.

[370] Deshpande, M. D., et al., "A new Approach to Estimate Complex Permittivity of Dielectric Materials at Microwave Frequencies Using Waveguide Measurements," *IEEE Transactions on Microwave Theory and Techniques*, Vol. 45, pp. 359-366, 1997.

[371] Coccioli, R., Pelosi, G., and Selleri, S., "Characterization of Dielectric Materials with the Finite-Element Method," *IEEE Transactions on Microwave Theory and Techniques*, Vol. 47(7), pp. 1106–1112, July 1999.

[372] Rekanos, I. T., Yioultsis, T. V., and Tsiboukis, T. D., "Inverse Scattering Using the Finite-Element Method and a Nonlinear Optimization Technique," *IEEE Transactions on Antennas and Propagation*, Vol. 47(3), pp. 336–344, Mar. 1999.

[373] Thakur, K. P. and Holmes, W. S., "An Inverse Technique to Evaluate Permittivity of Material in a Cavity," *IEEE Transactions on Antennas and Propagation*, Vol. 49(6), pp. 1129–1132, June 2001.

About the Authors

Giuseppe Pelosi was born in Pisa, Italy on December 25, 1952. He received a laurea (doctorate) degree in physics (summa cum laude) from the University of Florence in 1976. Since 1979, he has been with the Department of Electronics and Telecommunications of the University of Florence, where he is currently a full professor of electromagnetic fields. Dr. Pelosi was a visiting scientist at McGill University, Montreal, Quebec from 1993 to 1995 and a professor at the University of Nice-Sophie Antipolis (France) in 2001.

Dr. Pelosi is mainly involved in research in the field of numerical and asymptotic techniques for electromagnetic engineering, with particular interest in antennas, circuits, microwave, and millimeter-wave devices and scattering problems. He is also very active in the divulgation of electromagnetic engineering and telecommunications history.

He is a coauthor of over 300 scientific publications on the aforementioned topics, has appeared in international referred journals, and has spoke at national and international conferences. He has been a guest editor of several special issues of international journals: *IEEE Transactions on Antennas and Propagation*, 2001 (with V. Grikunov and J.L. Volakis); *International Journal of Numerical Modelling: Electronic Networks, Devices and Fields*, 2000 (with P. Guillon and T. Itoh); *Electromagnetics*, 1998 (with J.L. Volakis); *Annales des Telecommunications* (with J.L. Bernard and P.Y. Ufimtsev), 1995; *COMPEL*, 1994 (with P.P. Silvester); *COMPEL*, 2002; and *Alta Frequenza-Rivista di Elettronica*, 1992.

He is also coauthor of three books: *Finite-Element for Wave Electromagnetics* (with P.P. Silvester, IEEE Press, 1994), *Finite Element Software for Microwave Engineering* (with T. Itoh and P.P. Silvester, Wiley, 1996) and the first edition of *Quick Finite Elements for Electromagnetic Fields* (with R. Coccioli and S. Selleri, Artech House, 1998).

Dr. Pelosi is a fellow of the IEEE "for contributions to computational electromagnetics." He has been a member of the board of directors of the Applied Computational Electromagnetics Society (ACES) (1999–2001), a member of the board of directors of the IEEE Central and South Italy Section (1992–1995 and 1995–1998) and chairman of the IEEE Magnetics Chapter of the same section (1996–1999).

Roberto Coccioli received his degree (laurea), cum laude, in electronic engineering and his Ph.D. in computer science and telecommunications from the University of Florence in 1991 and 1995, respectively. He is currently a principal engineer at the Inphi Corporation in Westlake Village, California, where he is responsible for electrical, thermal, and mechanical design aspects of electronic packaging for high-speed digital and microwave products.

In 1992 and 1994 he was a visiting scholar at McGill University, Montreal, Canada, doing research on finite element methods. From September 1996 to December 1999 he was at the Electrical Engineering Department of UCLA, first as a visiting scholar and then as a postdoctoral fellow, working on the application of FEM to microwave circuits, antennas, and characterization of electromagnetic bandgap materials. From January 2000 to April 2001 he was a staff engineer with the Advanced Packaging Development Group in Conexant Systems. He joined Inphi Corporation as a senior engineer in May 2001.

He is a coauthor of over 30 publications in peer-reviewed international scientific journals and over 60 conference presentations, an instructor of several professional development courses on electronic packaging, and holds seven U.S. patents.

Roberto Coccioli is a senior member of the IEEE.

Stefano Selleri was born in Viareggio, Italy, on December 9, 1968. He obtained his degree (laurea), cum laude, in electronic engineering and his Ph.D. in computer science and telecommunications from the University of Florence in 1992 and 1997, respectively. He is currently an assistant professor at the University of Florence, where he conducts research on numerical modelling of microwave devices and circuits with particular attention to numerical optimization.

He was a visiting scholar at the University of Michigan, Ann Arbor, MI in 1992; at McGill University, Montreal, Canada in 1994; and at the Laboratoire d'Electronique of the University of Nice–Sophia Antipolis in 1997. From February to July 1998 he was a research engineer at the Centre National d'Etudes Telecommunications (CNET) France Telecom. In July 2007 he was a visiting professor at the Escuela Técnica Superior de Ingenieros de Telecomunicación de Madrid of the Polytechnic University of Madrid (Spain) and has been a visiting scholar several times in 2007 and 2008 at the Mecatronics Engineering Department of the University of Saarland.

He is coauthor of nearly 100 scientific publications on the aforementioned topics, and has appeared in international referred journals. He is the associate editor of the *International Journal of Antennas and Propagation* where he also worked with G. Toso on a special issue focusing on active antennas for space application that came out in March 2009. He is also on the editorial board of the *International Journal of RF and Microwave Computer-Aided Engineering.*

He is coauthor of one international book (the first edition of this book) *Quick Finite Elements for Electromagnetic Waves* (with G. Pelosi and R. Coccioli, Artech House, 1998).

Dr. Selleri is a senior member of the IEEE and a member of the Italian Order of Engineers.

About the Contributor

Giacomo Guarnieri was born in Florence, Italy, on June 5, 1978. He received his laurea degree (cum laude) in electronic engineering in 2003, and his Ph.D. in RF microwave and electromagnetics in 2007, both from the University of Florence.

In 2007 he joined the Selex Galileo Co., Radar and Advanced Targeting Department in Florence, where he works on radar antennas design and electromagnetic applications research and development.

His research interests also span numerical techniques for electromagnetics with particular attention to time domain methods and advanced finite element techniques. Dr. Guarnieri contributed to the second part of this book.

Index

L

M

Recent Titles in the Artech House Electromagnetic Analysis Series

Tapan K. Sarkar, Series Editor

Advances in Computational Electrodynamics: The Finite-Difference Time-Domain Method, Allen Taflove, editor

Analysis Methods for Electromagnetic Wave Problems, Volume 2, Eikichi Yamashita, editor

Analytical and Computational Methods in Electromagnetics, Ramesh Garg

Analytical Modeling in Applied Electromagnetics, Sergei Tretyakov

Applications of Neural Networks in Electromagnetics, Christos Christodoulou and Michael Georgiopoulos

CFDTD: Conformal Finite-Difference Time-Domain Maxwell's Equations Solver, Software and User's Guide, Wenhua Yu and Raj Mittra

The CG-FFT Method: Application of Signal Processing Techniques to Electromagnetics, Manuel F. Cátedra, et al.

Computational Electrodynamics: The Finite-Difference Time-Domain Method, Second Edition, Allen Taflove and Susan C. Hagness

Electromagnetic Waves in Chiral and Bi-Isotropic Media, I. V. Lindell, et al.

Engineering Applications of the Modulated Scatterer Technique, Jean-Charles Bolomey and Fred E. Gardiol

Fast and Efficient Algorithms in Computational Electromagnetics, Weng Cho Chew, et al., editors

Fresnel Zones in Wireless Links, Zone Plate Lenses and Antennas, Hristo D. Hristov

Grid Computing for Electromagnetics, Luciano Tarricone and Alessandra Esposito

High Frequency Electromagnetic Dosimetry, David A. Sánchez-Hernández, Editor

Iterative and Self-Adaptive Finite-Elements in Electromagnetic Modeling, Magdalena Salazar-Palma, et al.

Numerical Analysis for Electromagnetic Integral Equations, Karl F. Warnick

Parallel Finite-Difference Time-Domain Method, Wenhua Yu, et al.

Quick Finite Elements for Electromagnetic Waves, Giuseppe Pelosi, Roberto Coccioli, and Stefano Selleri

Understanding Electromagnetic Scattering Using the Moment Method: A Practical Approach, Randy Bancroft

Wavelet Applications in Engineering Electromagnetics, Tapan K. Sarkar, Magdalena Salazar-Palma, and Michael C. Wicks

For further information on these and other Artech House titles, including previously considered out-of-print books now available through our In-Print-Forever® (IPF®) program, contact:

Artech House
685 Canton Street
Norwood, MA 02062
Phone: 781-769-9750
Fax: 781-769-6334
e-mail: artech@artechhouse.com

Artech House
16 Sussex Street
London SW1V 4RW UK
Phone: +44 (0)20 7596-8750
Fax: +44 (0)20 7630 0166
e-mail: artech-uk@artechhouse.com

Find us on the World Wide Web at: www.artechhouse.com